Making Sense of Mining History

This book draws together international contributors to analyse a wide range of aspects of mining history across the globe including mining archaeology, technologies of mining, migration and mining, the everyday life of the miner, the state and mining, industrial relations in mining, gender and mining, environment and mining, mining accidents, the visual history of mining, and mining heritage. The result is a counterbalance to more common national and regional case study perspectives.

Stefan Berger is Professor of Social History and Director of the Institute for Social Movements, Ruhr University Bochum, Germany, Chairman of the Foundation History of the Ruhr, and Honorary Professor at Cardiff University, UK.

Peter Alexander is the Director of the Centre for Social Change, University of Johannesburg, South Africa, and holds the South African Research Chair in Social Change.

Routledge Studies in Modern History

www.routledge.com/history/series/MODHIST

For a full list of titles, please visit: www.routledge.com/history/series/MODHIST

Making Sense of Mining History

Themes and Agendas

**Edited by Stefan Berger and
Peter Alexander**

Routledge
Taylor & Francis Group

LONDON AND NEW YORK

First published 2020
by Routledge
2 Park Square, Milton Park, Abingdon, Oxon OX14 4RN

and by Routledge
711 Third Avenue, New York, NY 10017

Routledge is an imprint of the Taylor & Francis Group, an informa business

First issued in paperback 2021

British Library Cataloguing in Publication Data
A catalogue record for this book is available from the British Library

Library of Congress Cataloging in Publication Data
Names: Berger, Stefan, editor. | Alexander, Peter, 1953 October 29- editor.
Title: Making sense of mining history / [edited by] Stefan Berger (director
   of the Institute for Social Movements, Ruhr University Bochum,
   Germany, and chairman of the Foundation History of the Ruhr), Peter
   Alexander (director of the Centre for Social Change, University of
   Johannesburg, South Africa).
Description: Abingdon, Oxon ; New York, NY : Routledge, 2019. | Includes
   bibliographical references and index.
Identifiers: LCCN 2019007833 (print) | LCCN 2019009143 (ebook) |
   ISBN 9780429513527 (adobe) | ISBN 9780429520389 (mobi) | ISBN
   9780429516955 (epub) | ISBN 9780367198688 (hardback) | ISBN
   9780429243806 (ebook)
Subjects: LCSH: Mines and mineral resources—History. | Mines and
   mineral resources—Social aspects. | Coal mines and mining—History. |
   Mine accidents—History.
Classification: LCC TN15 (ebook) | LCC TN15 .M275 2019 (print) | DDC
   622.09—dc23
LC record available at https://lccn.loc.gov/2019007833

ISBN: 978-0-367-19868-8 (hbk)
ISBN: 978-1-03-208860-0 (pbk)
ISBN: 978-0-429-24380-6 (ebk)

Typeset in Times New Roman
by Swales & Willis Ltd, Exeter, Devon, UK

Contents

Foreword

Stefan Berger and Peter Alexander

This volume has been a long time in the making. The first idea of it goes back to the International Mining History Congress in Johannesburg in 2011. Since then we have invited a range of authors to submit, rework and resubmit the chapters that are now at long last assembled here. On the way, we had another international conference that took place at the Ruhr museum in Essen in 2013, where versions of these articles were also presented as papers. All of the authors of this volume are united by their desire to move to more transnational and comparative perspectives on the history of mining. Mining history, in our view, has the potential to be an important lens for global history. After all, it is very international and very longue durée. Technologies and markets provide strong commonalities, whilst at the same time there remain manifold contradictions and ambiguities between the local and the transnational. Transnational language communities, such as the Anglophone language community, provided strong global ties. So far, the dominant local, regional and national imagination in mining history is preventing these transnational perspectives from breaking through.

At the same time as this volume wishes to put forward an argument for more transnational mining history, it also presents an argument for a longer historical perspective. The developments in world markets have often taken place over long periods of time and our historiographical caesuras around ancient, medieval, early modern, modern and contemporary histories might not be the most adequate to get into focus those long-term continuities in the development of capitalist practices in mining. The significance of mining in different areas of the world has been shifting over long periods. Looking at mining history over longer periods of time would also allow us to work more with diachronic comparisons that allow for different chronologies in different places. As many of the chapters in this volume indicate, we are unlikely to move to a universal periodisation in mining history. Instead we can identify patterns and tendencies that are global – in sub-fields of mining history such as the technologies of mining, the gender of mining, histories of industrial relations in mining and others that are the subject of the following pages.

Although we have assembled here 11 themes and sub-fields in mining history, there are still a number of gaps. Thus, we would have liked to have an additional chapter on small-scale mining and one on the economics and finance regimes in mining. Geographically, the historiography of mining is still overall heavily

Western-centric, and although many of the chapters below have tried to refer also to non-Western literatures, there simply remain too many gaps and open questions which new research hopefully will address over the next decades. There is also a surprising gap within the West: Russia and Eastern Europe is heavily underrepresented here, which is also partly to do with the non-availability of literature in the English language and partly with the non-availability of English as lingua franca at least among an older generation of scholars in Russia and Eastern Europe.

The subsequent chapters are also aimed at providing readers with a range of theoretical perspectives that have been and continue to be fruitful in mining history. These include theories around the new materialism discussed in LeCain's chapter and gender theories, discussed in Lahiri-Dutt's chapter. The use of various forms of social theory, from classical Weberian and Marxian frames to the ideas of Bourdieu and Foucault are also present here. Miners' lives and mining communities can be studied from a great variety of different theoretical and methodological viewpoints. There is no right or wrong kind of theory; instead the latter fit particular research questions that are being posed to the history of mining.

A number of concepts have been key to mining history for many years. These include the concepts of class, gender, race, ethnicity, religion and culture. These are all characterised by a great degree of intersectionality, and in our view mining historians still need to take more account of such inter-sectionality. Overall, we hope that the present volume will help mining historians to move towards a more intersectional, a more theoretically informed, a more long-term and a more transnational form of mining history in the future.

Bochum and Johannesburg
January 2019

Contributors

Peter Alexander is Director of the Centre for Social Change and a Professor of Sociology at the University of Johannesburg, where she holds the South African Research Chair in Social Change. Most of his work is in the fields of comparative labour history, social movement studies, and the sociology of class, and his books include *Marikana: A View from the Mountain and a Case to Answer* (2013) and *Class in Soweto* (2013). Recent publications include a chapter in a festschrift for Marcel van der Linden entitled 'On the Road to Global History – via Comparison' (2018).

Stefan Berger is Professor of Social History and Director of the Institute for Social Movements at Ruhr University Bochum. He is also Executive Chair of the Foundation History of the Ruhr and an Honorary Professor at Cardiff University in the UK. He has published widely on the history of deindustrialization, industrial heritage, memory studies, the history of historiography, nationalism, and labour movement history. His most recent publications are a special issue, co-edited with Steven High, on de-industrialization by the North American journal *Labor* 19:1 (2019) as well as a special issue on German labour history by the British journal *German History* 32:2 (2019).

Michael Farrenkopf is head of the Montanhistorisches Dokumentationszentrum (Mining History Document Centre) at the Deutsches Bergbau-Museum Bochum (German Mining Museum Bochum), to whose board he belongs. As a historian with lectureships at the Ruhr University Bochum and the TU Bergakademie Freiberg; he is equally active in culture and research. He has realized numerous exhibition projects, most recently a large joint exhibition with the Ruhr Museum entitled 'The Age of Coal. A European History' on the World Heritage Site Zollverein in Essen. In addition to the catalog (co-edited by Franz-Josef Brüggemeier and Heinrich Theodor Grütter, Klartext Verlag, Essen 2018), he most recently presented a *History of Mining* (with Lars Bluma and Stefan Przigoda, L & H Verlag, Berlin 2018).

Dagmar Kift is Head of the research department of the LWL-Industriemuseum and its Deputy Director. She is working and publishing on the social and cultural history of coal-mining, leisure, gender and migration. Recent publications are On 'Events Heard' – Researching and Re-Using Industrial Soundscapes. The EU-Project 'Work with Sounds', in: *Moving the Social* 56 (2016), with Konrad Gutkowski, and *Bergbaukulturen in interdisziplinärer Perspektive. Diskurse und Imaginationen*, co-edited with Eckhard Schinkel, Stefan Berger and Hanneliese Palm (2018).

Ad Knotter is Honorary Professor at Maastricht University and Research Fellow at the International Institute of Social History in Amsterdam. He has published widely on (global) mining history, most recently in Karin Hofmeester and Marcel van der Linden (eds.), *Handbook Global History of Work* (De Gruyter Oldenbourg, 2017). His latest book *Transformations of Trade Unionism: Comparative and Transnational Perspectives on Workers Organizing in Europe and the United States, Eighteenth to Twenty-first Centuries* is available in Open Access at Amsterdam University Press.

Kuntala Lahiri-Dutt is a Professor at the Resource, Environment and Development group at the Crawford School of Public Policy, College of Asia and the Pacific, The Australian National University. Kuntala's extensive research on gendered labour in the mines and quarries of India, Indonesia, Lao PDR, and Mongolia has made her a global authority on the subject. She is also researcher of precarious livelihoods in environmental resource-dependent communities – whether in the transient chars (river islands) or in the mineral-rich tracts – extracting a tentative living. More about Kuntala's work can be gleaned from her staffpage at https://crawford.anu.edu.au/people/academic/kuntala-lahiri-dutt

Timothy James LeCain is Professor of History at Montana State University. His first book, *Mass Destruction: The Men and Giant Mines That Wired America and Scarred the Planet* (Rutgers, 2009), was awarded the 2010 best book award by the American Society for Environmental History. LeCain's most recent book is *The Matter of History: How Things Create the Past* (Cambridge, 2017), which develops a neo-materialist theory and method of history. LeCain has been a fellow at the Rachel Carson Center in Munich, Germany, and the Center for Advanced Study in Oslo, Norway.

T. Dunbar Moodie is Professor Emeritus of Anthropology and Sociology at Hobart and William Smith Colleges and a Research Associate at the Institute for Society, Work and Politics (SWOP) at the University of the Witwatersrand. He is the author of two books, *The Rise of Afrikanerdom: Power Apartheid and the Afrikaner Civil Religion* and *Going for Gold: Men, Mines and Migration*, both published by the University of California Press, as well as numerous articles in scholarly journals and academic collections dealing with various aspects of South African society.

Jeremy Mouat is a Professor Emeritus with the University of Alberta, where he taught from 2005 until retiring in 2017. He has a longstanding interest in the history of mining and metallurgy, on which topic he has published numerous articles and two books. He also served as guest editor of two journals for theme issues on mining history. He was awarded the Mining History Association's Rodman W. Paul Award For Outstanding Contributions to Mining History in 1997, and later served as President of the Association. He has held visiting fellowships in Australia, New Zealand, the UK, and the USA, including as a Fulbright Visiting Research Chair at Arizona State University in 2011. He is currently planning the twelfth meeting of the International Mining History Congress, to be held at Laurentian University in Sudbury in June 2020.

Pavithra Narayanan is Associate Professor of English and affiliated faculty of the Collective for Social and Environmental Justice at Washington State University Vancouver, USA. She is the author of *What Are You Reading? The World Market and Indian Literary Production* (Routledge, 2012). Her research areas include the history of the book, processes of decolonization, Indigenous land rights and movements, civilian resistance movements, political and economic policies, social and environmental justice, nationalism, and postcolonial theory. She is also a documentary filmmaker; her most recent work is a co-directed film on *Chicago's Black Arts Movement* (2019).

Alma Parra Campos is a full-time researcher at the National Institute of Anthropology and History, Mexico. She obtained an MSc from the London School of Economics and is completing her doctoral studies at the National University of Mexico. Her work centres on Mexican mining history, entrepreneurial activity, and the international trade networks for the distribution of supplies, technology, and commodities for mining. She has co-authored books on Mexican mining and edited a dossier on international mining published by CIDE, Mexico, as well as published numerous articles.

Stefan Siemer is a historian and researcher at the Montanhistorisches Dokumentationszentrum (Mining History Document Centre) at the Deutsches Bergbau-Museum Bochum (German Mining Museum Bochum). He has published in the field of the history of technology, environmental studies and museology. Recent publications include studies on mining museums and collections in the Ruhr area and, co-edited with Michael Farrenkopf, a forthcoming volume on mining museums in Germany.

Paul Stewart was recently appointed Associate Professor in the Department of Sociology at the University of Zululand after teaching for 20 years at the University of the Witwatersrand. He has published both locally and internationally on mining labour with specific reference to labour time, the struggles of rock drill operators, mechanisation and mine safety. His most recent co-authored publication is on the 'Right to Refuse Dangerous Work' in the *Journal of the Southern African Institute of Mining and Metallurgy* 119:1 (2019). A further co-authored article is to appear in the *Journal of International Development Policy* – Special Edition 10th Anniversary of the ILO (2019).

Simon Timberlake is a geologist with experience of working in the museum sector in the UK, and then between 1996 and 2000 on geo-archaeological (i.e. Leverhulme Trust-funded) research projects studying various aspects of prehistoric-Roman mining on Alderley Edge (University of Manchester) and Bronze Age copper mining in Wales (Bangor University and Coventry University). Following this he was employed as Senior Researcher with the Cambridge Archaeological Unit (University of Cambridge), and now works freelance as an archaeological specialist on stone, clay, metalworking, and palaeo-environmental analysis. He is the author of two books and more than 50 published papers on the study of ancient mining.

Chris Wrigley is Emeritus Professor of History at Nottingham University. His books include new (2018) editions of *David Lloyd George and the British Labour Movement* (1976) and *Lloyd George and the Challenge of Labour* (1990), *British Trade Unions Since 1933* (2002) as well as biographies of Lloyd George, Arthur Henderson, Winston Churchill and A.J.P. Taylor. His edited books include three volumes of *A History of British Industrial Relations* (1982–96), *Challenges of Labour: Central and Western Europe* (1993) and *The First World War and the International Economy* (2000). He was President of the (UK) Historical Association, 1996–99. He was awarded an honorary doctorate by the University of East Anglia in 1998. *Labour and Working-Class Lives: Essays to celebrate the Life and Work of Chris Wrigley*, edited by K. Laybourn and J. Shepherd was published in 2017.

Abbreviations

AGMS	Adivasi Gothre Mahasabha, the
AGMS	Grand Assembly of Adivasi, the
AIM	American Indian Movement
AMCU	Association of Mineworkers and Construction Union, the
AMS	Accelerator Mass Spectroscopy
AMWU	African Mine Workers' Union
ANC	African National Congress, the
BHP	Broken Hill Proprietary
CBI	Central Bureau of Investigation
CFMEU	Construction, Forestry, Mining and Energy Union, the
CIL	Coal India Limited
CIMI	Indigenous Missionary Council, the
CMRCC	Coalmining Research Controlling Council, the
COIAB	Coordination of the Indigenous Organizations of the Brazilian Amazon, the
CRIC	Regional Indigenous Council of Cauca, the
CRL	Collieries Research Laboratory, the
CWU	Colliery Workers' Union, the
DGMS	Directorate General of Mines Safety, the
ECSC	European Coal and Steel Community, the
FIFO	Fly In Fly Out
GAD	Gender and Development theory
GPR	Ground Penetrating Radar
IBA	International Building Exhibition, the

ICLU	Indian Collieries Labour Union, the
IGBCE	Industriegewerkschaft Bergbau, Chemie, Energie, the
ILO	International Labour Organization, the
IPCC	Intergovernmental Panel on Climate Change, the
IPN	Indigenous Peoples' Network, the
JATAM	Jaringan Adovaksi Tambang, the
LIDAR	Light Detection and Ranging
MFGB	Miners' Federation of Great Britain, the
MHSA	Mine Health and Safety Act, the
MHSI	Mine Health and Safety Inspectorate, the
MOU	memorandum of understanding
NAFTA	North American Free Trade Agreement, the
NCW	National Commission for Women, the
NHRC	National Human Rights Commission, the
NRC	Native Recruiting Corporation
NUM	National Union of Mineworkers
PJEC	People's Judicial Enquiry Commission, the
RAG	Ruhrkohle Aktiengesellschaft
SA	South Africa
SC/ST Commission	Scheduled Caste and Scheduled Tribes Commission, the
SOE	state-owned enterprise
SPD	Social Democratic Party, the
SWMF	South Wales Miners' Federation, the
TUC	Trades Union Congress, the
UDM	Union of Mineworkers
UMWA	United Mine Workers of America, the
WBK	Westfälische Berggewerkschaftskasse, the
WCIP	World Council of Indigenous People, the
WDM	World Development Movement, the
WNLA	Witwatersrand Native Labour Association, the

1   Mining history

Sub-fields and agendas

Stefan Berger

Introduction

Mining history has been a thriving sub-field of historical writing for many decades. This volume attempts to provide a survey of key aspects of 11 important fields in mining history that the editors believe have been particularly vibrant and important over the years and where we can expect innovative new work to emerge in the future. In any undertaking like this, there is always the danger of overlooking sub-fields and being wrong about those subjects that really matter within mining history. What we present is therefore preliminary, tentative and, as all historical writing, subject to criticism and revision. Nevertheless, I hope that the pages of this volume will be helpful to mining historians to consider where the subject is at and how it may best progress into the future.

In this first chapter, I shall endeavour to introduce the 11 themes, subsequently to be elaborated at greater length by experts in those fields of mining history. I have connected each theme to a particular date. In other words, each of the subsequent 11 dates symbolizes an event in mining history that stands for a particular theme of great significance for mining and its history. Within the themes chosen, other historians may well regard different dates to be of much greater significance than the ones I have chosen, but I hope that we can at least agree on the importance of the themes that are, of course, overlapping and interconnected – like a system of communicating tubes.

One of the giants of mining history, Klaus Tenfelde, established a comparative framework for the study of mining societies, which was based on the assumption of a great commonality of social conditions. Mining, he argued, was ultimately dependent on geology; it needed huge investments and was often subject to early state legislation. The work processes in mining were quite similar and there was a strong connection between the work environment and the wider lifeworld, making for densely knit mining communities.[1] More than ten years ago I attempted to develop this idea further in the hope of providing a basis on which mining histories could be compared.[2] By identifying what I regard as the most important themes in mining history I am also hoping to contribute to the foundations of a more developed transnational and comparative history of mining. Many of the subsequent chapters attempt to survey their respective themes with a view to transnational and comparative dimensions underlining the fruitfulness of looking beyond one specific locality, region or nation.

Let me introduce the 11 dates and the themes they are associated with. First, there is the archaeology of mining and its attempts to tell us more about how the extraction of raw materials from the earth has been an activity shaping human existence for tens of thousands of years. About 40000 BC the first traces of underground mining can be found in Africa and ever since, archaeologists have contributed in vital ways to our knowledge of mining. Simon Timberlake's chapter in this volume introduces the research landscape of pre-industrial mining in particular.

The second theme is about technological progress that has played, again from early on, a central part in the development of mining. The date I have chosen here lies somewhere between AD 960 and 1127, when the blast furnace was invented under the Sung dynasty in China. The development of technological progress is subsequently at the heart of Jeremy Mouat's chapter on the importance of engineering changes to the development of modern mining in the US, South Africa and Australia. It should, however, also be noted that mining was one of those industries where technological progress was at times limited by natural geological conditions. This was nowhere more the case than in gold mining, which always remained incredibly labour-intensive precisely because of the limits of technological change. Dunbar Moodie's chapter compares the consequences of this for workers' lives in gold mines in South Africa and India.

Third, there is the science of mining that has had such important repercussions on the organization of work processes in the industry. An iconic early publication for the increasing scientificity of mining is Georg Agricola's *De Rerum Metallica*, first published in 1556. The knowledge of mining and the science of mining was institutionalized in mining academies in early modern Europe, from where they travelled across the globe – together with Western imperialism. In her chapter, Alma Parra looks at the importance of knowledge transfer in the silver mining of Mexico, highlighting the importance of local knowledge adapting the more universal scientific recipes that were imported from abroad.

The fourth theme to be explored in this volume is migration. The iconic date chosen to illustrate that theme is the discovery of the first major diamond in South Africa, the 'Eureka Diamond' in 1866. It was found in river gravels from the Orange River near Hopetown in the Cape Colony. It initiated a massive migration of miners to the diamond-fields of South Africa and is illustrative of the importance that migration played in the history of mining. Ad Knotter's chapter speaks to the global ramifications of migration to a history of mining.

Fifth, mining produced particular forms of working-class culture. The leisure pursuits of miners were part and parcel of miners' culture, which has been an extremely fruitful area for research for many years. The date I have chosen here is 1900, when Cornish miners in Mexico played, for the first time in Mexico, a game of football. Peter Alexander's chapter in this volume compares the impact of miners' culture on what he calls 'classed identities' and the emergence of a trade union consciousness in the Ruhr, South Wales, West Virginia, Alabama,

Jharia (India), Enugu (Nigeria), the Zambian Copperbelt and the Transvaal (South Africa). His chapter provides tantalizing glimpses onto a vast tableau of research that opens up in the field of culture.

The sixth theme is gender. Of course, all the other themes identified here as key themes also benefit from using gender as a category of historical analysis.[3] But it is true that few areas have seen such exciting and innovative work being produced as is the case with the gender history of mining. The iconic date chosen here is 1933, when underground work for women was officially prohibited in Japan. Kuntala Lahiri Dutt surveys the diverse ways in which gender perspectives have, over recent years, enriched the field of mining history.

The seventh theme focuses on one of the saddest aspects of mining – mining accidents. It has long been recognized that mining is a particularly dangerous form of economic activity and bad accidents litter its history. The iconic date chosen is 1942, when 1,954 miners lost their lives at the Benxihu colliery in China in what is to date the biggest accident in mining history – in terms of loss of life in one incident. Mine accidents devastated whole communities and the ever-present threat of accidents is often seen as one of the reasons for the legendary solidarity of miners – they had to rely on each other if they wanted to be safe. This volume has two contributions focusing on mining accidents. Michael Farrenkopf's contribution focuses on global mining accidents caused by explosions and how the international scientific community over time sought to reduce this risk to the miners' lives. Dagmar Kift and Paul Stewart's chapter looks at the miners' experiences of accidents through the lens of oral history – also taking into account the positions of the coal companies and the state in their comparison of South Africa and Germany.

The eighth theme returns to the importance of the state to mining history. State action proved vital to the miners' existence at several key junctures in history. The iconic moment chosen here is the nationalization of the coal mines in Britain in 1946 – the realization of long-held aspirations of many miners in Britain. Chris Wrigley's chapter provides a comparative argument around the role of the state in labour conflicts in mining which highlights the benefits of 'bringing the state back in'[4] when dealing with mining history. Almost like gender, the state comes into almost every aspect of mining history. Another area where it has been prominent is the area of land rights. They have been key to the expansion of mining companies, and Pavithra Narayanan's chapter reminds us of the many struggles of indigenous populations against mine companies' insatiable hunger for land and for expansion.

Ninth, there is the visual history of mining. The visual turn in history writing[5] has also reached mining history. More and more historians are paying attention to the diverse ways in which mining has been pictorialized and in which images of mining have played a key role in shaping and understanding the mining industry. The iconic date chosen here is here is 1978 – the death of Australian miner and artist Samuel Michael Byrne, whose naïve paintings, in particular of mining in Broken Hill, are reminiscent of a wider trend that connected naïve painting to mining, not only in Broken Hill, but also in Upper Silesia and the

Ruhr, among other places. Stefan Siemer's chapter focuses on how diverse visualizations of coal have framed diverse memory cultures of mining in different parts of the world.

The tenth and penultimate theme deals with the environmental consequences of mining which have, of course, been manifold and the source for strong moral condemnation of the industry. The iconic date chosen here is 1992, when the first Earth Summit was held in Rio de Janeiro. Here carbon dioxide emissions were blamed for destroying the ozone layer and contributing to global warming, threatening the planet with self-destruction. Tim Le Cain's chapter introduces a new materialist reading of the relationship between mining and the environment, which has attracted much attention in recent years and promises more fruitful research in years to come.

The eleventh and final theme has to do with mining heritage. Globally there is no end in sight for mining. Yet in many parts of the world mining has come to an end – often after decades of an active history which determined the physical and mental outlook of whole communities. The fate of the remnants of a defunct mining industry has been extremely diverse in different parts of the world. On one end of the spectrum everything has been demolished and almost nothing remains of a once proud industry; on the other end, whole industrial landscapes have been preserved, arguably nowhere more comprehensively than in the Ruhr area of Germany – once Europe's biggest coalfield. Hence the iconic date chosen is 2001, when Zeche Zollverein in Essen became a UNESCO world heritage site. Stefan Berger in his chapter provides a comparative perspective on the heritage of the coal industry in different parts of the world. In the remainder of this chapter, I would like to introduce the 11 dates and themes a bit more.

Extracting raw materials from the earth as an integral part of human economic activity

With the beginning of the New Stone Age, around 40000 BC, come the first traces of underground mining in Africa. Like human civilization itself, mining originated in Africa.[6] The Lion Cavern in Swaziland is the oldest mine on archaeological record, with radiocarbon dating putting it at 43,000 years of age.[7] The mineral hematite, containing iron, was mined here, and from it, Africans made the red pigment ochre. During the Neolithic period ochres, ambers, flints and salt were among the most important objects of trade. Gold also came to dominate human attention from very early on (certainly before 6000 BC, when we think about the ancient Nubian and Egyptian empires).[8] Copper, bronze and iron were vital for making weapons and changing or maintaining power structures and empires from the copper age onwards. The rise of Sweden from a tributary state of Denmark to a major European power in the seventeenth century would be inexplicable without its mineral wealth,[9] and similar stories can be found elsewhere in the world. Mining, from its early beginnings, was associated with trade and trading networks and it has become a global activity critical to wealth creation, power structures and to what the French philosophes called human progress. It is not by chance that whole ages are associated with minerals or their derivates, which involve mining – from 'bronze age' to 'iron age' to 'nuclear age'. Gold, silver, bronze, copper, lead, zinc, coal, salt, tin, iron, uranium, oil and potash are among the most important of those minerals that had to be extracted from the earth through mining. Mining is a foundational activity and accompanies human history throughout the ages.

Much of the mining has been and continues to be surface mining (open-pit mining, quarrying, strip-pit mining), although a considerable part has also been sub-surface mining (drift mining, slope mining, shaft mining). Interestingly, to the popular imagination of mining, sub-surface mining has arguably been much more important than surface mining, regardless of the fact that it has been far less important overall.

By the fifth and fourth centuries BC, thousands of miners, largely slaves, were working in the silver-bearing lead ore mines near Athens (Mount Laurion), where more than 2,000 shafts, some of them up to 425 feet deep, were dug.[10] In the Roman Empire mining was an important activity in Spain, Gaul, Britain and the Alpine regions south of the river Danube. Some historians have argued that without the reserves of lead, silver, mercury, copper and iron ores from Spain, the Roman Empire could not have lasted. Elsewhere empires also depended on mining – the famous Iron Pillar in Delhi dates back to AD 310; iron coins were minted in China as early as AD 525, and the gem stones of the Orient were mined from at least 1000 BC onwards. Throughout the Middle Ages, mining remained of vital importance in various parts of the world. The ancient kingdom of Ghana, for example, produced some of the world's most skilled metalworkers during its Golden Age between 920 and 1050 BC.[11]

In ancient and especially prehistoric mining, archaeology has been the key historical discipline for many years. While industrial archaeology, of course, has also much to contribute to the modern and contemporary period, Simon Timberlake here reviews examples from premodern, mainly Palaeolithic and Iron Age mining, extending to the Roman and medieval periods. He not only includes European but also African and pre-Columbian mining in the Americas. Landscape analysis, fieldwalking, geophysical surveys, the development of mining tools and their role in dating mining activities – they are all extensively discussed in a chapter that highlights the ongoing relevance of primitive mining practices in the developing world until today. Timberlake introduces us to the fascinating world of the working lives of ancient miners and discusses the meaning of the Bronze Age industrial revolution. And he provides intriguing glimpses into the representation of ancient mining in the modern world – especially the nationalization of ancient mining activities and attempts to relate them to nation-building attempts. Overall, Timberlake's chapter amounts to a fascinating *tour d'horizon* of the possibilities and promises of the prehistoric archaeology of mining.

Technological progress and mining

From those earliest days of prehistoric mining onwards, the technological problems of extracting precious things from the earth abounded. How to deal with water in mines and how to bring light into the darkness of the mine were problems waiting to be solved through technological change. One of the early centres of technological innovation in mining was China. Here, under the Sung dynasty, around AD 960–1127, we saw the invention of the blast furnace about 500 years before it was reinvented in Europe.[12] Mining activity in China, related to copper, can be traced back to the third millennium BC, when copper was already used for bronze and brass production. The blast furnace allowed for the reduction of coal to coke, which formed the basis for profitable coal mining all over the world. Marco Polo, who visited China towards the end of the thirteenth century, reported about the 'burning of black stone' in China, something entirely unknown in his native Italian lands.[13] The history of mining, this date reminds us, is to a large extent the history of how man dealt with geology and how human beings managed, with the help of technological progress, to extract minerals from the bowels of the earth and process them into something useful and valuable. Largely unconnected to earlier developments in China, some of the major technological advances in relation to mining came about in sixteenth- and seventeenth-century England.[14] During the eighteenth and nineteenth centuries, Britain led the way in coal mining. The number of coke-fired blast furnaces increased from 17 in 1760 to 81 in 1791.[15] No similar expansion could be found on the continent at that early time, indicating that Britain was on its way of becoming the first workshop of the world.[16] Some of the key technological problems of mining had to do with how to sink shafts, how to drain off water (from water wheels to the steam engine) and how to haul the minerals to the surface (from the conveyor belt to hydraulic hoisting), how to break hard rock, how to ventilate mines and how to illuminate them. The ancient Romans had already developed hydraulic mining via aqueducts.[17] Explosives were used for the first time in the Banská Štiavnica mines in Hungary in 1627.[18] The development of steam-powered lifts, drills and pumps during the Industrial Revolution brought further major technological advances in mining. The improvements made to trucks, diggers, hauling equipment, cranes, lifts, crushers and other tools were as vital for the progress of mining as the rapid development of the science of metallurgy, so important for the processing of many minerals once they were out of the earth.[19] All of this, of course, puts engineers and scientists into the spotlight of the historians' gaze. Regimes of knowledge and their application to industry have been vital to the development of mining worldwide.

Jeremy Mouat, in his chapter, highlights the transnational aspects of mining engineering – discussing the importance of American mining engineers for the transition to deep-level mining in places such as South Africa and Australia. Analysing issues surrounding new technology and labour costs in Butte, Montana, Johannesburg and Broken Hill, New South Wales, he finds very similar challenges during the 1890s and 1900s, which allowed American engineers to develop generic processes that could be used independently from place and local conditions. Such economies of scale, according to Mouat, explain the transnational success of American mining engineering in this period.

While technological development was key to many forms of mining in many different parts of the world over a very long time, it also had its limits – nowhere more so than in gold mining. Dunbar Moodie's comparison of gold mining in South Africa and India points to the remarkable similarities in the geological conditions and production processes in both countries, even if gold mining happens in South Africa on a much larger scale than in India. Yet in gold mining, what was always more important than technological innovation was cheap labour costs. The proletarianization of labour was seen far more negatively in South Africa than in India, where workers escaped from rural bondage and could experience mine work as liberation, despite starvation wages and horrible working conditions. As Moodie also shows, both groups of miners had clear ideas about 'a moral economy' which set standards of wages and working conditions. If management did not satisfy them, it sparked forms of resistance. Moodie's chapter is already strongly related to our next theme, work processes in mining and their grounding in scientific research about mining.

The science of mining and work processes in mining

Technological progress was intimately related to the progress in science. One of the early landmarks in the history of sciences related to mining was the 1556 publication of Georg Agricola's *De Rerum Metallica*.[20] Agricola, a typical representative of humanism and friend of Erasmus of Rotterdam, spent a lifetime studying mining and his book was the most comprehensive depiction of methods and techniques used in mining.[21] It remained the standard work on this topic for the best part of 200 years. The knowledge about mining was, of course, fostered by the early mining academies.[22] In Europe the first was in Kongsberg/ Norway, founded in 1757. The most famous in the eighteenth and much of the nineteenth century was arguably Freiberg, founded in 1765. It is rumoured that Potosi in Bolivia had a mining academy as early as the sixteenth century.[23] Mining engineers and their 'science' became scientific leaders and mining studies became a lead science of major impact on many neighbouring disciplines. Mining engineers saw themselves, and were widely seen by outsiders, as constructors of a better world and as pathfinders of industrialization and the Industrial Revolution, something which coincided with the quasi-religious adoration of technology. As a profession, they were right from the very beginning a very transnational group of people. As early as 1786, a meeting of 154 scientists from 15 countries interested in mining and mining technology that included Lavoisier from France, Watt from England and Goethe from Germany took place in Schemnitz (today Slovakia), where they discussed the process known as amalgamation. This meeting led to the formation of the first mining society as a learned society in Europe.[24] Major inventions and the technological progress of mining were propelled forward by those academies and the engineers they produced.

This also had a major impact on work processes. Technological inventions and improvements, such as the Davy lamp, invented in 1815, have been vital for determining work processes practices in mining. In all forms of sub-surface mining, it was vitally important to see in the dark, yet open methane gas flames were incredibly dangerous. Hence the importance of Sir Humphrey Davy's invention – the fine mesh that enclosed the flame left enough air through to allow combustion, meaning the flame would not go out. At the same time the mesh also avoided ignitions and explosions.[25] Many of the work processes focused on questions of work safety. Mining work was not only, for much of its history, incredibly hard physical labour, it was also dangerous and unpredictable. It was this danger to life and health associated with the work process that helped to produce the strong sense of solidarity in mining communities. While we clearly do well to avoid any occupational determinism in constructing ideal typical figures of 'the miner', it still seems to me worthwhile to ask how work processes made the experiences of miners contribute to common lifeworlds and how these very same lifeworlds were indeed segmented by those experiences.

Throughout much of the history of mining, accident rates were high and a workers' life counted for little. One of the deadly threats was associated with dust. As the South Walian miners' writer Bert Coombes wrote:

> Dust. If the fog drops on these areas a large number of men are in torture. At one period one man every week was permanently disabled by dust at one colliery – out of less than a thousand men. I have known men in whole families wiped out by dust disease and dying in agony.[26]

Work processes determined to a large extent the everyday life of the miners in mining communities and common work experiences united those communities. However, workplace solidarities should not be overemphasized. After all, we know all too well that hauliers could be played off against hewers, the old could be pitted against the young, the employed against the unemployed, and the emergence of new technologies, such as long-wall mining, led to the individualization of pay, which undermined solidarities.

Alma Parra's chapter looks at the diffusion of knowledge and the transfer of technology from the West to the silver mines of Mexico. She emphasizes that much was imported and foreign intervention, in the context of colonialism and imperialism, and certainly shaped the technological progress of mining. Yet she also underlines the crucial role of local knowledge mediating this transfer and having a major impact on the way that imported knowledge was put into practice locally. This constant interplay between local and transnational forms of knowledge is at the heart of knowledge transfer, of which Mexican silver mining is a good example.

Migration and mining

Another factor that often divided rather than united miners was ethnicity. In many mines across the world the workforce was ethnically segmented, which had to do with the pull of mining. Migration to minefields was common and the bigger minefields attracted tens and even hundreds of thousands of miners, coming from near and far – attracted by the economic opportunities afforded by mining. The various nineteenth-century rushes associated with precious gems and metals are just particularly spectacular examples of the general importance of migration to mining activities.[27] When, in 1866, the first major diamond was discovered in South Africa, the so-called 'Eureka Diamond' triggered just such a rush. The gold rushes in California (1848), on the Klondike and in Ballarat, Australia (1851) would be other famous instances. Gold and diamond rushes played vital roles in the industrialization of the US, South Africa and Australia, as they included a transport revolution and the building of railways which opened up much of the land of those vast countries.[28] They also attracted capital (often from abroad) and involved the massive migration of peoples, as they were hunting for minerals and the wealth that minerals promised. In South Africa, within 20 years from the discovery of this first diamond, diamonds were exported to the value of GB£3 million annually and made up the value of half of all exports of the Cape colony.[29] Kimberley, one of the epicenters of diamond mining, rapidly developed into the second city of the Cape colony. The great majority of the 18,000 black workers employed in the Kimberley mines were migrants from the Transvaal, Lesotho and Southern Mozambique. The strong racial division of labour in the Kimberley mines set the scene for the gold mines of South Africa.[30]

Racial and ethnic tensions and segregation were common in mining communities across the globe, although nothing quite seems to symbolize the potential racism of miners as starkly as the rallying cry of the white miners of South Africa in the midst of their epic 1922 struggle: 'Workers of the World Unite and Fight for a White South Africa'.[31] White miners in the US not only waged wars on native Indians, they were also famously hostile to Blacks. And there were important rifts between 'whites', for Protestant English, Welsh and German miners found it hard to accept the Irish and Slav miners as their equals.[32] In Australia the mining communities were often bastions of racism,[33] and even between culturally strongly related nations, such as Spain and France, French miners in Southern France were deeply hostile to Spanish miners.[34]

However, there were also famous instances of cross-ethnic solidarity, as in the Ruhr strikes of 1905 and 1912, when Polish and German miners came together,[35] or in the 1926 General Strike in Britain, when conflicts between English and Irish and workers of different ethnic origins were often secondary,[36] or as in the US Knights of Labor's attempts to unite different racial and ethnic groups in one union.[37] And, of course, we also have multiple examples of global ethnic networks of miners which originated from one tiny place on the earth but kept their solidarity to that place everywhere on the globe. Welsh miners, for example, settled in Australia, Patagonia and North America, among other places. Wherever they went they took with them notions of a Welsh *gwerin*, a Welsh people to which they showed fierce loyalty and commitment.[38] Of course, the South Wales coalfield had seen massive in-migration in the nineteenth century, largely from England (with a sprinkling of Spanish miners), again pointing to the basic fact that booming mining industries, throughout the history of mining, had an insatiable appetite for labour.[39]

Ad Knotter's chapter in this volume reminds us how vital migrant labour was for mining worldwide. Ethnic differences in the labour force were often related to skill differences, but the ethnic segmentation of the labour force could at times also result in cross-ethnic and cross-skill solidarities. Knotter analyses patterns of recruitment of migrants, where a variety of different recruitment systems can be distinguished. These could include systems of coercion, including convict and indentured labour. Knotter also finds various diasporic workers who were often highly skilled and migrated all over the world in order to make use of their sought-after skills. And he discusses various forms of state-regulated migration to the coalfields of Western Europe between the end of the First World War and the 1970s. Finally, Knotter asks about the relationship of migrants to the miners who were already working in the coalfields. To what extent were they integrated into a pre-existing host culture and did they try to retain cultural ties to their regions of origin?

Mining and leisure

The workers who migrated into mining came to work and to earn good money. However, they also wanted to have fun, and the little leisure time they often had they spent in pursuit of things that gave them pleasure. Sport was one of those things. In 1900, the first game of football was played in Mexico by Cornish miners, who also formed Mexico's first football club.[40] Once again we can observe the transnational side of mining, where migratory processes encouraged an exchange of social and cultural practices – also in the field of leisure. There is, of course, a big overlap but also a significant difference between working-class leisure cultures, labour movement cultures and miners' leisure cultures. Whereas labour movement cultures tended to propagate a civilizing process of workers and often adopted the middle-class values and understandings of culture, working-class cultures emerged more from within the milieu of workers in their factories and neighbourhoods. Miners' cultures are a specific form of working-class culture.[41] The establishment of reading clubs, of bursaries to attend universities, of the promotion of cultural events mimicking middle-class cultural festivals – they all belong to a labour movement culture whose aim it was to 'educate' workers to better themselves and gain access to forms of 'high culture'. This labour movement culture, prominent in many mining communities, could produce tensions with a working-class culture that rejected middle-class and labour movement tutelage alike, avoided cultural paternalism and was rooted in a kind of 'Eigensinn' of working-class life.[42] Drinking, prostitution, gambling, the music hall, fun fairs, specific forms of (often male) camaraderie, and a host of sports, such as boxing and football (both in egg-shape and round form), were associated with a rougher side of working-class and miners' cultures.

Forms of popular culture depended on things such as housing that was available to miners, and influences of company paternalism or local or state governments. Thus, it has, for example, been demonstrated for the copper miners of El Teniente that attempts to rid the mining community of rampant prostitution backfired badly and produced considerable resistance from within the male mining community, while at the same time pitting miners' wives and employers against miners.[43]

Was the culture of miners then an alternative culture to the bourgeois culture that dominated elite culture, including labour movement culture, in the nineteenth and twentieth centuries? I think we would do well to treat ideas of 'counter culture' or 'alternative culture' with some caution; at best we can perhaps speak of a non-bourgeois cultural world, but it rarely was a conscious alternative to bourgeois culture. Furthermore, when we talk about mining culture, we should not forget that it cannot be restricted to modern industrial workers and their organizations. In Central Europe, for example, the estates-based culture of mining had deep roots and produced much 'high culture', e.g. golden cups/tankards or church windows.[44] This cannot be ignored when we talk about mining culture in the *longue durée*. From time immemorial we can in fact see a strong correlation between mining activities and cultural renaissances. The strong development of mining in Central Europe during the twelfth and thirteenth centuries was the basis for a wonderful wealth of religious art made largely for churches. The mining booms of the fifteenth century played a major role in laying the material foundations for the Renaissance art and culture emanating from Italy throughout Europe.[45]

Peter Alexander's chapter provides insights into the linkages between culture, 'classed identities' and trade union consciousness in a range of African, Asian, American and European countries. He demonstrates that forms of working-class culture contributed, in diverse circumstances, to the emergence of 'classed identities' which in turn sustained trade unionism – sometimes across periods of massive repression of these unions. Depending on the urban/rural context in which these working-class cultures developed, they could sustain more or less internationalist class orientations. Especially in colonial settings, in Asia and Africa, retaining links to the land and to an agricultural way of life could be an important resource for miners which had a deep impact on how they came to understand issues of class. Religious and ethnic divisions were vital in determining how cultural organizations were functioning and how they could or could not contribute to the emergence of a unifying class discourse bridging religious and ethnic particularities. In South Africa, for example, dance societies and mutual-aid societies were vital aspects of working-class culture, out of which forms of trade unionism could develop. The positioning of employers vis-à-vis forms of working-class culture was also vital in establishing the degree to which class identities were to emerge – where employers were antagonistic and oppressive, common class consciousness finding expression through cultural forms was more widespread. Interracialism, as Alexander shows, could work within trade unionism to protect wages and jobs but in broader working-class cultures it was not necessarily a guarantee that everyday racism was absent from relations between black and white miners.

The gender of mining

Gender was another aspect of mining culture that often, but not always undermined and divided class identities. Until 1933 Japanese women had worked underground alongside men in the mines – producing sometimes tension and sometimes solidarity. When the Japanese state introduced legislation that banned women from working underground, it was the last developed country that followed the bourgeois gender order, which had been established in the West during the nineteenth century.[46] The idea of 'separate spheres' for men and women, public for men and private for women, proved to be a powerful normative code structuring public and private lives to an extraordinary extent, although there always remained exceptions to the rule.

In mining, this idea meant that centuries-old practices of women working alongside men came to an end – at least in the global West.[47] Especially where work underground was involved, the heat meant that male and female workers alike wore very little clothing. In Japan they even shared coalpit baths. All of this deeply offended bourgeois morality.[48] In Europe, Britain set the pace in abolishing female labour underground with the Mines Act of 1842.[49] Many Western European countries also instigated legislation in the nineteenth century. Mining thus became an overwhelmingly male activity, identified one-sidedly (and often incorrectly) with male labour. The more the new bourgeois work codex found roots, the less employment opportunities mining offered for women. Some mining towns were overwhelmingly male, with hardly any women present, but most operated on a segregated labour market, where male miners were breadwinners, enjoying male forms of camaraderie at work and male forms of associationalism in their free time, while women looked after the homes and children and developed their own forms of sociability. Yet, as always, such strict differentiations tended to break down on close examination, especially if we look at mining in developing countries.[50] If we stick with the example of Japan, even after underground women's work was officially banned in 1933, as Donald Smith has shown, many women continued to work underground especially in rarely inspected smaller mines, where the percentage of women workers actually increased between 1932 and 1934. During the Second World War women were again sent underground to replace men who went off to fight the Emperor's war. As late as 1961, women were found to work underground in some Japanese coal mines. Moreover, women throughout the history of mining in Japan played an important role in labour conflicts, backing the demands of men for better pay and working conditions with effective and radical direct action, thereby performing a very public role and not restricting their activities to the private sphere.[51]

Research on mining communities in Europe and North America has recently emphasized how voluntary organizations, including friendly societies and temperance societies, offered opportunities for women in mining communities to become active outside the narrow confines of family life.[52] And women frequently played active and supportive roles during strikes, as well as leading consumer protests in mining communities.[53] The gender of mining has been a very fruitful form of exploration in other places as well. Thus, Carolyn Brown's study of the Red Enugu in the Nigerian Iva Valley has shown how metropolitan management practices constructed the African miner as feminine, childlike and immature, thereby justifying withholding from African workers the wages and rights enjoyed by metropolitan white workers. As a response, African workers reenforced their masculine identities and foregrounded them in their demands for a 'family wage'.[54]

Kuntala Lahiri-Dutt's chapter provides a masterly survey over mining as a gendered activity. She demonstrates that mining cannot be properly understood without gender. Investigating the rich gendered symbolism of mining, she explains how, despite the overall maleness of the industry, women were also economic agents, even if their work was often unpaid and rendered invisible. Lahiri-Dutt also discusses women as victims of mining whose dispossession meant loss of livelihood and home. She underlines how gender was strongly interrelated to issues of race and class. Finally, she also stresses the many instances where women took on an explicitly political role – in strikes and in labour movements.

Mining accidents

Women were also deeply affected by mining accidents, as they often meant injury or death of husbands and sons and the associated emotional turmoil and loss of income. Mining was a dangerous activity and accidents were a frequent occurrence. The largest recorded accident in history was the one at Benxihu Colliery in China in 1942, were 1,954 miners lost their lives. Part of the Japanese-controlled puppet state of Manchukuo at the time, the mine was operated by the Japanese who maltreated the Chinese miners, using them effectively as slave labourers. When a gas and coal dust explosion hit the mine on 26 April 1942, the Japanese, in trying to control the underground fires, closed the ventilation of the mines and sealed the pithead without as much as trying to evacuate any surviving Chinese miners trapped underground. A Soviet investigation of the incident at the end of the Second World War found that most miners had died of carbon monoxide poisoning due to this lack of ventilation. Hence the high death toll in this incident was directly related to the racist treatment of Chinese labourers by the Japanese occupants during the Second World War.[55]

The history of mining is littered with terrible mining accidents costing hundreds of lives in one single accident – even without racism having a hand in it. The Courrières mine disaster, Europe's worst mining accident, caused the death of 1,099 miners (including many children) in Northern France on 10 March 1906.[56] It was again caused by a coal dust explosion. This French mine disaster was not just notorious for the number of dead involved, but also for the huge national and international media attention it engendered. Accidents like these brought to the attention of nationalized publics the dangers faced by miners and made miners into national icons for whom appeals to social justice seemed more justified than to other social groups. Jean Jaurès, the French socialist leader, wrote after Courrières:

It is a call for social justice that comes to the nation's representatives from the depths of the burning mines. It is the harsh and suffering destiny of work that, once more, manifests itself to all. And would political action be something else than the sad game of ambitions and vanities if it didn't propose to itself the liberation of the workers, the organisation of a better life for those who toil?[57]

This appeal to social justice in the name of the nation has remained a lasting feature of mining accidents. Perhaps the last prominent example of this was the 2010 Chilean miners' rescue, when 33 miners were trapped underground for 68 days.[58] When they were hauled to safety, the Chilean president and the Roman Catholic Church were prominently represented in front of a watching global audience. It had become a national and international media event, exploited for political reasons and nationalist purposes alike. The Catholicism of the Chilean miners was also a pertinent reminder of the enduring strong link between mining and religion in many parts of the world.[59] The cult of St. Barbara in many Christian mining communities across the world is a very visible sign of this religiosity.[60] The special dangers connected to the job seem to have exacerbated the need for a belief system that transcends the banality of life in the here and now.

The Chinese mining accident under Japanese occupation also is a pertinent reminder of the importance of colonialism and imperialism to mining and vice versa. The early modern colonial expansion of Europe in the fifteenth and sixteenth centuries would have been unthinkable without mining.[61] Iron, steel, copper, lead and other metals were the material basis on which European superiority and power in the new world came to rest. In addition, the colonial expansion itself was soon to lead to a great boost in mining activities in the colonies itself, for example the beginnings of silver mining in Potosi, Bolivia.[62] Mexico became the mineral treasure house of early modern Spain, without which it arguably could not have become one of the foremost powers in Europe.[63] The link between power and mining can also be seen in war more generally. Of course, war often was extremely detrimental to mining activities. One only has to think about the effects of the Thirty Years' War in Europe which interrupted much Central European mining, often for a very long time.[64] However, mining was at the same time of crucial importance in warfare – iron ore, coal and other metals connected to mining were the material basis of war economies.[65] After all, the ruthlessness of the exploitative practices of the Japanese in China cannot only be attributed to racism but also to the need of the Japanese war economy for coal.[66]

Michael Farrenkopf's chapter in this volume reminds us of the difficulties of generating comprehensive and reliable comparative data for different countries on a global scale. He focuses on accidents caused by explosions, explaining the various strategies of protecting miners from this hazard. An improved scientific understanding of the cause of explosions reformed safety procedures, thereby significantly reducing the risk of explosion, especially in Europe and the Western world after 1945. Elsewhere the costs of introducing new safety technology often meant considerable delays or they prevented it from being adopted altogether – in some case until today.

Whilst Farrenkopf focuses on accidents and accident prevention, Dagmar Kift and Paul Stewart analyse what accounts of accidents in Witbank, South Africa and the Ruhr, Germany tell us about miners' experiences. Workers' safety representatives, awareness training and accident simulation exercises were all appointed and introduced in Germany much earlier than in South Africa, which had much to do with the racialized character of mining in South

Africa and the strength of mining trade unions in West Germany after 1945. Kift and Stewart stress that the workers themselves had rather matter-of-fact attitudes towards accidents, and that their voices were rarely heard in both cases when it came to accident prevention strategies.

The ownership of mines

Seeking a voice in their industry was a struggle that miners faced not just vis-à-vis accident protection. Over the course of the nineteenth and first half of the twentieth centuries, many miners came to believe that their interests were best served if the industry was nationalized. Private ownership came to be identified by many with exploitation. The nationalization of the coal mines in Britain in 1946 was an iconic moment in that respect – a moment that many British miners had already sought to achieve after the end of the First World War.[67] The nationalization of the coal mines in Britain was celebrated by unions and unionized miners as a new dawn and precisely the promise of the kind of greater social justice demanded by Jaurès and many other labour and miners' leaders across the world. Many now hoped for fairer wage levels, greater work protection and better working conditions.

Nationalization seemed to many the end point of a century-old history of exploitation of miners. The ancient Greek historian Diodorus Siculus reported that miners near Athens were slaves who toiled for 24 hours a day and were basically working themselves to death.[68] Forms of unfree labour continued in mining for a very long time to come, with one of the most notorious being practiced in early modern Peru, the so-called mita.[69] But even free labour often toiled under intolerable conditions. Hence, it is not surprising that the social status of miners was often low. From ancient Egypt to the absolutist monarchs of early modern Europe, the 'regalian system' meant that mines were by law seen as property of the monarch. It was rivaled, since Roman times, by the 'accession system', in which those who owned the land also owned the mines. Various forms of state and private ownership of the mines developed from these different systems. Seeking to protect themselves through association, petitions and combination, the miners formed craft guilds in the Middle Ages and associations, like the Knappschaften in Germany.[70] Later trade unions and the labour movement more generally, found much support among miners, and certainly their status and material position improved considerably.

But were the hopes invested in nationalization justified? After all, in Britain the National Coal Board oversaw the decline and ultimately demise of the post-war mining industry. Under the government of Margaret Thatcher, intent on waging a class war from above on the unions, the National Coal Board became the main adversary of the miners in the legendary 1984/5 miners' strike that all but ended coal mining in the UK.[71] The experience of miners in Eastern Europe under communism was also hardly evidence for the benefits of nationalization.[72] In parts of Western Europe, in particular Scandinavia, West Germany and Austria, models of social partnership and tripartism seemed to achieve better deals for mining industries and their employees than was the case under nationalization in Britain.[73] By contrast, nationalization per se had not guaranteed miners a powerful voice.

Chris Wrigley's chapter provides a strong comparative perspective on the impact of the state on industrial relations. Reviewing labour conflicts in mining in many parts of the world, he concludes that Britain had been somewhat of an exception in the nineteenth and twentieth centuries. The earlier and greater neutrality of the state and the greater power of mining unions allowed for a relatively early system of autonomous industrial bargaining. A constitutional and liberal state did not perceive the labour movement as an existential threat and instead co-opted both unionism and working-class parties into the structures of the state without much difficulty. Labour conflict in Britain, already in the nineteenth century, produced hardly any fatalities. Miners could feel properly represented in such a system, industrially and politically, which was clearly not the case in many other systems around the world.

Wrigley highlights the importance of the state for industrial relations and labour relations more generally. The state is also often a key actor when it comes to land rights that mining companies seek in order to establish the industry in a particular area. Pavithra Narayanan in her chapter looks at the struggle of indigenous populations against mining companies in India who seek to expropriate them. Indigenous land rights clash with economic interest of mining in many parts of the world, and in most cases, state power is with the economic interest of mining rather than with the indigenous population. Yet such collusion sparks resistance, in India and elsewhere, and the struggle against displacement and discrimination is an ongoing one with a very long history. In such struggles over land rights, visualizations often play an important role. The mining companies, often in collusion with the state, use powerful public relations departments to paint rosy pictures in film and photography about the future of those areas designated for mining. And the indigenous population counters this with artwork where the destruction of their natural habitat through mining is bemoaned and criticized.

Visual turn in mining history

Visualizing mining companies, resistance to mining companies and mining unions are just some aspects of a much wider visual turn in mining history that is promising important research results in years to come. The visual history of mining is incredibly rich and ranges across geological and geographical images, photography and film, but also painting. In many places of the world, miners themselves were active artists who painted the world around them. Naïve painting was particularly favoured as a style by miner painters across a variety of different cultures, including the Ruhr, Upper Silesia and Australia.[74] Our iconic date here is 1978, when the well-known Australian miner and artist Samuel Michael Byrne died. His naïve paintings, in particular of mining in Broken Hill, spoke to his own socialist inclinations and ambitions and sought to represent those vis-à-vis a wider world.[75] But it was not just painting that was generated by mining.

Photography and film are other areas of visual culture in which mining has played a considerable role. Bert Hogenkamp has pointed to the popularity of mining in film in Germany, France, Belgium and the USSR. Pit disasters, strikes and the family unit, including strong mother figures, were all popular themes of miners' films in the four countries investigated by Hogenkamp.[76] As Eric Margolis has pointed out, the meaning of mining photography can be ambiguous and it has certainly been used by employers and unions alike to make particular points about the industry that suited their respective aims and ambitions.[77]

Stefan Siemer's chapter reviews the potential of a visual history of coal focusing on the pictorial world of mining which is indeed impressive. He highlights that the pictures associated with mining were being generated by the reality of mining, while they in turn also contributed to generating that reality. A whole load of different experiences in and with the industry produced very different pictures, some nostalgic, some condemnatory, some accusatory, some matter-of-fact, some upbeat and triumphalist. Different pictures signaled different memory cultures, thus making coal emblematic for a variety of different things in different parts of the world. Siemer looks at pictures related to engineering, to child labour, to class images, to national images and to environmental disasters in order to analyse different levels of meaning produced through diverse forms of visualization.

Mining and the environment

Images to do with the environmental destruction caused by mining have led to a boom in histories focusing on mining's relationship to the environment. The awareness of mining having a detrimental effect on the environment has grown exponentially over recent decades. In 1992 the first Earth Summit in Rio de Janeiro blamed carbon dioxide emissions for destroying the ozone layer and contributing to global warming.[78] It put the replacement of fossil fuels firmly on the global political agenda.

Mining has been problematic for the environment for many reasons. Opencast mining destroyed the natural and built environment and left nasty scars in the landscape. Whole villages and towns were destroyed and their population had to be resettled. Erosion, deforestation, sink holes, the loss of biodiversity, contamination of soil and groundwater – they all were the nasty side effects of mining.[79] The industry reacted vigorously to such charges of environmental destruction, which grew in the context of the rise of the environmental protection movement from the 1970s onwards.[80] Under the guidance of the International Council on Mining and Metals, a conglomeration of nine leading international mining companies, it developed environmental codes of conduct and various forms of environmental monitoring.[81] It set social and environmental standards and began tackling major problems such as waste disposal. Nevertheless the examination of the environmental side effects on mining at local, national and international levels firmly belongs to the central concerns of mining historians and has in my view been one of the most vigorous, innovative and exciting sub-fields of mining history over recent years.

Tim LeCain's chapter investigates how environmental history has, over recent years, engaged with the history mining (and vice versa). He focuses in particular on the impact of new material theory on this sub-field of mining history, which allows the mining historian to go beyond forms of representation and investigations of cultural constructivism. Instead, it asks about how the material world, in this case mining, has constructed human beings. But arguably, according to LeCain, the new material history goes even further than this by breaking down the binary divisions between 'humans' and 'things'. Instead, everything becomes material flows and processes. Humans are part and parcel of the material world around them. Social power, for example, cannot be thought of separately from energy power. The history of minerals is intimately related to the history of the human body. Doing the environmental history of mining, Le Cain concludes, is pointing us to the materiality of mining in its different forms.

The heritage of mining

The materiality of the mining industry is also at the centre of many attempts to save the heritage of mining in places where it belongs to the past. In the biggest coalfield in Europe, the Ruhr, the last coal mine closed in 2018. In 2001, the biggest coal mine in interwar Europe, Zeche Zollverein in Essen, became a UNESCO world heritage site. Tens of thousands of miners worked here in the 1950s and 1960s. When the mine closed in 1986, the huge area that the mine covered lay dormant for years, protected by a preservation act of North-Rhine Westphalia. Eventually it was decided to move the most important regional history museum of the Ruhr, the Ruhr museum, into the coal washery of the former mine – a daring aesthetic and architectural feature, and a Foundation Zollverein developed the entire area into a touristic attractive place.[82]

Today, hundreds of thousands of tourists from all over the world visit the site. New companies have relocated to Zollverein, occupying buildings that once were mine buildings or building anew. Other museums, including a design museum, have also moved to the site, as has one of the most reputable German universities for design and architecture, the Folkwang university in Essen. Thousands of people have found a new occupation on the territory of the former mine, which can be seen, on the one hand, as an extremely successful example of structural social change. On the other hand, however, it has also been pointed out that the neighbourhood of Zeche Zollverein, the Essen suburb of Katernberg, has no part in this regeneration. The tourists don't go there, the people who now work at Zollverein, don't live there and the residents of Katernberg are not among the museum visitors at Zollverein either. Social problems in the neighbourhood, including unemployment, poverty, poor infrastructure, lack of educational opportunities, crime etc. abound. Katernberg and Zeche Zollverein, once symbiotically connected through the mine, have become parts of different worlds.[83] Nevertheless, during 2010, when Essen on behalf of the Ruhr was European capital of culture, the images of industrial heritage played their part in promoting the Ruhr throughout Europe.[84]

The importance of mining heritage is underlined by the mere fact that several mining history-related world heritage sites exist across the world.[85] Mining and its heritage have been crucial to forms of identity constructions in regions of heavy industry across the world. In Europe we now even have a European Route of Industrial Heritage promoted by the European Union as one way of promoting European consciousness.[86] Mining regions, such as the Ruhr, had an important national significance as the economic powerhouse of the German nation during much of the nineteenth and twentieth centuries. Hence, mining heritage is connected to various forms of spatial identity, but it is equally linked to forms of non-spatial identity formation, including questions of religion, ethnicity, race, class and gender. And of course, as ever, spatial and non-spatial identity constructions are interrelated in manifold and complex ways that need further exploration, especially in comparative and transnational dimensions.

But the question of heritage is not only connected to questions of identity. It also plays a role in economic regeneration and the branding of regions (often for touristic purposes). The preservation of mining heritage has arguably been more popular than the preservation of heritage to do with a great number of other industries. What is it about mining, one may be tempted to ask, that seemed to call for the preservation of its heritage and the representation of its everyday experience? There has clearly been something about the mythology of mining that captured the imagination of many artists and ordinary people alike and contributed to this popularity of mining heritage, which seems to be a universal phenomenon, although it is clearly more strongly developed in some regions than in others.[87]

My own chapter in this volume surveys the economic, political and cultural preconditions that have to exist for heritage initiatives in mining to thrive. It also asks about different narrative strategies that have emerged in mining heritage in order to connect the past with the present and the future. What impact does mining heritage have on contemporary communities and their identities? To what extent does it contribute to the structural transformation of former regions of mining? What actors can be located in heritage and what are their motivations? All of these questions have produced much valuable research in different parts of the world. Yet much of this research remains focused on single cases, and I will try in my chapter to provide some tentative comparative perspective on this literature.

Conclusion

We have seen over the last decades a welcome development towards ever more complex stories of mining and mining societies, many of which are rooted in local studies. This has greatly enriched our understanding of the complexity of mining societies and their historical development. But our respective geographical specialisms should not lead us to go into our respective corners and only talk to those with similar geographical specialisms. Instead we need a sustained reflection about what binds us together as mining historians across the globe and what was, has been and continues to be distinctive about mining and its history in different parts of the world. How can we integrate the many stories into an overarching story of the development and characteristics of mining societies throughout the ages without losing that precious complexity that has been the hallmark of mining history over the past decades? My plea is to avoid local parochialisms and use transnational comparative and cultural transfer history to shed light on both commonalities and differences of mining communities. From that comparative and global perspective, I think it is vitally important to reflect on what constitutes our subject and how we measure its importance. Of course, Kerr and Siegel's isolated mass hypothesis from the 1950s is not a fit framework for comparison today.[88] And the 'archetypal proletarian' thesis, put forward by Peter Laslett in 1974, has also not stood the test of time and was, of course, already comprehensively critiqued by Royden Harrison's 1978 book, *Independent Collier*.[89] The comparative concern traditionally was with strikes and forms of solidarity – understandable in the heydays of labour history in the 1960s and 1970s.[90] But since then, the concerns have shifted and I would argue that it is high time to broaden our understanding of mining communities and integrate labour struggles into a much larger picture of mining societies.[91] The themes I have chosen here – human progress, technological change, work processes, migration and ethnicity/race, religion, popular culture, the gender of mining, accidents and issues of mining safety, labour struggles vis-à-vis states and company owners, visualizations of mining, its environmental impact, and the question of the heritage of mining – are among the crucial themes, I would argue, that we need to explore as interlinked circles in order to move closer to a fuller understanding of mining societies and provide a more complex comparative framework for the study of those societies. The subsequent chapters are meant to explicate this in greater detail.

Notes

1 Klaus Tenfelde, 'Comparative Research on the History of Mining Workers: some Problems and Perspectives', in: Gustav Schmidt (ed.), *Bergbau in Grossbritannien und im Ruhrgebiet: Studien zur vergleichenden Geschichte des Bergbaus 1850–1930* (Bochum: Brockmeyer, 1985), pp. 18–35; see also Klaus Tenfelde (ed.), *Towards a Social History of Mining in the Nineteenth and Twentieth Centuries* (Munich: C. H. Beck, 1992).
2 Stefan Berger, Andy Croll and Norman LaPorte (eds), *Towards a Comparative History of Coalfield Societies* (Aldershot: Ashgate, 2005).
3 Joan Wallach Scott, 'Gender: a Useful Category of Historical Analysis', in: idem, *Gender and the Politics of History*, new edn (New York: Columbia University Press, 2018).
4 Peter B. Evans, Dietrich Rueschemeyer and Theda Skocpol (eds), *Bringing the State Back in* (Cambridge: Cambridge University Press, 1985).
5 Peter Burke, *Eyewitnessing: The Uses of Images as Historical Evidence* (London: Reaktion Books, 2001).
6 R. J. Mason, 'Pre-Historic Mining in South Africa and Iron-Age Copper Mines in the Dwarsberg, Transvaal', *Journal of the South African Institute of Mining and Metallurgy* 82 (1982), pp. 135–142.
7 Fred. G. Bell and Laurance J. Donnelly, *Mining and its Impact on the Environment* (London: Taylor & Francis, 2006), p. 1.
8 Rosemarie and Dietrich Klemm, *Gold and Gold Mining in Ancient Egypt and Nubia: Geoarcheology of the Ancient Gold Mining Sites in the Egyptian and Sudanese Eastern Deserts* (Berlin: Springer, 2013).

9   Paul Douglas Lockhart, *Sweden in the Seventeenth Century* (Basingstoke: Palgrave Macmillan, 2004), p. 19.

10 R. Shepherd, *Ancient Mining* (London: Institute of Mining and Metallurgy, 1993), p. 225.

11 Augustin F. C. Holl, 'Metals and Precolonial African Society', in: Michael S. Bisson, S. Terry Childs, Philip de Barros and Augustin F. C. Holl (eds), *Ancient African Metallurgy: The Socio-Cultural Context* (Lanham, NY: Rowman & Littlefield, 2000), chapter 1.

12 Justin Y. Lin, 'The Needham Puzzle, Weber Question and China's Miracle: Long-Term Performance since the Sung Dynasty', *China Economic Journal 1*(1) (2008), pp. 63–95.

13 Laurence Bergreen, *Marco Polo: From Venice to Xanadu* (New York: Alfred A. Knopf, 2007), p. 156.

14 W. Rees, *Industry Before the Industrial Revolution*, 2 vols. (Cardiff: University of Wales Press, 1968).

15 Peter Lane, *The Industrial Revolution: The Birth of the Modern Age* (New York: Barne & Noble, 1978), p. 209.

16 Peter Mathias, *The First Industrial Nation: The Economic History of Britain, 1700–1914* (London: Routledge, 1969).

17 John Robert Travis, *Coal in Roman Britain* (London: John and Erica Hedges, 2008), p. 103.

18 Josef Henrych, *The Dynamics of Explosion and its Use* (Leiden: Elsevier, 1979), p. 30.

19 Michael Coulson, *The History of Mining: The Events, Technology and People Involved in the Industry that Forged the Modern World* (Petersfield: Harriman House, 2012).

20 *The Mining Magazine* provided a full English translation in 1912.

21 Friedrich Naumann, *Georgius Agricola: Berggelehrter, Naturforscher, Humanist* (Chemnitz: e-sights, 2015).

22 Christoph Bartels, 'The Administration of Mining in Late Medieval and Early Modern Europe', in: Nanny Kim and Keiko Nagase-Reimer (eds), *Mining, Monies and Culture in Early Modern Societies: East Asian and Global Perspectives* (Leiden: Brill, 2013), pp. 115–132.

23 On the science of mining compare Jakob Vogel, *Ein schillerndes Kristall. Eine Wissensgeschichte des Salzes zwischen früher Neuzeit und Moderne* (Cologne: Böhlau, 2008).

24 Michael Fessner and Christoph Bartels, 'Die Anfänge der Montanwissenschaften bis zur Gründung der Bergakademien' in: Christoph Bartels and Rainer Slotta (eds), *Geschichte des deutschen Bergbaus, vol. 1: Der alteuropäische Bergbau von den Anfängen bis zur Mitte des 18. Jahrhunderts* (Münster: Aschendorff, 2012), pp. 551–554.

25 Frank James, *The Davy Lamp: Inventing the Miners' Safety Lamp* (London: Unicorn Publishing, 2018).

26 Bill Jones and Chris Williams (eds), *With Dust Still in His Throat. A.B.L. Coombes Anthology* (Cardiff: University of Wales Press 1999), p. 22.

27 Angela Suzanne Hawk, *Madness, Mining and Migration in the US and the Pacific, 1848 to 1900* (Berkeley, CA: University of California Press, 2011).

28 Douglas Fetherling, *The Gold Crusades: A Social History of Gold Rushes 1849–1929*, rev. edn (Toronto: University of Toronto Press, 1997).

29 Alan Mabin, 'The Underdevelopment of the Western Cape, 1850–1900', in: Wilmot G. James and Mary Simons (eds), *The Angry Divide: Social and Economic History of the Western Cape* (Cape Town: David Philipp, 1989), pp. 82–94.

30 Robert Vicat Turrell, *Capital and Labour on the Kimberley Diamond Fields, 1870–1890* (Cambridge: Cambridge University Press, 1987).

31 Jeremy Krikler, *White Rising: The 1922 Insurrection and Racial Killing in South Africa* (Manchester: Manchester University Press, 2005).

32 Mildred Allen Beik, *The Miners of Windber: The Struggles of New Immigrants for Unionization, 1890–1930* (University Park: Pennsylvania State University Press, 1996).

33 David Carment, 'Mining, Race and Politics of the North Australian Frontier', in: Tenfelde (ed.), *Sozialgeschichte des Bergbaus* (Munich: C. H. Beck, 1992), pp. 242–253. D. Hollinsworth, *Race and Racism in Australia* (New Delhi: Social Science Press, 1998).

34 Donald Reid, *The Miners of Decazeville* (Cambridge, MA: Harvard University Press, 1985).

35 Brian McCook, *The Borders of Integration: Polish Migrants in Germany and the United States, 1870–1924* (Athens, OH: Ohio University Press, 2011).

36 Charles Ferrall and Douglas McNeil, *Writing the 1926 General Strike: Literature, Culture, Politic* (Cambridge: Cambridge University Press, 2015).

37 Carol A. Horton, *Race and the Making of American Liberalism* (Oxford: Oxford University Press, 2005), chapter 3: 'Race and the Emancipation of Labour', pp. 61–94.

38 Bill Jones, 'Welsh Identities in Ballarat, Australia During the Late Ninenteenth Century', *Welsh History Review 20* (2001), pp. 283–307. Ronald L. Lewis, *Welsh Americans: A History of Assimilation in the Coalfields* (Chapel Hill, NC: University of North Carolina Press, 2008).

39 Philip Nicholas Jones, *Mines, Migrants and Residents in the South Wales Steamcoal Valleys: The Ogmore and Garw Valleys in 1881* (Hull: University of Hull Press, 1987).

40 Peter Grego, *Cornwall's Strangest Tales* (London: Portico, 2013), 'A Cornish Wave in Mexico'.

41 Mary Murphy, *Mining Cultures: Men, Women and Leisure in Butte, 1914–1941* (Urbana, IL: University of Illinois Press, 1997).

42 For the concept of 'Eigensinn', see: Alf Lüdtke, 'Polymorphous Synchrony: German Industrial Workers and the Politics of Everyday Life', *International Review of Social History*, supp. 38 (1993), pp. 39–84.

43 Thomas Miller Klubbock, *Contested Communities: Class, Gender and Politics in Chile's El Teniente Copper Mine, 1904–1951* (Durham, MD: Duke University Press, 1998).

44 Angelika Westermann, 'Bergstadt und Montankultur, 1350–1850', in: Wolfhard Weber (ed.), *Geschichte des Deutschen Bergbaus, vol. 2: Salze, Erze und Kohlen. Der Aufbruch in die Moderne im 18. und frühen 19. Jahrhundert* (Münster: Aschendorff, 2015), pp. 409–560.

45 Rainer Slotta and Christoph Bartels (eds), *Meisterwerke bergbaulicher Kunst vom 13. Bis zum 19. Jahrhundert* (Bochum: Deutsches Bergbaumuseum, 1990).

46 Karen Hagemann, Sonya Michel and Gunilla Budde (eds), *Civil Society and Gender Justice: Historical and Comparative Perspectives* (Oxford: Berghahn Books, 2008).

47 Laura Mercier and J. Gier-Viskavotoff (eds), *Mining Women: Gender in the Development of a Global Industry, 1670–2005* (Basingstoke: Palgrave Macmillan, 2016).

48 W. Donald Burton, *Coal-Mining Women in Japan: Heavy Burdens* (London: Routledge, 2014).

49 Mike Sanders, 'Introduction', in: idem (ed.), *Women and Radicalism in the Nineteenth Century, vol. 1: Specific Controversies* (London: Routledge, 2001), p. 28.

50 Kuntala Lahiri Dutt and Martha Macintyre (eds), *Women Miners in Developing Countries: Pit Women and Others* (Aldershot: Ashgate, 2006).

51 W. Donald Smith, 'Gender and Ethnicity in Japan's Chikuho Coalfield', in: Stefan Berger, Andy Croll and Norman LaPorte (eds), *Towards a Comparative History of Coalfield Societies* (Aldershot: Ashgate, 2005), pp. 204–218.

52 Anita Ernst Watson, Jean E. Ford, and Linda White, ' "The Advantage of Ladies' Society": The Public Sphere of Women on the Comstock, in: Ronald M. James and C. Elizabeth Raymond (eds), *Comstock Women: The Making of a Mining Community* (Reno, NV: University of Nevada Press, 1998), pp. 179–202, esp. p. 192.

53 For Britain see Roy Church and Quentin Outram, *Strikes and Solidarity: Coalfield Conflict in Britain, 1889–1966* (Cambridge: Cambridge University Press, 1998), p. 262.
54 Carolyn A. Brown, *'We Were All Slaves': African Miners, Culture, and Resistance at the Enugu Government Colliery* (Santa Barbara, CA: Greenwood Publishing, 2003).
55 See the entry in the online version of Encyclopedia Britannica, www.britannica.com/event/Honkeiko-colliery-mining-disaster (accessed 6 August 2018).
56 Michael Farrenkopf and Peter Friedemann (eds), *Die Grubenkatastrophe von Courrières: Aspekte transnationaler Geschichte* (Bochum: ISB, 2008).
57 Quoted in 'Il y a cent ans, la catastrophe minière du 10 mars 1906', *L'Humanité* (10 March 2006).
58 Marcia Armidon Lusted, *The Chilean Miners' Rescue* (North Mankato: ABDO, 2012).
59 June Nash, 'Religious Rituals of Resistance and Class Consciousness in Bolivian Tin-Mining Communities', in: Christian Smith (ed.), *Disruptive Religion: the Force of Faith in Social-Movement Activism* (London: Routledge, 1996), pp. 87 ff.
60 Gerhard Heilfurth, *Der Bergbau und seine Kultur. Eine Welt zwischen Licht und Dunkel* (Stolberg: Atlantis, 1981).
61 Raymond E. Dumette (ed.), *Mining Tycoons in the Age of Empire, 1870–1945: Entrepreneurship, High Finance and Territorial Expansion* (Aldershot: Ashgate, 2009).
62 Enrique Tandeter, *Coercion and Market: Silver Mining in Colonial Potosi, 1692–1826* (Albuquerque: University of New Mexico Press, 1993).
63 James Mahoney, *Colonialism and Postcolonial Development: Spanish America in Comparative Perspective* (Cambridge: Cambridge University Press, 2010), p. 143.
64 Dirk Hoerder, *Cultures in Contact: World Migrations in the Second Millenium* (Durham, MD: Duke University Press, 2002), p. 86.
65 Michael Renner, 'Violent Mining Conflicts and Diamond Wars', in: Gavin M. Hilson (ed.), *The Socio-Economic Impacts of Artisanal and Small-Scale Mining in Developing Countries* (Lisse: Swets & Zeitlinger, 2003), pp. 75–95, esp. p. 76.
66 Erich Pauer (ed.), *Japan's War Economy* (London: Routledge, 1998).
67 Barry Supple, *The History of the British Coal Industry, vol. 4: 1913–1946: The Political Economy of Decline* (Oxford: Oxford University Press, 1988); W. A. Ashworth and Mark Pegg, *The History of the British Coal Industry, vol. 5: The Nationalized Industry* (Oxford: Oxford University Press, 1986).
68 Robert Shepherd, *Ancient Mining* (Michigan, IL: Institution of Mining and Metallurgy, 1993).
69 Jeffrey A. Cole, *The Potosi Mita, 1573–1700: Compulsory Indian Labor in the Andes* (Stanford, CA: Stanford University Press, 1985).
70 Martin H. Geyer, *Die Reichsknappschaft. Versicherungsreformen und Sozialpolitik im Bergbau, 1900–1945* (Munich: C. H. Beck, 1987).
71 Jim Philipps, *Collieries, Communities and the Miners' Strike in Scotland, 1984/5* (Manchester: Manchester University Press, 2012).
72 David Kideckel, 'The Unmaking of the East-Central European Working Class', in: C. M. Hann, *Postsocialism: Ideals, Ideologies and Practices in Eurasia* (London: Routledge, 2002), pp. 114–132.
73 Franz-Josef Brüggemeier, *Grubengold. Das Zeitalter der Kohle von 1750 bis heute* (Munich: C. H. Beck, 2018), pp. 326–348.
74 Dagmar Kift, 'Kultur, Kulturpolitik und Kulturgeschichte im Ruhrbergbau nach dem Zweiten Weltkrieg', in: Dagmar Kift, Eckhard Schinkel, Stefan Berger and Hanneliese Palm (eds), *Bergbaukulturen in interdisziplinärer Perspektive: Diskurse und Imaginationen* (Essen: Klartext, 2018), pp. 29–42.
75 Ross Moore, *Sam Byrne: Folk Painter of the Silver City* (New York: Viking, 1985).
76 Bert Hogenkamp, 'A Mining Film Without a Disaster is Like a Western Without a Shoot-Out: Representations of Coal Mining Communities in Feature Films', in: Berger, Croll and Laporte (eds), *Coalfield Societies*, pp. 86–98.

77 Eric Margolis, 'Mining Photographs: Unearthing the Meanings of Historical Photos', *Radical History Review 40* (1988), pp. 33–49.
78 Ranee K. L. Panjabi, *The Earth Summit at Rio: Politics, Economics and the Environment* (Lebanon, NH: Northeastern University Press, 1997).
79 Timothy J. LeCain, *Mass Destruction: The Men and Giant Mines that Wired America and Scarred the Planet* (New Brunswick, NJ: Rutgers University Press, 2009).
80 Drew Hutton and Libby Connors, *A History of the Australian Environment Movement* (Cambridge: Cambridge University Press, 1999), pp. 247–250. John R. McNeil and George Vritis (eds), *Mining North America: An Environmental History Since 1522* (Berkeley, CA: University of California Press, 2017).
81 International Council on Metals and the Environment (ed.), *Case Studies Illustrating Environmental Practices in Mining and Metallurgical Processes* (New York: United Nations Publications, 1996).
82 On the Foundation Zollverein see www.zollverein.de/stiftung-zollverein/ (accessed 7 August 2018); on the Ruhr museum see the catalogue of the permanent exhibition: Ulrich Borsdorf and Heinrich Theodor Grütter (eds), *Ruhr Museum – Natur. Kultur. Geschichte* (Essen: Klartext, 2001).
83 Michael Krummacher, Roderich Kulbach, Viktoria Waltz, and Norbert Wohlfahrt, *Soziale Stadt – Sozialraumentwicklung – Quartiersmanagement. Herausforderung für Politik, Raumplanung und soziale Arbeit* (Opladen: Leske & Budrich, 2003), p. 230.
84 Jürgen Mittag, '"Kulturhauptstadt Europas": eine Idee, viele Ziele, begrenzter Dialog – das Programm "Kulturhauptstadt Europas" und die Kulturhauptstädte des Jahre 2010 in diachroner und synchroner Perspektive', in: Thomas Ernst and Dieter Heimböckel (eds), *Verortungen der Interkulturalität: die europäischen Kulturhauptstädte Luxemburg und die Grossregion (2007), das Ruhrgebiet (2010) und Istanbul (2010)* (Bielefeld: transcript, 2012), pp. 59–94.
85 Helmuth Albrecht and Friederike Hansell (eds), *Industrial and Mining Landscapes within World Heritage Context* (Freiberg: Zweckverband Sächsisches Industriemuseum, 2014).
86 www.erih.net/ (accessed 7 August 2018).
87 Christian Wicke, Stefan Berger and Jana Golombek (eds), *Industrial Heritage and Regional Identities* (London: Routledge, 2017).
88 See already P. K. Edwards, 'A Critique of the Kerr-Siegel Hypothesis of Strikes and the Isolated Mass: A Study of the Falsification of Sociological Knowledge', *The Sociological Review 25*(3) (1977), pp. 551–574.
89 Royden Harrison, *Independent Collier: The Coal Miner as Archetypal Proletarian* (New York: St. Martin's, 1978).
90 See for example: L. H. Haimson and Charles Tilly (eds), *Strikes, Wars and Revolutions in an International Perspective: Strike Waves in the Late Nineteenth and Early Twentieth Centuries* (Cambridge: Cambridge University Press, 1989).
91 See, for example, Nimura Kazuo, *The Ashio Riot of 1907: A Social History of Mining in Japan* (Durham, MD: Duke University Press, 1997).

2 Archaeology of mining in the pre-industrial age

The recognition and interpretation of ancient mines

Simon Timberlake

Introduction

What do we mean by ancient mines? In the modern period this term has been used quite subjectively to describe any 'primitive looking' mine site for which there is no historic record. However, in the older literature we find the term 'ancient' substituted by other pre-modern identities common to local mythology and tradition, particularly when these are associated with former metal-using occupying cultures familiar in name, but beyond living memory, such as the 'Romans', 'Phoenicians', 'Danes', 'Saxons', 'Arabs', 'Jews' or even the 'Portuguese'.[1]

This account of the archaeology of mining is a personal view based on 25 years of investigation looking at some of the very earliest metal mining sites, therefore in terms of period it will mostly be looking at examples of prehistoric (Palaeolithic–Iron Age) mining, but will also cover, for example the Roman and Medieval periods, yet still including examples of pre-colonial mining in Africa and pre-Columbian mining in South and North America. It will look at some of the issues raised by the study of these sites, the archaeological controversies surrounding them, and also provide some ideas on a didactic approach to the archaeological study of mines and ancient mining landscapes. The main emphasis will be on metals, yet the range of this discussion will include early mining for pigments, salt, coal flint and stone. Furthermore, I believe that it is helpful sometimes to be able to transgress these strict chronological and subject boundaries, given that we might better understand some of the very oldest mining technologies and approaches to extraction by first familiarising ourselves with the earliest documented treatises on mining such as Georgius Agricola's *De Rerum Metallica*,[2] and perhaps more surprisingly, through an examination of contemporary 'primitive mining' practices being undertaken today within the unofficial or artisanal mining sector. For example, the extraction of gold, tin and semi-precious/precious stones by landless squatter miners throughout the developing world.

The world of mining today is indeed a world of contrasts, the depth of poverty with which some of it is associated being captured in the lens of a camera. We see this in the accompanying Sebastian Delgado (Magnum) photograph taken some 20 years ago showing the urban-poor working serf-like as gold miners at Serra Pelada in the Venezuelan Amazon (Figure 2.1a), and just as poignantly, the photograph of a child miner taken underground in a mine in Colombia (Figure 2.1b). This type of impoverished mining I have witnessed myself on the Cerro Rico, Potosi, Bolivia when I visited the mines back in 1985, an experience which has since inspired me to attempt to compare both primitive modern and recent ethnographic examples of mining and smelting with those from the ancient past. In fact, when conducting fieldwork in landscapes without any documented histories it is not always easy to tell these sorts of (old and new) mining apart.

Indeed, witnessing such scenes around the world today, one is sometimes tempted to refer to this as the beginning of an era of post-industrial mining – the point at which the future begins to head towards the distant past.

The archaeological study of ancient mines, excepting fortuitous discovery, begins with an analysis of the landscape; understanding the geology of it like a modern mineral prospector, the known distribution of archaeology, and also the tell-tale signs of surface disturbance.

Figure 2.1a Gold mining at Serra Pelada, Venezuela (copyright Sebastian Salgado/Magnum)

Figure 2.1b 20th-century child miner (permission DBM Bochum)

Recognising ancient mines: identifying mines within landscapes devoid of modern mining

We are now much more aware of the fact that many of the key sites of prehistoric mining for rich easily smeltable ores (or other quite different types of mineral) are likely to be found in those same places where large amounts of generally lower grade ore were mined in the historic period. As a result, there is often a presumption that later period(s) of mining will have removed most, if not all of the evidence for the earliest activity. Fortunately, this is not always the case. Some of these early-mined areas focused on locally restricted deposits of rich surface-oxidised ore. The absence of other ore minerals meant that they were of little interest later on. There are still other examples where the whole deposit was just too small to have been bothered with subsequently. Finally there are the materials which were never prized to the same degree in post-prehistoric times, such as flint and some types of stone used for the making of tools. The result is that there remain areas just outside of the modern economic extraction zones which have been left more or less untouched – and it is to these that any modern prospection for ancient mines should be directed.

Modern prospection for ancient mines should involve the use of high-quality vertical and oblique air photographs taken at different times of the year under different grass and crop growth conditions, and also at different times of the day under different light conditions for shadow and contrast. A good example of the sort of patterning one might be looking for to detect the presence of shallow infilled shafts dug to a stone bed or mineralised horizon can be seen in some of these examples of Neolithic flint mines, for example the Late Neolithic mines at Grime's Graves near Thetford in Norfolk, UK.[3] Another very useful tool which can be used to prospect large areas of landscape from the air is LIDAR (Light Detection and Ranging) – a remote sensing technology which illuminates the surface topography using a laser, and then analyses the reflected light. This technique helps to detect and model slight changes in topography due to earthwork construction or disturbance such as might be

Figure 2.2 LIDAR image of the prehistoric workings on the Hauptgang, Mitterberg
(T. Stollner, permission DBM Bochum)

caused by mining. The lineation of buried former mineral veins, associated shafts, spoil and leat channels can often be picked out, even within partially forested terrains, and we see good recent examples of the use of this within the Alps. On the Hauptgang Vein, Mitterberg (Figure 2.2) it has been possible to distinguish the characteristic workings as well as the heavily eroded processing areas associated with this Bronze Age mine.[4]

The evidence on the ground

Many of these ancient mines will only be found as a result of prospection on the ground, in particular through fieldwalking carried out within promising areas of surface mineralisation. Occasionally this can be difficult, such as was discovered during survey work carried out to locate mine entrances and earlier opencast working above the Early Bronze Age tin mine at Kestel near Camardi in the Taurus Mountains of Southern Anatolia.[5] Arid landscapes without covering vegetation normally make such field investigations easier, yet the very eroded rock sides and deep scree infill across these opencasts made them difficult to recognise as man-made features (Figure 2.3a). Only the evidence of rock-working close up, some traces of mineralisation, a few buried tools downslope and pounding hollows in the bedrock (Figure 2.3b) provided

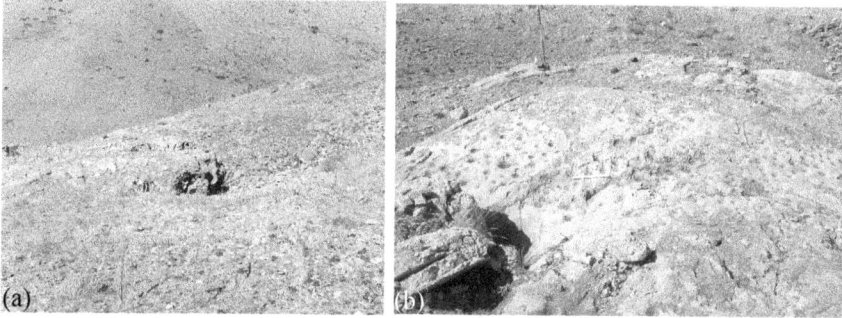

Figure 2.3 (a) Bronze Age opencast at Kestel Tin Mine, Anatolia; (b) ore pounding hollows within bedrock (photos by S. Timberlake)

the clues we were looking for. Within afforested and peat-covered landscapes such as we find in the Alps and in the UK (for example within mid-Wales, the Pennines and Scotland) it is sometimes necessary to look for the buried surfaces of ancient mine tips appearing from underneath eroding peat hags and within the sides of stream sections. It will also be necessary sometimes to resort to the use of soil augers for recovering samples rich in charcoal and finely crushed mineralised rock – the latter a likely indicator of prehistoric or early mining. A considerable thickness of peat cover over these tips and workings is also likely to be a good indicator of age.

Alluvial mining such as tin-streaming and gold washing involves the sifting through and sluicing of hundreds of tons of river gravel and sediments dug up from valley floors of both tin- or gold-bearing streams lying within the watershed below the eroding primary source of these placer minerals (i.e. native gold and cassiterite (tin oxide)). This appears to be a very ancient practice, the first alluvial gold probably having been extracted 8000 years ago, with the practice of tin-streaming following some 3000–4000 years later with the arrival of the European tin-bronze age. Given that tin and gold grounds are commonly reworked, it may be difficult now to find undisturbed prehistoric alluvial mining sites, the effects of periodic flooding also having removed many of the traces of early extraction. However, the techniques adopted and patterns of waste created by this activity have changed very little over time; these forming quite distinctive features within the areas of working. The parallel waste heaps dumped at right angles to the stream, such as we find on Dartmoor and some of the Cornish granite moorlands (Figure 2.4), reflect the use of a wooden sluice box (*tye*) and also a constant flow of water that is required to separate out the heavier tin from the lighter quartz and clay minerals.[6] [7] The latter waste material is usually dumped to one side as the tin-bearing gravel is worked upstream.

Yet another process called hushing was used throughout the upland metallogenic areas of central northwest Europe from Roman (or even pre-Roman times)

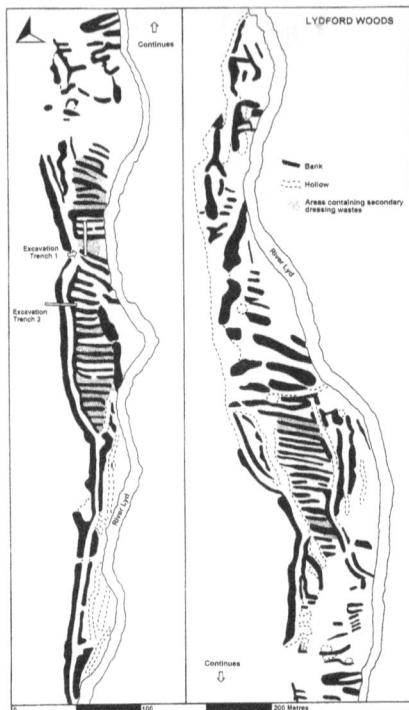

Figure 2.4 Map of waste heaps from Medieval tin-streaming, Lydford Wood, Devon (copyright P. Newman and RCAHM (English Heritage)

for the purposes of prospecting valley sides for minerals (tin, gold, copper, lead and iron ores). Today the best examples of this are to be found in northwest Spain, the Pyrenees and Britain (in particular Wales, the northern Pennines and southern Scotland). This same technique was also developed by the Romans on a much larger scale to extract gold, and very occasionally other heavy minerals such as lead, through repeated quarrying and undercutting of mineral vein outcrops on mountain slopes, then hydraulicing these with water torrents released from leat-fed earthwork tanks located on the scarp slopes above. This is a process graphically described by Pliny the Elder in AD 79 in his *Natural History*[8] as *ruinas montana*; the general process otherwise known as *arrugia* (gold washing). A good example of Roman or Medieval and later postmedieval prospection hushing can be seen on the slopes of Copa Hill, Cwmystwyth in mid-Wales (Figure 2.5). This is a hillside which preserves the record of some 3500 years of mining history.[9] Distinguishing in the field between natural gullying and erosion and artificially induced gullying and erosion is a skill that only comes with gradual familiarity and understanding of these sites, but once recognised, this can be the key to interpreting many mountainous mining landscapes.

Figure 2.5 Roman/Medieval and postmedieval hushing on Copa Hill, Cwmystwyth,
Wales, UK (photo by S. Timberlake)

Geological maps, geophysical and geochemical surveys

One of the first steps of an archaeological investigation following fieldwork and
the identification of a site is the undertaking of a geophysical survey. These are
carried out to try and determine the presence of sub-surface workings or voids, the
identification of buried stone or wooden structures associated with ore processing,
the presence of spoil sediment layers, and where they occur within the vicin-
ity of the mine itself, ore roasting or smelting hearths. Ground penetrating radar
(GPR) has been used successfully for the first of these tasks, while magnetom-
eter and resistivity surveys may reveal some trace of the layout of the working
areas. Magnetic susceptibility is another technique which has proved successful in
locating ancient smelting hearths, such as those confirmed by the author at Banc
Tynddol, Cwmystwyth in mid-Wales.[10]

Soil geochemical survey is now a commonplace technique used to help identify
concealed mineral vein outcrops, alongside the buried mining, processing (washing) and smelting remains surrounding some of these ancient mines. Using portable
X-Ray Fluorescence equipment such as the lightweight Innov-X alpha-6500
PXRF or Niton tools it is now possible to undertake rapid grid-based sampling of
large areas of ground. Take for instance the survey of the Great Orme's Head near
Llandudno in North Wales which we carried out in 1996.[11] The central anomaly
of >0.1% copper surrounds the Pyllau Valley Bronze Age Mine, yet the four other
smaller anomalies along the north coast of this headland don't match any of the
known mineral outcrops (Figure 2.6). These are the sites of natural springs to

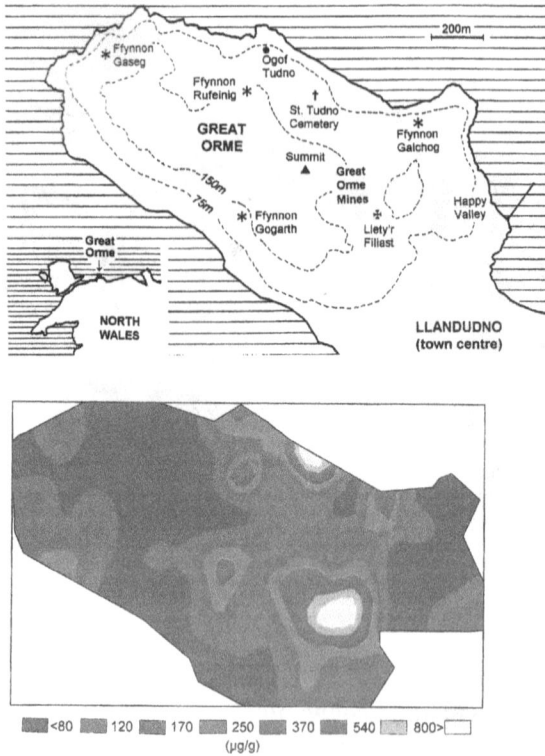

Figure 2.6 PXRF geochemical soils anomaly map of Great Orme's Head,
Landudno, UK (photo by D. Jenkins and S. Timberlake)

which many hundreds of tons of mine waste were brought at some point during
the distant past in order to wash, separate and concentrate the mined copper ores.

Understanding the ores, geology and structure of the host rock, and the formation and character of the mineral deposit, is in fact the next logical step to
understanding the early exploitation of a site. Because of this I think it essential
for mining archaeologists to work closely with geologists.

Some characteristic features of prehistoric mines

As soon as we put a spade into the ground, we begin to look for different bits of
evidence; trying to identify what is sealing what, the sequence of burial and the
tell-tale signs of later investigation or reworking.

Quite often we find the entrances to these ancient mines completely infilled by
slowly accumulated sediment, whilst the land surface above this has been buried by
later mining, or sometimes by agricultural or domestic structures. A good example
of this was the prehistoric house with its bread oven that we found at Kestel in
the Taurus Mountains, the one-time occupants of this dwelling presumably quite

unaware of the mineworking that lay beneath them. In another part of the same mine we found one of the passages reused as a mortuary chamber for burials. These Bronze Age people were interred here perhaps only 100 years after the mine's abandonment. Meanwhile, on Copa Hill in mid-Wales, we excavated into the top of the infilled Bronze Age opencast only to find the remains of a water tank and dam linked to later Medieval mining and hushing.[12][13]

It is not just the archaeological evidence that can be used to imply antiquity to a site. Another interesting exercise we undertook when investigating early mine sites was the study of place names, or rather the use of arcane Welsh words as descriptions for some of the oldest sites that we examined in Wales.[14] It seemed to us that some of these sites had acquired names a very long time ago, probably centuries after they were abandoned in the Bronze Age. Several of these mines in fact were referred to as caves, such as Ogof Llanymynech and Ogof Wyddon (Cave of the Wizard), while others whose identities were known seem to have acquired their mining names in the distant past, hence the use of the clearly old-fashioned descriptive term in Welsh *twll y mwyn* (mine pit).

Of course, some prehistoric mines are caves, or rather they are caves which were once mined, or mines which were once accessed through karstic (cave) passages. As a consequence these workings were free of water, enabling miners to reach greater depths and exploit a variety of different mineral veins. A good example of such was the Great Orme Bronze Age mine in North Wales. Within some areas of this mine we find backfilled prehistoric mine waste sealed by carbonate flowstone or speleothem.[15] The mine of Grotta della Monica in Calabria, Italy is another example of a mineworking in karst, but this time as a mined cave deposit dating from the Neolithic-Bronze Age. One part of the cave was used for human burial, but then this same area was also mined intermittently on several occasions throughout the Chalcolithic and Early Bronze Age for goethite (iron hydroxide) and malachite (copper carbonate) – both minerals being exploited for use as pigment.[16]

Firesetting

Firesetting is a characteristic technology associated with prehistoric mining. However, this technique is not in any way exclusive to prehistoric mining, being used in a calculated and semi-scientific way in just a few European mines (such as Rammelsberg (Harz Mountains) in Germany and Kongsberg in Norway) up to the end of the nineteenth century.[17] Much depended upon the geology of the host rock, its strength or brittleness, the degree of weathering, the presence of open joints or fractures, the type of deposit and the whereabouts of the mineral sought.

Most likely the majority of prehistoric fireset workings were just small mines or trials, like the abandoned adit on Mt Gabriel, County Cork in Ireland[18] which is recognisable as fireset on account of its characteristic concave arch-shaped profile formed as a result of firing across the direction of cleavage developed within this fine-grained old red sandstone rock (Figure 2.7). The prehistoric miners who did this were clearly skilled in the technique. They could mine along strike or down-dip by following small fractures and faults (slickensides), and where necessary cutting across these beds, chasing quartz veins or other mineralised structures. However, there is little

Figure 2.7 Fireset gallery, Bronze Age primitive mine, Mt Gabriel, Ireland (photo by S. Timberlake)

Figure 2.8 Elevation drawing of fireset chambers, Mahden Mine, Khetri and Zawar Mala, Rajasthan, India (by permission of B. Craddock)

evidence to suggest that they used water to quench the rocks, despite the fact that it would have helped; its use only contemplated perhaps where a supply of this became readily available. True perfection in prehistoric firesetting technology can be seen in the finely sketched scale elevation panoramas of fireset chambers and tunnels at the Mahden Mine, Khetri and Zawar Mala Mine in Rajasthan, India (Figure 2.8), as drawn by Brenda Craddock.[19] On the far right at the top of Figure 2.8 it is possible to see a stone platform which has been raised upwards towards the roof of the mine. This once supported a large brushwood fireplace used to extend the height of the chamber. Even within the Chalcolithic–Early Bronze Age gold mine at Sakdrissi in Georgia we find quite sophisticated 'tube-like' galleries fireset driven through some of the hardest rocks that it is possible to encounter.[20]

Dating by evidence of tool use

Archaeological work undertaken within mines rarely produces large numbers of cultural objects, and typically many of the artefacts (such as tools) are not chronologically or typologically distinctive. What we can do though is to distinguish between the use of stone and metal tools. This change from stone to metal mining tools is significant, as in Europe and the near East this commonly corresponds to the transition between the Early and Middle Bronze Ages. Locally however we may find that the currency of stone mining tools extended into the Late Bronze Age, or alternatively was completely replaced by metal during the Early Bronze

Figure 2.9 Batter marks on roof of Bronze Age gallery, Copa Hill Mine, Wales, UK (photo by S. Timberlake)

Age. In sub-Saharan Africa stone mining tools were used for much longer, only finally being superseded by the introduction of cheap good-quality wrought iron. In South America, stone mining tools were still being used during the Inca period, which was effectively a copper-bronze using culture.

In general the marks of stone tool use (such as the batter marks seen on the roof of the prehistoric gallery on Copa Hill in Figure 2.9) are a pretty good indication of Bronze Age (Early Bronze Age) date, this also being the case with the very slightly different pitting impressions seen upon the surface of the coarse-grained sandstones excavated for malachite at Engine Vein, Alderley Edge, Cheshire[21] and at Timna in Israel. At the latter site we see a combination of both stone and metal tools (copper chisels) being used at the beginning of the Bronze Age.[22]

Stone to iron: the evolution of mining tools

Grooved and ungrooved stone mining implements

While specific forms or local variants of these tool types can occasionally be recognised, suggesting individual 'styles of work', in general we find that these

Figure 2.10a Grooved mining maul from Engine Vein, Alderley Edge (photo by
S. Timberlake)

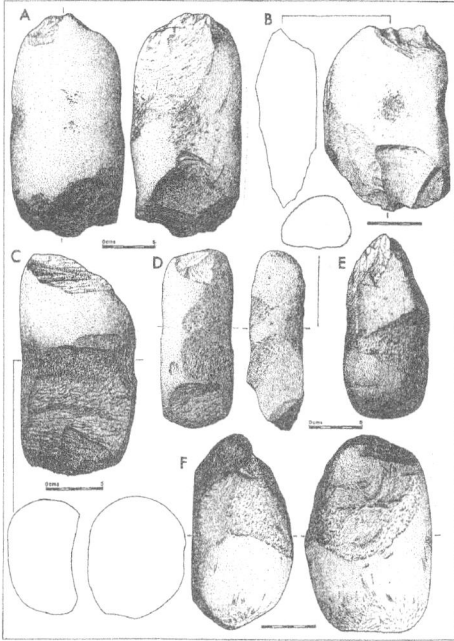

Figure 2.10b Drawings of cobble stone mining tools from Copa Hill, Cwmystwyth
(by B. Craddock in Timberlake & Craddock 2013)

cobble stone mining hammers are made and used in very similar ways the world over. We see this in examples studied from the Mesolithic, Neolithic or Chalcolithic, all of which share similar characteristics to those still being used in the Late Bronze Age. Nevertheless, some assemblages and sites produce more fully grooved (or modified) examples of these tools than others.

The decision to fashion a grooved (Figure 2.10a) rather than an ungrooved hammerstone might simply be influenced by the quality of the stone or the hardness of the host rock, both being factors influential in the longevity of the tool's use; the important point here being that both grooved and ungrooved (though sometimes still notched) hammerstones were hafted for use as rock-breaking tools, chisels and wedges. Needless to say, many of the smaller tools were handheld, and many of those which broke were then recycled, commonly as smaller implements, though sometimes as mortar stones and anvils when used upon their flat and broken surfaces. We tend to see the greatest amount of recycling at the most remote mining sites. For example at Copa Hill, Cwmystwyth located some 25 kilometres inland within the heart of the Cambrian Mountains of mid-Wales we find at least 41% reuse of the tools.[23] In fact the combined assemblages of stone tool fragments found at many of these sites equates to the presence of fairly complete toolkits. These well-selected and sometimes slightly modified but often little-artefacted beach or river pebbles offered the user the ultimate in a utilitarian and interchangeable mining tool (Figure 2.10b). Indeed the longevity of small-scale mining at some of these prehistoric copper mines is reflected in the actual numbers of tools which have survived. We see this for example at the Amerindian native copper mine on Isle Royale, Lake Michigan, where pits were found filled quite simply with thousands of different-sized basalt cobbles.[24] Similarly the Greenland Inuit of Cape York brought basalt hammerstones from as far away as 50 kilometres to mine and work the iron of the 3 ton Woman meteorite, the residue of this 'mining operation' consisting of more than 10,000 broken stone tools. [25]

The first metal mining tools: copper and bronze picks

In the use of stone tools we are in effect looking at a 'Stone Age' technology to mine metal before metal becomes abundant, and it is interesting therefore to speculate when this point of abundance and technological demand was reached. At Timna and other sites in the near East we may be looking at a transition to the use of copper tools such as picks and chisels for the sinking shafts by the Middle to Late Bronze Age,[26] as it was at Thorixos in Greece,[27] yet in the Alps this change to metal tool use seems to have taken place earlier, probably around 1500 BC.[28] The Alpine miners were using rather similar-looking copper and tin-bronze picks mounted in knee-length wooden hafts, the evidence for which we find in the salt mines of Hallstatt, and more rarely in the copper mines of the Mitterberg and Tyrol.[29]

The introduction of iron tools: Iron Age-Roman-Medieval mining

The introduction of iron tools for use as mining implements pretty much follows the earliest adoption of iron, a phenomenon occurring at the beginning of the La Tene period of the European Iron Age. In fact, the earliest iron mining picks in use at the Durnberg salt mine near Hallein, Austria mimic the style of the bronze picks, although generally they were smaller.[30]

Late Iron Age mining, as can be seen from the remains of the timberwork found within the gold mines of Limousin in Central France,[31] becomes increasingly sophisticated, such that by the beginning of the Roman period mining becomes standardised, with the type of pickwork and trapezoidal shape of the mine galleries reflecting a new style of mining. This is a style which can now be recognised in some form or other right across the Roman Empire from Britain, to Spain, to Roumania (Figure 2.11).

Early Medieval mining is characterised by a new and quite distinctive form of tool use known as hammer and gad. This was a method of hand-driving a level using a combination of a hand-held pick with an interchangeable point and hammer end and a sledge to drive this home. The levels themselves were often narrower than the Roman examples, and commonly coffin-shaped, the shape chosen perhaps to accommodate miners with loads of ore, or maybe just with tools upon their shoulders.

Figure 2.11 Roman mine gallery of second to third centuries AD: Catalina Monulesti, Rosia Montana, Romania (photo by S.Timberlake)

Some problems of interpretation: the continuation of primitive mining practices

The rare but continuing use of some the most simple and supposedly 'primitive' stone tools for gold beneficiation within parts of Uganda and Mali offers us an interesting insight into the perpetuation of supposedly prehistoric mining technologies. It is difficult to draw any clear conclusion from this, given that the scale of this activity was minimal. Nevertheless, it was a great leveller for us to find people out there still working with such tools, given that at the same time we were undertaking this as an archaeological experiment.

The example quoted above from Uganda[32] was interesting in that the ore grinding stone was still part of somebody's living and working space, an arrangement which seems likely to have been the same in prehistoric times, as in the vast majority of pre-modern small-scale mining. However, a much more commonly observed traditional practice seen within the artisanal mining sector is the crushing of ore on large stone anvils using iron hammers, such as Paul Craddock witnessed at the lead mine of Ishiagu in Nigeria[33] (Figure 2.12a), this being exactly the same process illustrated by Agricola in *De Re Metallica*[34] (Figure 2.12b), and probably also at late eighteenth/early nineteenth century dressing floors at mines in Britain, much of this work being carried out then (as it still is today within the unofficial artisanal mining sector) by women.

Figure 2.12a Women crushing galena on knockstones, Ishiagu, Nigeria, late twentieth century (copyright P. Craddock)

Figure 2.12b Crushing ore with an iron hammer on a stone anvil c.1550 (woodcut from G. Agricola's *De Re Metallica*)

Another traditional mining practices used until about 60 years ago was the technique of gold panning and washing practised in Transylvania, Romania.[35] This included the recovery of alluvial gold on a sheepskin – the so-called 'Golden Fleece' of Jason and the Argonauts; a likely process practised in prehistory, but needless to say one that leaves no trace of itself in the archaeological record.

There are numerous other examples of 'primitive' mining practices still being carried out across the developing world. I was particularly interested in one such technique witnessed by Paul Craddock in southern India about 20 years ago, perhaps the very last use of firesetting to extract stone.[36] This was being used in a slightly different way to that employed in the driving of a level, yet the principle of using fire to break up the rock was the same. In the Byrath granite quarry at Kothur near Bangalore the quarry workers were using small controlled brushwood fires to carefully split thin sheets of granite; they were lighting these fires over the top of iron wedges previously hammered into joints, the expansion of these wedges opening up a number of well-chosen cracks in the rock. This traditional quarrying process, now probably ceased, may be over a thousand years old.

Alluvial tin mining as still practised by tribal peoples in the jungles of Chattisgarh and Orissa might shed some light on the sort of organization and technology witnessed on the Indian sub-continent almost 3000 to 4000 years ago. The age-old practice of tin-streaming using bamboo baskets combined with local smelting of cassiterite appears to be linked to a vast unwritten body of knowledge concerning prospection. For instance, the locals look for signs of tin by examining the leaves of the Sarai tree (*Shoria robusta*). When growing on tin-rich ground these leaves often appear to be covered in yellow spots, as if suffering from a disease (in fact the leaves were found to contain some 700 ppm of tin on analysis!)[37] This type of knowledge regarding the observation plants and geology is exactly the sort of dimension that is missing from our current understanding of prehistoric and ancient mining.

The antiquity of 'ancient' mines: the use of absolute dating techniques

Radiocarbon dating (typically using Accelerator Mass Spectrosopy (AMS)) is probably now the standard absolute dating tool of archaeology, and is a technique which has had particular relevance to the discipline of mining archaeology, given that most ancient mining sediments rarely contain much in the way of dateable cultural artefacts.

Although there remains the danger that any charcoal recovered from these mines could have been redeposited, a much better understanding of the taphonomy of this material as well as the technical processes of mining should help to distinguish this from the burnt wood associated with contemporary firesetting. At the same time short-lived material (i.e. one to two year-old branchwood) may preferentially be chosen for more accurate C14 dating. The dating of *in situ* mine timbers is a better approach as regards the integrity of the material with the mining event, although the age effect of dating 50 to even 100 year-old wood must always be taken into account. Dating layers of short-lived organic material such as *in situ* peat formed within the infill of mines can also be helpful in providing an end date for the working as well as its abandonment – what is commonly referred to as a *terminus ante quem*. It remains to say that by far the best objects to date the mining activity are the actual tools that the miners used – either wooden ones, or in the context of British prehistoric mines, the bone and antler chisels, scrapers and antler picks. Even where these are found redeposited, their recognition as mining tools is enough to justify their use as dating material. New techniques now being applied to ancient mining investigations include the uranium series dating of stalagmite speleothem sealing mine spoil (such as we find on the Great Orme[38]), optically stimulated luminescence (OSL) dating of contemporarily buried near-surface sediments and rock walls, and finally archaeomagnetic dating of fine waterlain sediments or *in situ* firesetting and smelting hearths.

Radiocarbon has helped to push back the dates of the very earliest mining operations undertaken by mankind. Stefan Berger in this volume nominated the c. 41,000–43,000 year-old hematite mine at Lion Cave, Bomvu Ridge in Swaziland as one of Africa's contributions to this list of honour. Part of this ancient mining site has now been preserved inside of what is arguably one of the most modern of large iron mines, yet here at the very beginning of the Upper Palaeolithic period people were mining this deposit of specular hematite with dolerite hammerstones in order to use it as a red pigment, possibly as a body paint, but perhaps also for rock art.[39]

Man's first collection of iron ochre for use as a pigment extends back another 500,000 years, if not more. But if we are talking of actual mining of raw materials, then Lion Cave is not the oldest site. Palaeolithic chert mining and the digging of shafts for its extraction at Quena and Nazlet in Egypt have been dated to around 50,000 years BP.[40]

Mining metallic minerals before the metal age: the mining and use of copper and iron pigments

While Lion Cave may be the oldest confirmed example, the mining of metallic pigments in prehistory is clearly an important, and almost certainly a neglected, subject in the field of mining archaeology.

It is important to say at this point that many of our grandest statements concerning the rise of metallurgy and the earliest exploitation of oxidised copper minerals within Chalcolithic–Bronze Age may be false, given the absence in many cases of associated smelting furnaces. In fact, we are making a big assumption that these mines were always for the extraction of copper, given that the use of green or blue minerals for pigment or for jewellery may have been equally prized. Needless to say, the actual evidence for this is unlikely to be confirmed by the archaeological record.

As part of our examination of the archaeology of the prehistoric copper mines of Alderley Edge in Cheshire I was interested in the association between the locations of Mesolithic hunting camps and the outcrops of copper ore.[41] These included the brightly coloured malachite and azurite, brown and yellow ochres, and black

Figure 2.13 Crushing azurite nodules to make pigment, Engine Vein, Alderley Edge, UK
(photo by S. Timberlake)

manganese minerals present within some of the softer easily scrapeable sandstone
beds. As an exercise in experimental archaeology we mined and crushed some of
the azurite nodules to make a blue paint-like pigment (Figure 2.13). We also won-
dered whether the nearby Roman mine on Engine Vein[42] might have been working
a mudstone rich in nodules of azurite nodules. These could have been used to make
the blue wall plaster pigment known as 'Egyptian Blue'. In fact there is a very
strong similarity between this site and the Roman pigment mine dug for azurite at
Wallerfangen, Saar in Germany.[43]

Some years back a number of small hammerstones were found in the part-
natural cave/part-iron ochre mine of Clearwell Caves within the Forest of Dean
in the UK. Alongside other examples of mortar stones found in the Forest, these
tools could be many thousands of years old.[44]

In 2010 I visited Taltal on the coast of northern Chile where they were excavating
a 12,000 year old prehistoric trench mine worked by firesetting and hammerstones.
This mine had been dug to recover specular hematite which was then crushed on-
site using small hammers and mortar stones to make a red pigment.[45]

The first copper mines

Currently some of the earliest mines worked for the extraction of copper are to be
found within the Balkans. Even at that end of the Neolithic/beginning of the Copper
Age we are witnessing the dual use of malachite to smelt copper and to make

beads, as has recently been shown at Belovode in Serbia.[46] Nevertheless some 7000 years ago at Rudna Glava in Serbia and Ai Bunar in Bulgaria, malachite and other oxidised minerals were being mined using stone tools, and then ores transported to settlement sites and smelted.[47] This period of exploitation also seems to coincide with the earliest copper extraction at sites in Anatolia and Mesopotamia. However, the first working of native copper would have been earlier than this, and in Anatolia both copper and lead slags have been found at Yarim Tepe[48] and Catal Huyuk,[49] both dating to the early sixth millennium BC.

The use of copper and the mining and smelting of copper ores then spread west into Italy, France and Iberia during the fourth to third millennia BC,[50] and south into the Middle East where we find significant mining of oxidised copper ores at Timna in Israel[51] and Wadi Feinan in Jordan.[52]

The Bronze Age 'industrial revolution'

In terms of the evolution of copper metallurgy, oxidised copper ores were worked first, then copper-arsenic sulphide (fahlerz) minerals, and finally the primary copper-iron-sulphide ores such as chalcopyrite. The smelting of the latter required a better design of shaft furnace, greater use of fuel, and proper slagging techniques.

Figure 2.14 Experimental matte production using an Alpine-type copper smelting furnace, Mitterberg, Austria 2012 (photo by S. Timberlake)

This developed smelting process was probably perfected in the Alps during the Middle Bronze Age (and probably elsewhere) around 1500–1400 BC. By this time large amounts of chalcopyrite were being deep-mined on the Mitterberg and at numerous other Alpine sites. The remains of some of these workings, such as those which survive underground on the Arthurstollen (Mitterberg) are currently being investigated by the Department of Montanarchaeologie at the Deutsches Bergbau Museum in Bochum.[53] Within the area of the Mitterberg alone some 200–300 ore roasting and smelting sites have been identified, most of which would have been located close to the then available fuel supplies, but also to sources of water.[54] For a number of years now Erica Hanning has been carrying out ore roasting, smelting, matte roasting and reduction experiments (see Figure 2.14) close to the former smelting sites.[55] From this she has tried to elucidate answers as to the nature of the furnace technologies associated with this revolution in copper production, the impact of which must have been felt on the shores of Britain following the first arrival of scrap metal from Europe.

Understanding prehistoric mining through experiment: processes and predictions

It has been a particular focus of my work over the last 10–15 years to try and understand ancient mining (in particular prehistoric mining) through undertaking experimental archaeology. In fact this approach has had a four-fold role: first to understand the construction of the mining tools and the nature and difficulties of the mining process; second to try and estimate the raw material requirements of a mining operation; third to look at the sorts of remains we might be leaving behind us; and fourth to try and predict what other sorts of tools might once have been used.

Firesetting experiments at Cwmystwyth, Wales

Firesetting experiments were carried out between 1987 and 1995 at our experimental site at Cwmystwyth in mid-Wales.[56] These were all undertaken against the same mineral-veined rock face of slaty rocks using between 100–250 kilos of firewood for each fireset, wherever possible including oak branchwood, as this was the wood species most frequently identified within the prehistoric mine. By the end of this period the wood fuel to stone extraction ratio had improved substantially from about 1:1 to 1:2; this improved yield of mined material reflecting an increase in our manual skills, a more economic use of fuel, and a much better understanding of the properties of the rock. All of this mining was undertaken using a combination of experimentally reconstructed hafted stone mining tools and antler picks, one of which had removed up to 1.5 tons of mined rock (Figure 2.15a).

The predictive results of this experimentation concluded that picks, most probably antler picks (because of their lightweight and robustness) would have been used in conjunction with the stone tools in the Bronze Age mine. Sure enough, examples of one of these picks was discovered the following year on Copa Hill

Figure 2.15a Experiment in prehistoric hard-rock mining using an antler pick, Cwmystwyth, UK (photo by S. Timberlake)

(Figure 2.15b). This had been used at the pick end and then reused on the crown as a mallet, in exactly the same way as we had been using them in the experiments. Furthermore, on observing the way that the experimentally hafted hammerstone handles made of twisted hazel (*Corylus avellana*) broke in use, this prompted us to look in more detail at the organic material we were recovering from the excavations. As a result, we found the remains of one of these withy handles which had broken in two in exactly the same way, and in the same place, as those we had made, thus confirming the probable type of hafting we had been experimenting with, and also the inherent shortcomings of these tools. After undertaking this short campaign of mining work we began to realise the essential nature of carrying out repairs, and also the importance of having somebody skilled to carry out this task.

Gold mining experiments at Sakridissi, Kazreti, Georgia

Similar experiments were carried out during the archaeological excavations of a late fourth to early third millennium BC gold mine of the Kura-Axes culture at Sakdrissi near Kazreti in Georgia, a site currently being excavated by a joint German-Georgian team of archaeologists as part of the 2007–2013 Bochum

Figure 2.15b Pick/hammer of red deer antler found within Bronze Age mine,
Cwmystwyth, UK (photo by S.Timberlake)

Caucasus project.[57] Our role in this was to try and reconstruct the gold mining,
milling and washing process as suggested by the archaeological evidence.

The Sakridissi Mine has been claimed, with some justification, as the earliest
example of (hard rock) gold mining in the ancient world. Its greatest enigma is
that the gold grains present within the quartz-hematite veins are so tiny (<0.5mm)
as to be invisible to the naked eye. While it is possible that the source of this
gold was traced upstream, it remains difficult to comprehend how this particular
deposit which consists of only of a small group of quartz veins carrying upwards
of 10–20 ppm gold in finely disseminated form was identified from amongst hundreds of other less-enriched deposits. Moreover, how did they come up with an
effective strategy to work it?

Mining

As in Wales, we began our experiments by finding suitable river cobbles for use
as tools, which were then hafted in handles made from local hazel (*Corylus* sp),
and secured by tying strips of rawhide in the style of the pre-Columbian stone
mining tools found at Chuquicamata in Chile. The results during the course of
repeated firesets carried out on several of the quartz veins were similar to those
we achieved in Wales. In one experiment a total 255kg of wood was burnt with
temperatures reaching a maximum of 800°C during two hours of firing, the subsequent mining of this rock with the stone hammers (Figure 2.16) and an antler pick

Figure 2.16 Mining with hafted stone hammers, Sakdrisi, Kazreti, Georgia (photo by
S. Timberlake)

producing some 279kg of rock.[58] It seems likely therefore that the fuel required to
excavate this size of prehistoric mine through some of the hardest of rhyodacitic
rocks would have been truly enormous, requiring thousands rather than just hundreds of tons of wood.

Crushing and separating

To process the potentially gold-bearing rock the quartz-hematite contact material
was carefully separated out from the waste and then crushed upon anvil stones,
or within the original small-hollowed Bronze Age mortars, using flat-sided and
hand-held pounding stones, the aim of this being to reduce this down to a grit-
size consistency (2–3mm) (Figure 2.17a). We decided then to try assaying each
metre of the vein, through milling and panning samples, just to see how practical a method it was to help determine where the richest gold values lay. The
crushed, cleaned and concentrated ore was then fine-ground to a powder on large
'saddle-quern-type' grind stones using suitably flat or slightly convex rubbing
stones. Our washings of the residues showed that a grain size of between 0.25
and 0.5mm was probably the best fraction for gold recovery.

Figure 2.17a Processing gold-hematite ore at Sakdrisi using a Bronze Age mortar stone (photo by S. Timberlake)

Gold washing and recovery

Samples of pulverised ore weighing between 0.5 and 1kg were panned with a small gold pan using water taken from an ancient-cut rock cistern perched on the edge of the opencast, a cistern which may have been used by the miners for this very purpose. Pan washing these samples for 10–15 minutes removed the quartz and produced a dark concentrate of hematite. This was much harder to pan away, yet a number of the samples with significant iron oxide contents yielded some the best heads of gold – the latter composed of fine yellow flakes, the largest grains being only 0.5mm in diameter (Figure 2.17b). The best result we achieved (perhaps >20–40 ppm gold average) came from the milling of a sample taken from the hematite-rich portion of a quartz vein exposed within a boulder recovered from the excavations – almost certainly one of the veins worked in prehistory. We deduced from this that it would have been feasible for the miners to assay each part of a vein for its gold content, and hence determine in which direction they were, or were not, going to mine.[59]

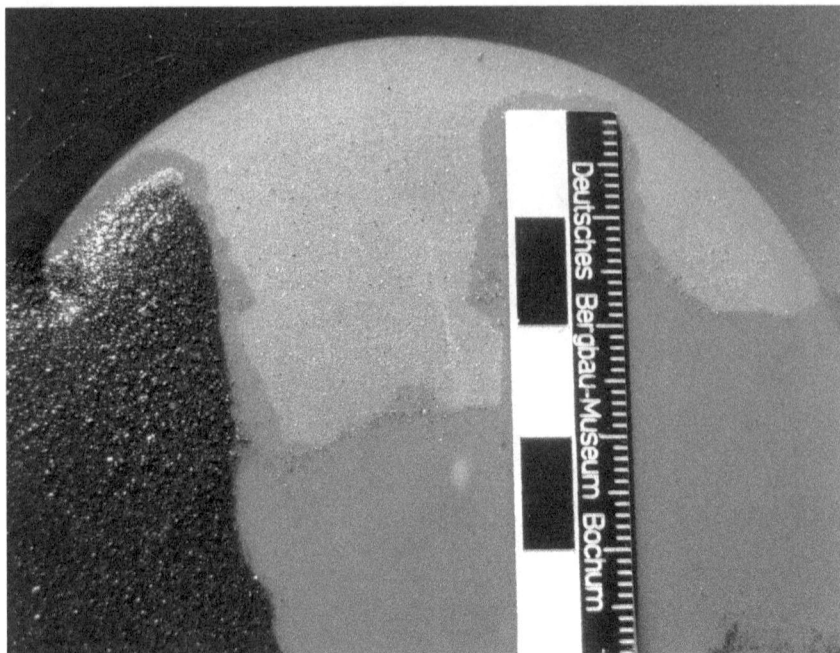

Figure 2.17b Gold flakes as 'head' within fine-grained hematite in pan, Sakdrisi (photo
by S. Timberlake)

Prehistoric tin mining: Kestel/ Göltepe – a tin source in Anatolia?

Debates on how prehistoric miners managed to work low-grade or seemingly complex mineral deposits are not new to this subject. One only has to look at the dispute between Muhly *et al.*[60] and Yener *et al.*[61] over the Early Bronze Age tin mine at Kestel in southwest Anatolia.

Working at this site in 1996, we were faced with a rather similar sort of enigma. The mine looked like a hematite deposit, which it might be argued was worked for pigment, yet in places this same orebody contained up to 1% cassiterite (tin oxide). Archaeology carried out here has since demonstrated that the finely dispersed tin within this hematite was extracted from the pulverised mineral within a series workshops at the nearby settlement of Goltepe; this process employed a complex sequence of enrichment which involved winnowing (or washing) as well as blowpipe smelting of the tin into beads upon the surface of large flat crucibles. The important ingredient here (apart from their obvious metallurgical skills) was the ingenuity of the processors, alongside a quite different approach to timescale, economy and the effort of collective labour.

Mining and the environment: a sense of scale

Records of atmospheric metal pollution dating back to ancient times have been preserved in the polar ice caps. In their pioneering work in 1969, Murozunni, Chow and Patterson[62] used lead levels in the ice from northwest Greenland to deduce that extensive contamination of the Arctic atmosphere with lead began before the Industrial Revolution. In fact the lead content of ice layers deposited between 500 BC and AD 300 was about four times that of background levels, implying widespread pollution of the Northern Hemisphere by emissions from Roman mines and smelters. Hong, Candelone and Patterson, by contrast, looked at the evidence for widespread atmospheric contamination with copper during the same period, matching the increases in pollution with the amount of copper mined.[63] For instance, a rise in copper contamination during the Late Bronze Age seems to have been linked to the exploitation of copper in Cyprus, while in Roman times there appears to be a signal, which could relate to production at Rio Tinto and other sites in Iberia, as well as to the Cypriot mines.

At Kargaly in the southern Ural steppes of Russia, Evegny Chernych estimates a level of production amounting to 2–5 million tons of copper ore during the Bronze Age (mostly during the Late Bronze Age). While the most significant documented production in this region took place during the nineteenth century, he considered the prehistoric evidence alone (in the form of a landscape full of infilled buried shafts) equivalent to 130–140 square km.[64] If this were so, and all the workings belonged to the one prehistoric period, the implications for the sourcing of wood for use as pit props and as fuel for smelting would have been enormous within this treeless area. In fact he admits that all this timber would have had to be brought in over distances of hundreds of kilometres. I am left wondering whether these still largely unexcavated 'prehistoric' remains represent multi-period rather than single period exploitation.

Ancient mining and environmental disaster

When I first saw quartz scree-covered mountainside at Puerto del Palo in the mountains of Asturias, northwest Spain I was under the impression that I was looking for a Roman gold mine. What I didn't know was that everything I could see was the mine, and I was standing on one tiny little part of it. In fact Puerto del Palo is an enormous opencast, one of a dozen such gold mining sites which cross the mountainous tracts of Asturias, Cantabria and Leon, all of them worked hydraulically by hushing during the Roman period.[65]

Hydraulic gold mining in northwest Spain undertaken during the first–second centuries AD caused an immense environmental impact – the removal of many millions of tons of soil and rock which resulted in the silting up of rivers and estuaries more than 150km away. This blocking of the waterways prevented access for shipping and impeded trade, and as a result caused a downturn in the fortunes of the area. The mining process here was described in some detail by Pliny the Elder in AD 79.[66]

Investigating the environmental record and examining the impact of ancient mining

In Britain I believe that we have found a small-scale version of one of the *ruinas montana* mines of northwest Spain[67] at Craig y mwyn in the Berwyn Mountains of northeast Wales (Figure 2.18). Unlike the Spanish mines, this was mined for lead, not for gold. A palaeoenvironmental investigation of the blanket peat bog above the mine undertaken by Tim Mighall and myself revealed three prob-able phases of working identified by their lead pollution peaks in the core.[68] This suggests mining activity during the Late Roman (1720+/–40 BP (AD 230–410)), Early Medieval (estimated 1127 years BP (nineth century AD)), and Late Medieval (estimated at 620 years BP (fourteenth century AD)) periods. The last of these phases of mining and hushing (perhaps undertaken by the Cistercians) was suggested by the excavation and radiocarbon dating of the peat fill of an aban-doned leat.[69] Concerning the suggested Roman working of the mine, four levels of water supply (coming in along the 480m, 500m, 520m and 525m contours) associated with small (10–15m long) rectangular holding and hushing tanks can be identified, although none have been archaeologically examined. The lowest of these hydraulic systems have since been lost due to the widening of the open-cast. This gradual working back of the opencast face meant that water had to be brought in at a higher level each time, just as described by Pliny.

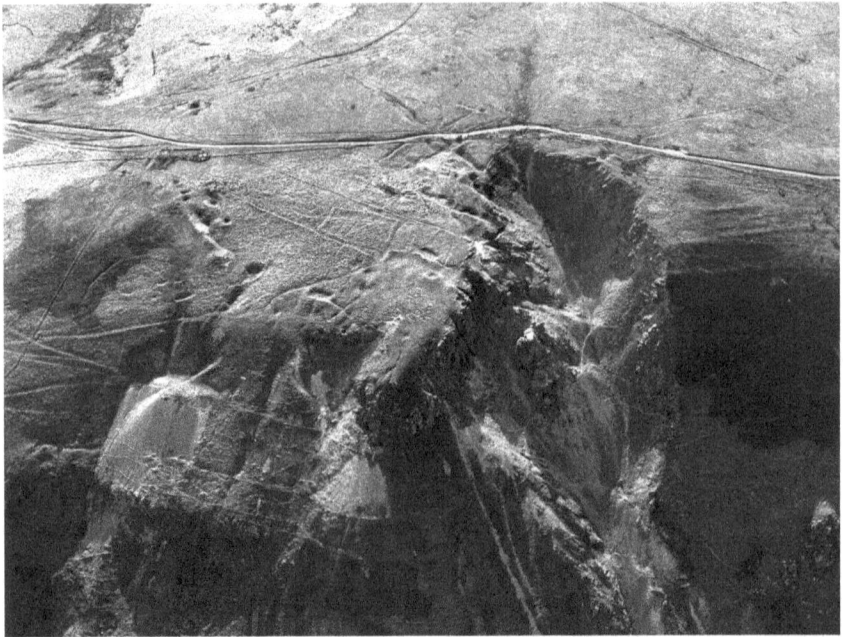

Figure 2.18 Craig y mwyn opencast, Berwyn Mountains, Wales, UK (Crown copyright RCAHMWales no.935097-53)

Preservation of objects

Conditions for the preservation for objects within the underground environment may be good or indifferent depending upon the type of material, the degree of waterlogging or desiccation, its acidity (pH) and degree of oxidation (Eh). In fact the underground environment within a metal mine can be among the most extreme of environments to be found anywhere on earth, with high acidity and metal toxicity due to the rapid oxidation of iron sulphide in deposits worked for copper ore. Two examples are the mines of Rio Tinto in Spain and Parys Mountain on Anglesey,[70] both sites linked to Bronze Age as well as Roman mining.

The waterlogging of mining deposits has permitted the survival of numerous organic artefacts such as the wooden tools found at Mt Gabriel in Ireland[71] and also the wooden handles, rope, basketry and red deer antler mining picks (see Figure 2.15b) preserved within some parts of the Copa Hill mine in mid-Wales.[72]

Bone and antler mining tools are better preserved wherever the groundwater conditions are more alkaline, thus we find good assemblages of these within limestone-hosted mines such as the Great Orme (which has produced many thousands of bone artefacts[73]) and Ecton in Staffordshire.[74]

Wood is sometimes well preserved within these heavy metal environments, partly on account of the waterlogging, but more often due to the absorption of lead, arsenic and copper, the high levels of which can stop any sort of biological activity. A study of the Copa Hill drainage launder revealed the growth of lead minerals within the cellular structure of the wood,[75] while the exceptional preservation of the 3700-year-old Alderley Edge shovel can be explained by the extremely high levels of copper and arsenic it contains (Figure 2.19). A recent

Figure 2.19 The Alderley Edge Bronze Age oak shovel (59.4 cm long) (permission of The Manchester Museum)

whole artefact study of this object using the Daresbury synchrotron has revealed a very interesting patterning of metal absorption.[76] It would seem that the highest concentration of copper and arsenic is to be found upon the handle and on the tip of the blade, and as such this contamination has been interpreted as being contemporary with its use.

Needless to say, some of the most favourable conditions for the preservation of organic objects are to be found in salt mines.

The working life of a miner 3000–4000 years ago: conditions inside the mine

Conditions suitable for the preservation of organic objects have provided us with a unique insight into the daily working life of the prehistoric or Roman miner. One example of this are the many hundreds of pine lighting splints or tapers found within the Middle Bronze Age copper mines on the Mitterberg,[77] the Bronze Age– Iron Age salt mines at Hallstatt and Durnberg,[78] and the second to third century AD Roman gold mines in Dacia (Romania).[79] This evidence provides a very clear picture of how these mines were illuminated underground more than two and a half millennia ago. The evidence from Hallstatt and Rosia Montana suggests that the miners took bunches of these tapers underground, some of which were tied together as torches at the work place.

A model for the ventilation required while firesetting underground in some of the Alpine mines was suggested by Zoschke and Preuschen as long ago as 1933.[80] However, it now appears that most of the deeper workings never used fire at all, all the mining being undertaken with the aide of copper or bronze picks.

Water would have been a perennial problem in any mine that reached the water table. At Copa Hill in mid-Wales we believe that the hollowed-out log launders or drainage pipes were arranged in series in order to try and tap water coming into the mine from a spring located on the wall of the opencast working.[81]

The people: archaeological identities of the miners themselves

What do we know of them? At Hallstatt the level of personal detail gleaned from the archaeological evidence left within the mines is phenomenal. This includes the recovery of the skin carrysacks, items of clothing, caps, shoes and even piles of human excrement preserved in the salt, some of these with the traces still preserved of the intestinal parasites such as roundworm and whip worm which infested their guts.[82] Outside of the mine many of the graves associated with the mining settlement have now been excavated. Almost the only thing left to find inside of the mine is the mummified remains of a miner buried in salt. Such a find is not inconceivable given that we now know that the mine was completely inundated by mudflows on several different occasions; yet to date no ancient human remains have (with any certainty) been found.

However, at the Cherhabad salt mine near Zanjan in northwest Iran a team from the Deutsches Bergbau Museum have been working with Iranian archaeologists

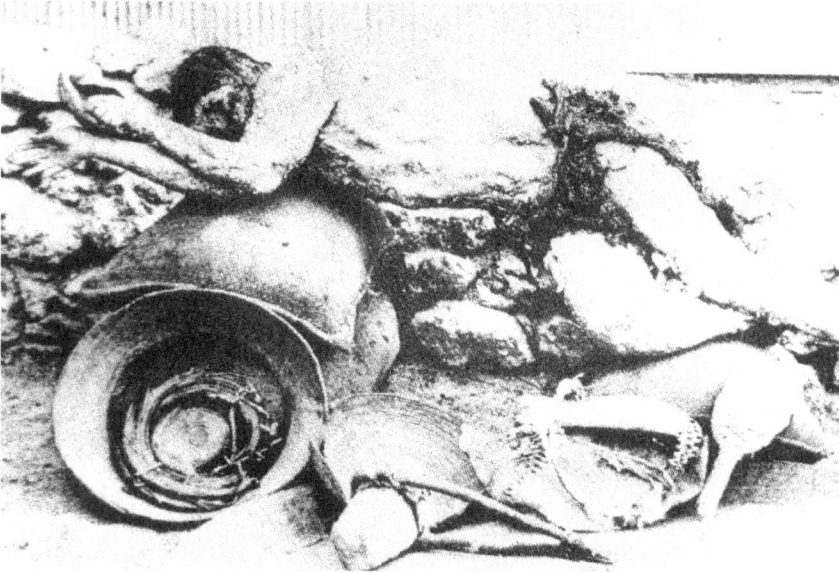

Figure 2.20 The Chuquicamata 'Copper man': a 1500-year old mummified miner from the Atacama, northern Chile (permission P. Craddock)

since 2008.[83] Here the remains of at least six mummified bodies, most probably those of miners, have been found at those points where the edges of the salt workings have collapsed in upon them. DNA analysis carried out on the fourth and best preserved of the mummies revealed that he was a young man who had evidently travelled some distance from Mazandaran Province in northern Iran some 2245 years ago to work as a miner at Zanjan.

Just as astonishing a discovery, yet preserved under quite different conditions of aridity in the Atacama Desert of northern Chile, was the near-perfectly preserved 1500-year-old mummy of an indigenous copper miner found buried beneath an ancient roof fall at the Restauradora Mine, Chuquicamata in 1899.[84] Surrounding him within the narrow gallery were his baskets, shovel and hafted stone hammers (Figure 2.20).

Modern depictions of ancient mines

To some extent artistic reconstructions of prehistoric or ancient mining scenes, as with other archaeological reconstructions, reflect a combination of different ideas and influences. Dominating this of course is the artist's own style, yet the way 'ancient people' or miners are portrayed often reflects current perceptions and fashion, and sometimes also national identity. I say this because there are now a lot of images of ancient flint and copper miners, and almost all of them look like

Figure 2.21a A 1911 painting of prehistoric miners from Cabrieres, southwest France (permission Esperou)

Figure 2.21b Ross Island Bronze Age miners, Killarney, Ireland (permission W. O'Brien)

Figure 2.21c Hallstatt salt miners (copyright NHM Vienna)

Figure 2.21d Alderley Edge miners, Engine Vein (permission The Manchester Museum)

Figure 2.21e Iron Age miners in Limousin, France (permission B. Cauet)

they come from quite different periods and worlds (see Figures 21a–e). I find the 1911 painting of the Bronze Age copper miners from Cabrières interesting in that it portrays them as being Palaeolithic-type cave men,[85] while the illustration of Bronze Age copper miners from Ross Island in County Kerry clearly makes them look Irish.[86] Indeed, I can't imagine them being from any other place. Meanwhile, deep underground within the Hallstatt mines one might be forgiven for thinking that these depicted ranks of armed miners are in fact orcs,[87] while the mining scene at Alderley Edge is rural and quite simply 'English'.[88] The best of them however has got to be Asterix the Gaul in this depiction of Iron Age gold mining in France.[89] I don't think this one was meant to be a joke, yet it certainly does reflect national identity.

Comparison with contemporary images of mining

So how do former civilizations see mining in their contemporary world? We have almost no images of Roman miners that I am aware of, apart from the Linares bas-relief. This of course gives very little away – and we are none the wiser about the Roman mining industry for it. By contrast, the Late Medieval wood-cut images, which liberally illustrate Agricola, are full of information and detail,

Figure 2.22 Annaberger Bergalter of 1521 (photo by S.Timberlake)

and are necessarily over-didactic in nature. Much more interesting is this central painting of a mining scene on the Erzgebirge as portrayed with artistic license at the very beginning of the sixteenth century in Hans Hesse's triptych *Annaberger Bergalter* of 1521 (Figure 2.22). I grew up with this painting which was hung in my parents' house. Nobody was really certain how it got there, yet it greatly influenced my life. Even now, every day I can still find something new in this painting to look at, and when staring into it, it feels like I am looking back into time, immersed in the detail of this busy 500-year-old silver mining scene.

Conclusion

Today the study of ancient mining can no longer be accused of being a dry and academic subject. Much is going on in a world accelerated by economic development, and in the search for new metal sources, traces of some of the earliest evidence for mining are being uncovered. However, systematic survey and an appropriate mining archaeological approach is essential to the recovery of meaningful information, alongside an understanding of 'primitive' mining technology and a willingness to engage in experimental archaeological research. Understanding prehistoric and ancient mining means that we can identify the same commonality of approach which links Early Bronze Age copper mining in Britain with gold mining in Georgia, or for that matter pre-Columbian copper mining in the Atacama Desert with pre-colonial copper and gold mining in Zimbabwe. In almost all cases we see a combination in the use of some of the most primitive toolsets alongside what appears to be quite complex modelling or processing sequences. This invites some interesting suggestions as regards the role that mining and raw material acquisition have played in our social and intellectual development as modern humans.

Acknowledgements

I would like to thank Brenda Craddock in particular for her contribution to the experimental projects described in this chapter. The Early Mines Research Group (Brenda Craddock, Phil Andrews, John Pickin and Anthony Gilmour) undertook most of the Welsh and English prehistoric mining excavations mentioned, while the National Museum of Wales, The Leverhulme Trust, The Manchester Museum, The Peak District National Park and English Heritage all helped to fund or support excavation work in the UK. I am very grateful to Thomas Stöllner of the Deutsches Bergbau Museum (DBM), Bochum and Irina Gambaschidze of the National Museum of Georgia for the invitation to work at Sakridissi, Georgia, and for permission also to use some data and photos in advance of full publication. I would also like to thank Beatrice Cauet at the TRACERS Laboratory, University of Toulouse for the opportunity to work at Rosia Montana in 2012.

Notes

1 Simon Timberlake, 'How old is the "Owld Man"?: Discovering ancient mines', *Mining History: Bulletin of the Peak District Mines Historical Society 14*(6) (2001), pp. 54–60: p. 59.
2 Georg Agricola, *De Rerum Metallica* (1556) (H. Hoover & L. Hoover (eds)).
3 Martyn Barber, David Field and Peter Topping (eds), *The Neolithic Flint Mines of England* (Swindon: English Heritage, 1999), p. 49.
4 Thomas Stöllner, et al. 'The Mitterberg as a large scale producer of copper', in: Gert Goldenberg, Ulrike Töchterle, Klaus Oeggl and Alexandra Krenn-Leeb (eds), *Research Programme HiMAT – News from the Mining History of the Eastern Alps, Archäologie Österreichs Spezial 4* (Vienna: Österreichische Gesellschaft für Ur- und Frühgeschichte, 2011), chapter 4, p. 130.
5 Lynn Willies, 'Kestel Tin Mine, Turkey: interim report', *Bulletin of the Peak District Mines Historical Society 12*(5) (1995), pp. 1–11.
6 Phil Newman, *The Field Archaeology of Dartmoor* (Swindon: English Heritage, 2011).
7 Sandy Gerrard, 'The early south-western tin industry: an archaeological view', in: Phil Newman (ed.), *The Archaeology of Mining & Metallurgy in SW Britain* (Matlock: Peak District Mines Historical Society, 1996), pp. 67–83: p. 71.
8 Pliny (the Elder), *Natural History* XXXIII, AD 79, pp. 67–78.
9 Simon Timberlake, 'Cwmystwyth, Wales: 3500 years of mining history – some problems of conservation and recording', *Journal of the Russell Society 5*(1) (1993), pp. 49–53.

10 Simon Timberlake, 'Medieval lead smelting boles near Penguelan, Cwmystwyth, Ceredigion', *Archaeology in Wales 42* (2002), pp. 45–59.
11 David Jenkins and Simon Timberlake, 'Geo-archaeological research into prehistoric mining for copper in Wales', unpublished report to the Leverhulme Trust, University of Wales, Bangor (1997).
12 Simon Timberlake, 'Excavations on Copa Hill, Cwmystwyth (1986–1999): an Early Bronze Age copper mine within the uplands of Central Wales', *BAR British Series* 348 (Oxford: Archaeopress, 2003).
13 Simon Timberlake, 'An archaeological examination of some early mining leats and hushing remains in upland Wales', *Archaeology in Wales 43* (2003), pp. 33–44.
14 David Jenkins and Simon Timberlake, *ibid.* (1997), p. 12.
15 Andy Lewis, 'Underground exploration of the Great Orme copper mines', in: Peter Crew and Susan Crew (eds), *Early Mining in the British Isles* (Plas Tan y Bwlch: Snowdonia Study Centre, Maentwrog, Snowdonia National Park, 1990).
16 Felice Larocca and Chiara Levato, 'From the imprint to the tool: the identification of prehistoric mining implements through the study of digging traces. The case of Grotta della Monaca in Calabria (Italy)', in: Peter Anreiter *et al.* (eds), *Mining in European History and its Impact on Environment and Human Societies, Proceedings for the 2nd Mining in European History Conference of the FZ HiMAT (Innsbruck, 7–10 November 2012)* (Innsbruck: Innsbruck University Press, 2013), pp. 21–26.
17 Simon Timberlake, 'Review of the historical evidence for firesetting', in Peter Crew and Susan Crew (eds), *Early Mining in the British Isles* (Plas Tan y Bwlch: Snowdonia Study Centre, Maentwrog, Snowdonia National Park, 1990), pp. 49–52.
18 William O'Brien, *Mount Gabriel – Bronze Age Mining in Ireland* (Galway: Galway University Press, 1994).
19 Brenda Craddock, 'Drawing Ancient Mines', in: Trevor Ford and Lynn Willies (eds), *Mining Before Powder, Bulletin of the Peak District Mines Historical Society 12*(3) (1994), pp. 9–12: p. 11.
20 Thomas Stöllner, Irina Gambaschidze and Andreas Hauptmann, 'The earliest gold mining of the Ancient World? Research on an Early Bronze Age gold mine in Georgia', in: Ünsal Yalcin, Hadi Özbal and A. Günhan Pasamehmetoglu (eds), *Ancient Mining in Turkey and the Eastern Mediterranean – Proceedings of an International Conference, Ankara* (Ankara: Atilim University, 2008), pp. 271–288.
21 Simon Timberlake and Carolanne King, 'Archaeological excavations at Engine Vein, Alderley Edge 1997', in: Simon Timberlake and John Prag (eds), *The Archaeology of Alderley Edge – Survey, Excavation and Experiment in an Ancient Mining Landscape* (Ann Arbor, MI: University of Michigan, 2005), pp. 33–57.
22 Paul Craddock, *Early Metal Mining and Production* (Edinburgh: Edinburgh University Press, 1995), p. 68.
23 Simon Timberlake and Brenda Craddock, 'Prehistoric metal mining in Britain: the study of cobble stone mining tools based on artefact study, ethnography and experiment', *Chungara Revista de Antropología Chilena 45*(1) (2013), pp. 33–59.
24 Michael L. Wayman, 'Native copper: humanity's introduction to metallurgy?', in: Michael Wayman (ed.), *All that Glisters: Readings from Metallurgical History* (Montreal: Canadian Institute of Mining and Metallurgy, 1989).
25 Vagn Fabritius Buchwald and Gert Mosdal (eds), *Meteoritic Iron, Telluric Iron and Wrought Iron in Greenland* (Copenhagen: Museum Tusculanums Forlag, 1985).
26 Paul, Craddock, *ibid.* (1995), p. 65.
27 Herman F. Mussche, Jean Bingen and J.E. Jones (eds) *Thorixos. IX 1977/1982 Rapport préliminaire* (Gent: Comité des Fouilles Belges en Grèce, 1990), pp. 115–143: p. 136.
28 Thomas Stöllner, 'Copper and salt – mining communities in the Alpine metal ages', in: Peter Anreiter et al. (eds), *ibid.* (2010), p. 297.
29 Anton Kern and Kerstin Kowarik, *Kingdom of Salt: 7000 years of Hallstatt* (Vienna: Museum of Hallstatt, 2009), p. 54.

30 Anton Kern and Kerstin Kowarik, *ibid.* (2009), p. 87.
31 Beatrice Cauet, *L'or des Celtes du Limousin* (Limousin: Archéologie Culure & Patrimoine en Limousin, 2004), p. 57.
32 Terry Worthington and Brenda Craddock, 'Modern stone tools', *Mining History: Bulletin of the Peak District Mines Historical Society 13*(1) (1996), p. 58.
33 Paul Craddock, *ibid.* (1995), p. 162.
34 Georgius Agricola, *ibid.* (1556), p. 270.
35 *Cãlãtorie în Tara Aurului* (Rosia Montana Gold, 2011), p. 12 and 46.
36 Paul Craddock, 'The use of firesetting in the granite quarries of South India', *Mining History: Bulletin of the Peak District Mines Historical Society 13*(1) (1996), pp. 7–11.
37 T.M. Babu, 'Advent of the Bronze Age in the Indian subcontinent', in: Paul Craddock and Janet Lang (eds), *Mining and Metal Production through the Ages* (London: British Museum Press, 2003), pp. 174–180: p. 179.
38 A. Lewis, 'Prehistoric Mining at the Great Orme: Criteria for the Identification of Early Mining', unpublished MPhil Thesis (Bangor: University of Wales, 1996).
39 Adrien Boshier, 'Ancient mining of Bomvu Ridge', *Scientific South Africa 2* (1965), pp. 317–320.
40 Pierre Vermeersch and Etienne Paulissen, 'Palaeolithic chert quarries and mines in Egypt', *Sahara 2* (1989), pp. 95–98.
41 Simon Timberlake, 'Geological, mineralogical and environmental controls on the extraction of copper ores in the British Bronze Age', in: Paul Anreiter, et al. (eds), *Mining in European History and its Impact on Environment and Human Societies* (Innsbruch: Innsbruck University Press, 2010), pp. 289–296.
42 Simon Timberlake and A. Douglas Kidd, 'The archaeological excavation of a Roman mine shaft and gallery ('pot shaft') at Engine Vein, Alderley Edge', in: Simon Timberlake and John Prag (eds), *The Archaeology of Alderley Edge – Survey, Excavation and Experiment in an Ancient Mining Landscape* (Ann Arbor, MI: University of Michigan, 2005), pp. 79–97.
43 Gabriele Körlin, 'Luxusgut Blau – Romischer Azuritbergbau in Wallerfangen/Saar', *Der Anschnitt 62* (2010), pp. 174–188.
44 Martin Strasburger, 'Early Ochre Mining in the Royal Forest of Dean (Gloucestershire)', unpublished manuscript (1999).
45 Diego Salazar, Donald Jackson., Jean Louis Guendon, Hernán Salinas et al., 'Early evidence (ca.12,000 BP) for iron oxide mining in the Pacific coast of South America', *Current Anthropology 52* (2011), pp. 463–475.
46 Milijana Radivojević, Thilo Rehren, Ernst Pernicka, Dusan Šljivar, Michael Brauns and Dusan Borić, 'On the origins of extractive metallurgy: new evidence from Europe', *Journal of Archaeological Science 30* (2010), pp. 1–13.
47 Evgenil Nikolaevich Černych, 'Aibunar – a Balkan copper mine of the fourth millennium BC', *Proceedings of the Prehistoric Society 44* (1978), pp. 203–218.
48 Nikolai Merpert, Rauf Munchaev and N. Bader, 'The earliest metallurgy of Mesopotamia', *Sovtskaya Arkheologiya 3* (1977), pp. 154–165.
49 James D. Muhly, 'The beginnings of metallurgy in the Old World', in: Robert Maddin (ed.), *The Beginning of the Use of Metals and Alloys* (Cambridge, MA: MIT Press, 1988), pp. 2–20.
50 Paul Craddock, *ibid.* (1995).
51 Beno Rothenberg, *Timna* (London: Thames and Hudson, 1972).
52 Andreas Hauptmann, 'The copper ore deposit of Feinan, Wadi Arabah: early mining and metallurgy', in: Susanne Kerner (ed.), *The Near East in Antiquity* 1 (Amman, Goethe-Institut, 1990), pp. 53–62.
53 Thomas Stöllner, Elisabeth Breitenlechner, Clemens Eibner, Rainer Herd, Tobias Kienlin, Joachim Lutz et al., '(4) Der Mitterberg – Der Grosproduzent für Kupfer im östlichen Alpenraum während der Bronzezeit', in: Gert Goldenberg, Ulrike Töchterle, Klaus Oeggl and Alexandra Krenn-Leeb (eds), *Neues zur Bergbaugeschichte der*

Ostalpen (Forschungsprogramm HiMAT), Archäologie Österreichs Spezial 4 (2011), pp. 113–144 and pp. 117–123.

54 Erica Hanning, Thomas Stöllner, Annette Hornschuch and Beate Sikorski, 'Quantifying Bronze Age smelting sites in the Mitterberg Mining District', in: *Mining in European History and its Impact on Environment and Human Societies: Proceedings of 2nd Mining in European History Conference*, Innsbruck (November 2012) (Innsbruck: Universität Innsbruck, 2013), pp. 67–72.

55 Erica, Hanning, 'Smelting of sulfidic ore during the Bronze Age in the Eastern Alpine region: ma mining, archaeological and experimental approach' (unpublished PhD thesis) (University of Bochum, 2014).

56 Simon Timberlake, 'The use of experimental archaeology/archaeometallurgy for the understanding and reconstruction of Early Bronze Age mining and smelting technologies', in: Susan La Niece, Duncan Hook and Paul Craddock (eds), *Metals and Mining: Studies in Archaeometallurgy* (London: Archetype/The British Museum, 2007), pp. 27–36: p. 28.

57 Thomas Stöllner, Irina Gambaschidze and Andreas Hauptmann, *ibid.* (2008).

58 Simon Timberlake, 'The study of cobble stone, bone, antler and wooden mining tools in prehistoric metal mining: new evidence from the British Isles and beyond', in: Jacquo Silvertant (ed.), *Research and Preservation of Ancient Mining Areas: Yearbook of the Institute Europa Subterranea* (Valkenburg: Trento, 2014), pp. 28–57, figure 12.

59 Thomas Stöllner, Brenda Craddock, Simon Timberlake and Irina Gambaschidze, 'Feuersetzen im frühesten Metallerzbergbau und ein Experiment im fruhbronzezeitlichen Goldbergbau von Sakridissi, Georgien', in: Klaus Oeggl and Veronika Schaffer (eds), *Die Geschichte des Bergbaus in Tirol und seinen angrenzenden Gebieten – Proceedings zum 6 Milestone-Meeting des SFB HiMAT 2011* (Innsbruck: Universität Innsbruck, 2012), pp. 65–76.

60 James D. Muhly, Friedrich Begemann, Ö. Öztunali et al., 'The Bronze Age metallurgy of Anatolia and the question of local tin sources', in: Ernst Pernicka and Gunher A. Wagner (eds), *Archaeometry 90* (Basel: Birkhauser, 1991), pp. 209–220.

61 Alishan Yener, Hadi Özbal, Ergun Kaptan, A.N. Pehlivan and Martha Goodway, 'Kestel: an Early Bronze Age source of tin ore in the Taurus Mountains, Turkey', *Science 244* (1989), pp. 200–203.

62 M.T. Murozuni, Tsaihwa J. Chow and C.C. Patterson, 'Chemical concentration of pollutant aerosols, terrestrial dusts and sea salts in Greenland and Antarctic snow strata', *Geochim. Cosmochim. Acta. 33* (1969), pp. 1247–1294.

63 Sungming Hong, Jean-Pierre Candelone, Clair C. Patterson and Claude F. Boutron, 'Greenland ice evidence of hemispheric lead pollution two millennia ago by Greek and Roman civilizations', *Science 265*(5180) (1994), pp. 1841–1843.

64 E.N. Chernych, 'Kargaly: the largest and most ancient metallurgical complex on the border of Europe and Asia', in: K. Linduff (ed.) *Metallurgy in Ancient Eastern Eurasia from the Urals to the Yellow River* (Lewiston, ME: The Edwin Mellen Press, 2004), pp. 223–237.

65 P.R. Lewis and G.D.B. Jones, 'Roman gold-mining in north-west Spain', *Journal of Roman Studies 60* (1970), pp. 169–185.

66 Pliny (the Elder), *Natural History* XXXIII, AD 79, pp. 67–78.

67 David Bird, 'Pliny's Arrugia: water power in Roman gold-mining', in: Peter Claughton (ed.), *Water Power in Mining A Special Issue of Mining History 15*(4/5) (2004), pp. 58–63.

68 Timothy Mighall, Simon Timberlake, S. Singh and M. Bateman, 'Records of palaeopollution from mining and metallurgy as recorded by three ombrotrophic peat bogs in Wales, UK', in: Susan La Niece, Duncan Hook and Paul Craddock (eds), *Metals and Mines: Studies in Archaeometallurgy* (London: Archetype/British Museum), pp. 56–66: p. 60.

69 S. Timberlake, 'Early leats and hushing remains: suggestions and disputes of Roman mining and prospection fopr lead', *Mining History* 15 (4/5) and *Bull. Peak District Mines Historical Soc.* (2004), pp. 64–76.

70 David Barrie Johnson, 'Geomicrobiology of extremely acidic subsurface environments', *Microbiol. Ecol. 81* (2012), pp. 2–12.

71 William O'Brien, *Mount Gabriel – Bronze Age Mining in Ireland* (Galway: Galway University Press, 1994).

72 Simon Timberlake, *Excavations on Copa Hill, Cwmystwyth (1986–1999): An Early Bronze Age Copper Mine within the uplands of Central Wales* (Oxford: British Archaeological Reports, 2003).

73 S. James, 'The economic, social and environmental implications of faunal remains from the Bronze Age Copper Mines at Great Orme, North Wales', (unpublished PhD thesis) (University of Liverpool, 2011).

74 Simon Timberlake, 'Prehistoric copper extraction in Britain: Ecton Hill, Staffordshire', *Proceedings of the Prehistoric Society 80* (2014), pp. 159–206, figure 16, p. 181.

75 Xoram Ever-Hadani, 'Metals investigation on the Cwmystwyth launders' (unpublished MA thesis in Conservation of Historic Objects (Archaeology)) (University of Durham, 2000).

76 Andrew D. Smith, D.I. Green, J.M. Charnock, E. Pantos, Simon Timberlake and John Prag, 'Natural preservation mechanisms at play in a Bronze Age wooden shovel found in the copper mines of Alderley Edge', *Journal of Archaeological Science 30* (2011), pp. 1–9.

77 Thomas Stöllner, Elisabeth Breitenlechner, Clemens Eibner, Rainer Herd, Thomas Kienlin, Joachim Lutz et al., *ibid.* (2011).

78 Anton Kern and Kerstin Kowarik, *Kingdom of Salt: 7000 years of Hallstatt* (Vienna: Museum of Hallstatt, 2009), p. 92.

79 Beatrice Cauet et al., *Raport de Cercetare Archeologicã Preventivã Masivul Cârnic Rosia Montanã Vols. I & II* (Tolouse: University Toulouse Le Mirail, 2009).

80 Karl Zschocke and Ernst Preuschen, 'Das urzeitliche Bergbaugebiet von Mühlbach-Bischofshofen, Materialien zur Urgeschichte Österreichs 6' (Vienna: Selbstverlag der Anthropologischen Gesellschaft, 1932).

81 Simon Timberlake, *ibid.* (2003).

82 Anton Kern and Kerstin Kowarik, *ibid.* (2009), p. 100.

83 Abolfazl Aali, Thomas Stöllner, Aydin Abar and Frank Rühli, 'The salt men of Iran: the salt mine of Douzlākh, Chehrābād', *Archäologisches Korrespondenzblatt 42*(1) (2012), pp. 61–81.

84 J.B. Bird, '"The Copper Man": a prehistoric miner and his tools from northern Chile', in: Elizabeth P. Benson (ed.), *Pre-Columbian Metallurgy of South America* (Washington, DC: Dumbarton Oaks, 1979), pp. 105–132.

85 Jean Luc Esperou, *2008 Les Mines de Cabrières: L'exploitation des cuivres argentiferes depuis las fin du IIIe millénaire avant notre ère* (Cabrieres Ed. Pro Baeteris, 2008), p. 115.

86 William O'Brien, *Ross Island: Mining, Metal and Society in Early Ireland – Bronze Age Studies 6* (Galway: National University of Ireland, 2004), p. 457.

87 Anton Kern and Kerstin Kowarik, *ibid.* (2009), p. 100.

88 Simon Timberlake and Carolanna King, *ibid.* (2005), p. 52.

89 Beatrice Cauet, *ibid.* (2004), p. 36.

3   Engineering changes

The cause and consequence of modern
mining methods at Butte, Montana;
Johannesburg, South Africa; and Broken
Hill, New South Wales

Jeremy Mouat

Introduction

Mining has played a critical role in the modern world. From the eighteenth century, access to coal as well as to base and precious metals drove Europe's industrialization and consequently was vital to economic growth. By the mid-nineteenth century the production and use of iron and steel were widely regarded as indices of modernity, 'the social barometer by which to estimate the relative height of civilization among nations', as one American contemporary put it.[1] The mid-nineteenth century gold rushes were further proof of the mineral economy's importance. Mining historians have focused less on this larger narrative, tending instead to describe the role of mining in specific regions. In part, this narrower perspective reflects the weight of the national paradigm, an influence that has proved largely immune to frequent calls for a larger unit of analysis. Yet most mining regions display a similar trajectory. Graphs charting mineral production through the nineteenth and twentieth centuries show common patterns of dramatic growth.[2] Making sense of these patterns remains a challenge for mining historians.

One way to better understand these patterns is by adopting a comparative perspective. Such an approach offers a way not only to transcend the problems inherent with a national focus but also to clarify the ways in which mining and its associated work processes have changed over the past several centuries. What follows is an attempt to make sense of the growth of the mining industry by briefly describing a key feature of this trajectory in three prominent mining communities: Butte, Montana; Johannesburg, South Africa; and Broken Hill, New South Wales. My focus is principally on the way in which the mining industry in each of these three communities made the transition to relying on low-grade ore deposits, after dramatic early growth made possible by extracting relatively high-grade ore. Mining engineers – often from the United States – were the key actors in this transition.

The mining industry – like virtually all other forms of intense industrial activity – has changed fundamentally over the course of the last two centuries. Arguably, the greatest change came in the period from the gold rushes to the years following the First World War. One way to appreciate the extent of that change is by noting that participants in the gold rushes employed mining methods similar to those

Figure 3.1 Operating a windlass, woodcut from *De re metallica* (1556) (public domain)

Figure 3.2 Windlass at the Barker Claim, William's Creek (1868) (reproduced by permission of the Royal BC Museum and Archives)

illustrated by the woodcuts that appeared in the sixteenth century mining classic, Agricola's *De Re Metallica.*

By the early twentieth century this was no longer the case: new processes had fundamentally changed the work of mining. Mining engineers oversaw these changes and had assumed a new significance in mining, not least in determining how work would proceed. The following pages examine this transition, focusing on the period of greatest change.

The genesis of modern mining in three communities

'Modern mining' did not begin with the gold rushes but those events certainly had an impact on mineral production from the middle of the nineteenth century. Beyond the economic stimulus of a significant increase to the gold supply, however, at first not much changed. Rudimentary mining methods remained adequate so long as the mineral resource could be profitably recovered.

The expanding base and precious metals mining industry relied chiefly on the exploitation of high-grade ore deposits found in those regions of the world then being colonized by Europeans. This process – in particular, with British and American expansion into the Pacific basin – led to growing numbers of Europeans occupying lands where large and unexploited mineral deposits existed. In a number of instances Aboriginal peoples had previously mined these resources extensively but were no longer doing so.[3] A common barrier encountered by early miners was underground water.

By the mid-nineteenth century, engineers and mine managers were developing new methods which enabled mining to overcome traditional barriers associated with working at depth. The British development of steam technology – itself closely associated with mining and metallurgy – addressed two significant constraints to underground mining. Steam-driven pumps could keep mines relatively free of water and steam-driven haulage systems could raise ore from underground, enabling deeper mines to be more easily worked.[4] European smelting methods were also becoming more efficient and formed part of the intellectual tool kit that German- and British-trained metallurgists took with them as they ventured overseas.[5] However, relocating European technology overseas was not a straightforward process. Mining camps in Australia, South Africa, Canada and the United States were typically places where wages were higher and transportation was far more challenging than in Europe. Mine managers tended to regard European methods with considerable suspicion and they were often under intense pressure to produce profits quickly.[6] In such circumstances, experimenting with novel processes was a risky business.

Mining depended not only on the ability to bring ore to the surface but also to treat it successfully, efficiently separating the mineral content from the waste rock. The market price of the metal was another factor that could determine whether the operation was commercially viable, as the collapse of silver in 1891 demonstrated. The intense pressure to generate profits in the face of high costs and by extracting ore of fluctuating value encouraged mining engineers to embrace what would come to be called 'modern mining'. As a prominent mining engineer put it in 1907, 'the crux of the whole subject is business results, and modern mining requires that the balance shall be on the right side of the ledger at the close of the year'.[7] Twenty years later, another engineer spoke at greater length on the subject. 'Modern mining', he told his colleagues at a conference, had

> to deal with ore of lower grade and ever increasing complexity, and in order to make a mining enterprise on those lines profitable, it became necessary to handle very large tonnages, which, of course, involved large capital expenditure. Modern mining had hence become, to a great extent, a problem in mass production, and, on that account, mechanical methods had necessarily been introduced extensively. That meant, however, that increasing demands were being made upon the modern mining engineer, not only in respect of his knowledge of mining, metallurgical, and mechanical details, but more particularly upon his administrative and organising abilities, mining enterprises today being vast and complex concerns.[8]

Implicit in this account is the assumption that modern mining occurred in tandem with the employment of a modern mining engineer.[9]

During the late nineteenth century and the first years of the twentieth, this fundamental change affected many of the world's leading mining regions, not least Butte, Johannesburg and Broken Hill. However unique these three locales were – in terms of their racial and ethnic tensions, labour relations, climate, geography, accessibility, ore and much else – one can argue that the mining industry in each community confronted broadly similar challenges over roughly the same period, from the 1890s through to the early 1900s. At their most basic, the issues confronting the industry involved the mines' need to remain profitable as costs rose and ore values declined. The only viable solution was to rely increasingly on economies of scale, and such economies of scale could only be achieved by adopting new techniques and processes. That said, in Butte, corporate manoeuvring was as critical as the recovery and treatment of copper ore. Both Johannesburg and Broken Hill, by contrast, faced a common dilemma: as mines reached greater depth, their ore became more complex and consequently more difficult to treat. Virtually everyone familiar with the industry recognized that its continued success would depend upon acquiring the expertise to treat low-grade (and often refractory) ores.

The adoption of two key processes, cyanidation and flotation, was crucial in the shift to a new regime. Although these two innovative methods were in processing technology, they had a profound impact on underground work methods and enabled mining low-grade ore to become cost-effective. The development of both processes reflected a more sophisticated grasp of chemistry applied to metallurgy, fast becoming the new science of mining.[10] Their effectiveness also created new problems, notably by dramatically increasing mining's environmental impact. (Pollution was already becoming a factor in Butte, where litigation over environmental degradation had assumed a significant role.[11]) While the environmental impact of mining would continue to be an issue – and in many ways was the inevitable consequence of the mass mining of low-grade ore – the adoption of cyanidation and flotation ensured that mining could continue to be profitable. By the 1910s, the mining companies of Butte, Johannesburg and Broken Hill had resolved the challenge of profitably mining low-grade ores.

This resolution relied not only on the adoption of new technology but also by containing labour costs. The three communities were hardly alone in their concern with cost reduction. By the early 1900s, this was a widely publicized preoccupation of the mining industry. The leading technical journals devoted much space to the 'The Costs of Mining' and in 1905, the prominent mining journalist, T. A. Rickard, produced an edited collection, *The Economics of Mining*. It was presumably successful since a second edition appeared two years later, in 1907. Another book appeared in 1909, Finlay's *The Cost of Mining: An Exhibit of the Results of Important Mines Throughout the World*.[12] Tied to this concern with costs was a growing consensus that the mining engineer possessed privileged technical knowledge, someone whose key function was to ensure commercial success.[13]

These books and journals were almost always published in the United States and by the 1910s American-trained mining engineers were achieving prominence around the world. These individuals not only controlled much of the industry, they also participated in professional organizations and supported a range of technical journals. Indeed the pages of the leading journals – New York's *Engineering and Mining Journal* and San Francisco's *Mining and Scientific Press* – often ran articles on innovations developed on mining fields in countries thousands of miles apart, articles which then could spark lively discussions in the correspondence pages among geographically dispersed practitioners. Those conversations illustrate the workings of what can be best described as the engineers' epistemic community and suggest the willingness of its members to share information.

The growing dominance of American mining engineering methods reflected several developments.[14] The overseas success of individuals such as John Hays Hammond, Hamilton Smith, Hennen Jennings, Herbert Hoover and others suggested that lucrative opportunities existed abroad, and surviving correspondence indicates that many of these individuals actively recruited colleagues to assist them. In addition, American engineering relied heavily on notions of portability and replication, far more than European engineering practice. Just as American locomotives and bridges could be prefabricated and exported cheaply (to the considerable irritation of English colleagues), so too could American mining methods be successfully applied in South African or South American mining camps.[15] Finally, the experiences on the American mining frontier often demanded creative problem-solving. As James Douglas noted in his presidential address to the American Institute of Mining Engineers in 1899 – an address entitled 'The Characteristics and Conditions of the Technical Progress of the Nineteenth Century' – 'Where [American mining engineers] differed from their brethren abroad was in their inoculation by the spirit of adaptiveness which is so strong a feature of the national character'.[16] In addition to the rough and ready conditions on the frontier that demanded such adaptiveness, engineers could often find themselves confronting daunting geological conditions. This too could stimulate innovation; as a leading geologist pointed out in 1906,

> The interests of good mining are not always served by the finding of rich ore. True progress in the art is more apt to be recorded at the low-grade mines . . . where small economies may make the difference between profit and loss, and so it comes that the best practice and the most modern inventions may be found at our base-metal mines.[17]

The examples of Butte, Johannesburg and Broken Hill tend to support this analysis.

By the 1910s and continuing into the 1920s, the work of mining was becoming increasingly standardized. The widespread adoption of mass mining and milling techniques meant that at most large mining camps, people and machines were integrated into measurable and controllable systems.[18] With some exceptions, mining was now chiefly concerned with the extraction of massive tonnages of low-grade ore. Two consequences followed. Far less reliance was placed on the skill of the working miner; instead, the ability of the professional mining engineer was regarded as the essential element in successful mining operations. And the environmental footprint of mining grew exponentially as operations extracted and treated far greater amounts of leaner ore.

Butte, Montana: the copper crucible

Butte became famous for its copper production in the late nineteenth century and was one of the world's first successful low-grade copper camps. Its spectacular growth helped to make the US the world's leading copper producer:

> Between the years 1880 and 1894, the United States definitely attained world leadership as a producer of copper. In 1880, she mined less than 20 per cent of the world's supply, while in 1895 the American output exceeded 50 per cent of world production. During the next two decades, her relative contribution continued to increase, though at a much less rapid rate. It reached almost 60 per cent in 1909, but did not go higher until the war-stimulated year of 1916.[19]

Despite soaring copper production during the 1880s as more and more mines in the west opened, mining companies in such places as Butte confronted unique challenges. In an account of 'The Mines and Reduction Works of Butte City, Montana' written in the early 1880s, E. D. Peters described some of these challenges. He admitted that their mining and ore treatment techniques were not especially efficient and that considerable copper was lost in the tailings, but he pointed out that

> in Butte *the ore is the cheapest thing we have*, while fuel, labor, and machinery are extremely expensive, and that *the copper contained in the ore has cost so little to mine that it does not acquire any considerable value until a certain amount of labor has been expended upon it.*[20]

In this environment, mining companies were keen to adopt any process that would enable them to save on fuel or labour costs. Their smelters soon featured a technology first proposed in England and turned into a working process in France. The Manhès Converter – developed in France in 1880–1881 – successfully applied Bessemer's method of treating iron to copper. Within a few years, Manhès converters were in use at Butte, Montana and by 1890 'the Bessemer converter had come to be generally recognised [in the US] as the most economical method for making copper from matte'.[21]

Peters went on to become the world's leading authority on copper smelting. In 1884 he helped install Manhès converters at the Parrot works in Butte, and the following year began a series on 'Modern American Copper Smelting' for the *Engineering and Mining Journal*, with columns on the topic appearing regularly for the next 11 months. These were then collated and published as a book, *Modern American Methods of Copper Smelting*, in 1887.[22] Peters was at pains to describe how these modern American methods had evolved and why they came to differ from the traditional Welsh method. He explained that in the US it was typically the case that the ore supply comes from only one or two sources, constant in its composition, and usually in very large quantities. This, with the high wages and exceedingly expensive fuel, has caused the introduction of labor-saving machinery and appliances to an unprecedented extent, as well as a constant endeavor to lessen the proportion of fuel to ore smelted. The lack of steady and skilled furnacemen, and the high cost of refractory materials ['fire-brick, clay, siliceous sand, etc.'], have also had a powerful influence in shaping the processes of treatment, and have perfected the water-jacketed cupola, without which many of our most successful metallurgical enterprises could hardly exist. The same influences have concentrated the works for the refining of copper in a very few hands, and located them with the view to cheap coal and refractory materials and to a market for the finished product.

In Wales, by contrast, Peters noted that one

is confronted by the inextricable mingling of the commercial with the metallurgical that is so characteristic of the English system. Without a thorough understanding of the peculiar local conditions under which the ores are purchased at the Swansea ticketings, it is impossible fully to appreciate the fine points of the complex and ingenious system that time and circumstances have elaborated, or to realize the important influence exercised on the whole subsequent series of operations by the amount of judgment displayed in the purchase of the ores, and in the adaptation of the same to the immediate needs of the works.[*Ed., original note*: See Percy on Copper for a full description of the Swansea ore sales, together with quality and value of ore offered] The Swansea smelter receives his ore in numberless small parcels, differing not only in richness, but in purity and other qualities. To carry out the reverberatory process to the best advantage, he requires, in addition to the main supply of sulphide ores, a certain proportion of oxides and carbonates, all of which are obtainable in the public ore market. His coal is of the cheapest and most suitable quality, and the refractory material – fire-brick, clay, siliceous sand, etc. – is obtainable at prices far below American rates. He also has at his command a body of experienced and skillful workmen who have grown up at the furnaces, and who, at very low wages, are fully capable of executing all the difficult operations demanded by this system of treatment. In addition, he has a market for his product, where every variety of metal brings the highest justifiable price.

It is very evident that such a state of affairs cannot be compared with average American conditions.[23]

Twenty years later, in 1907, William Gowland added to this analysis in his presidential address at the Institution of Mining and Metallurgy in London.

Speaking on 'The chief Advances in Copper Smelting in Modern Times', Gowland noted both the dramatic shift in copper production over the last half century (from 1854 to 1905) and the fact that Europe was no longer at the centre of copper production. He suggested that

> the enormous output of the metal at the present time would have been quite impossible with the old methods, furnaces and appliances, some of which still survive in this country and in continental Europe.
>
> In the United States all the conditions favourable to the introduction of new processes or improvements of the old were present, as there were no old plants to be discarded and written off, neither were there any old smelting customs or traditions to retard progress; moreover, the ores were present in superabundant quantities.[24]

His point that custom and older plant could retard progress often featured in critiques of English metallurgical practice.

Johannesburg, South Africa and the mines of the Rand

If Butte was famous for its extensive copper deposits, South Africa would become one of the world's most productive gold producers. This came somewhat later in South Africa: it was an important producer by the mid-1890s but then lost ground during and after the South African War (1899–1902). By 1903, significant mining was again underway and the Rand began to reestablish itself as one of the world's leading gold producers, a position it held for many decades thereafter.[25]

The key discovery came in 1886, in what was then the South African Republic (often known as Transvaal), with the discovery of the Witwatersrand Main Reef.[26] Unlike some other gold discoveries, individual miners had little chance of winning quantities of alluvial or placer gold; the mining industry that followed the 1886 discovery required large amounts of capital to cover the costs of sinking shafts and other ancillary plant needed to recover gold from the ore. The discovery of diamonds in South Africa nearly 20 years earlier meant that the region was familiar with mining and that significant sums of capital were available locally to invest in the new gold field.

One feature of the emerging gold mining industry was the number of American mining engineers who came to work on the Rand. Hamilton Smith's journey from California to South Africa is illustrative of the ways in which such individuals could make the journey. As superintendent of the North Bloomfield mine in the Sierra Nevada, Smith had become involved in a lawsuit over the environmental impact of mining in California. The lengthy litigation was initiated by the state's farming community, who wanted to prevent the dumping of many tons of gravel into the river system upstream from their farms. The court upheld the farmers, a decision which Smith saw as a mortal blow to hydraulic mining in California. In 1883, he accepted an offer from the Rothschilds to manage a gold mine that they owned in Venezuela. He soon demonstrated the wisdom of hiring an expert engineer: under his management, the cost of treating a ton of ore dropped from over $45 to just over $15. When Smith described how he had been able to do this, the leading mining journal editorialized that this 'must be not only extremely gratifying to the fortunate shareholders, but also a source of very justifiable pride to Mr. Hamilton Smith himself, and to our American mining engineers generally'.[27] Smith surrounded himself with other American engineers at the mine, and when the Rothschilds became interested in South African mining a few years later, Smith and his colleagues crossed the Atlantic and helped to build the gold mining industry then developing on the Rand. Smith's nephew, Hennen Jennings, played a particularly important role, for it was largely on his recommendation that the cyanide process was adopted by South Africa's gold mining industry, a key innovation that ensured the long-term profitability of that country's low-grade gold mines.[28] In addition to Hamilton Smith and Hennen Jennings, such individuals as Gardner Williams, Thomas Leggett and John Hays Hammond rose to fame there. By the end of the nineteenth century, influential investors in London believed that American mining engineers possessed particular ability and thus were especially well equipped to oversee the work of mining.[29]

Like other mining camps, the Rand went through an initial period of fevered development followed by one in which its longevity depended on adjusting to changing geological conditions. The region's political volatility as well as the South African War made this adjustment particularly challenging. The first difficulties came in 1889–1890, as the free milling ore was worked out and replaced by refractory ores.[30] The industry's ultimate recovery rested both on fundamental changes to the labour regime as well as deploying technology that would enable the massive deposits of low-grade ore to be treated efficiently.[31] Key to the latter was the cyanide process, which could guarantee extraction rates of 80 per cent and better. The process rested upon gold's solubility in cyanide, something that chemists had known for most of the nineteenth century. However this principle was not translated into an economical treatment method until 1887.

Through the 1890s gold fields around the world adopted the cyanide process, nowhere more successfully than in South Africa.[32] When leading metallurgists published *A Text-book of Rand Metallurgical Practice* in 1911, they dedicated it to J. S. MacArthur, the man responsible for developing the process, acknowledging that his

> pioneer researches and introduction in 1890 of the cyanide process as an essential feature of Rand metallurgical practice have rendered possible the successful treatment on scientific principles of low-grade banket ore, and have been a prime factor in establishing the Witwatersrand goldfields as the premier gold producer of the world.[33]

By this point, as Van Helten acknowledged, the mining industry had been transformed:

This transformation did not only concern the familiar change from a collection of diggers into a highly centralised industry, but also involved the equally crucial transformation from mines working a mixture of high grade ore out-crop/deep levels into an industry consisting almost exclusively of low grade ore deep levels . . . the financial and labour crises which befell the industry during the post-War period take on an added meaning, if they are considered within the framework of the mines' attempts to change into a low grade ore industry, which they accomplished, more or less successfully, towards the end of the period under review [1886–1910].[34]

As in Butte, the mining industry relied on the expertise of leading engineers to plan and execute the necessary changes. These engineers participated in an international community of practise, one fostered by industry journals, engineering associations and informal relationships. This was a community in which knowledge was freely exchanged, thereby enabling a relatively quick diffusion of ideas about mining technology and best practices, quite literally around the world.

In his inaugural address in 1908, for example, the new president of the Chemical, Metallurgical and Mining Society of South Africa noted that 'Our Society is now widely represented by members in all parts of the world' and drew attention to the election of several Corresponding Members (for Mexico, the United States and Asia). He hoped that these members overseas 'will do their utmost to further our interests by bringing in new members, and by obtaining for our information and instruction, papers upon the many and varied subjects which come under the scope of our work'.[35]

Broken Hill, New South Wales and the mines of the Barrier

Like families, mining camps can be fundamentally alike but dissimilar in their own way. On the western edge of New South Wales, Broken Hill differed from Butte and Johannesburg in that its longevity and fame as a mining camp reflected its enormous silver/lead/zinc deposits. Although silver was the metal that the first miners sought, Broken Hill would come to rely on base metals, differentiating it in some important ways from Johannesburg, whose gold had a stable value as well as particular political role.[36] Like Butte, both labour and transportation costs were high; it was relatively remote from major centres and had few amenities to offer.[37] Broken Hill also confronted unique challenges in treating its ore.[38]

Initially this was not a problem, as the surface ore was relatively easy to treat. In common with many other mining camps in the late nineteenth century – not least, South Africa – Broken Hill's metallurgical expertise was drawn chiefly from the United States. For example, the dominant company, Broken Hill Proprietary (BHP), employed American technology when it built the area's first ore concentrator in 1889.[39] The concentrates produced by the plant carried high mineral values which could be readily recovered by smelting. The system was not perfect since once the stampers in BHP's concentrating mill had pulverized the ore, much silver was lost in the tailings.[40] The company built a leaching plant in 1890 to try and recover this silver, although it too was far from efficient.

In the early years, BHP and the other Broken Hill mining companies regarded such lost mineral content as an unavoidable consequence of concentration. Outside observers took a different view of such inefficiencies. Donald Clark, 'Special Commissioner to the *Australian Mining Standard*', referred in 1904 to 'the extravagant treatment of the early days' and derided the 'Concentration methods [which] give frightfully wasteful results'. In an article on 'Concentration by Gravity Processes', Clark claimed that 'The recognised losses have been, and are now, enormous, and very little if any effort has been made to materially reduce them . . . .more than half [the mineral] value is lost'.[41] When the mineral value of Broken Hill ore began to decline, coinciding with a fall in metal prices, the companies were forced to improve their concentrating techniques.

The drop in Broken Hill's ore values reflected the gradual depletion of its rich oxidized ore. Production increasingly relied on the deeper, lower grade sulphide ore, distinct from oxidized ore in several ways.[42] It was leaner, that is to say, its mineral content was considerably less, but also its structure was more complex. The ore still contained silver and lead but it also included zinc and other minerals. Typically, the zinc and lead values were roughly equivalent, running from 10 to 15 per cent each in a ton of sulphide ore.[43] The presence of zinc seriously hampered smelting lead-silver ores. It had to be removed to permit efficient smelting operations but the zinc sulphide (or blende) and the lead sulphide (or galena) were so closely associated in the sulphide ore that the Broken Hill concentrating mills were unable to separate the two. A mood of pessimism prevailed as the mining companies contemplated their future once the oxidized ore reserves were exhausted.[44]

Facing the same discouraging prospect, BHP and the other mining companies arranged for a leading German metallurgist to visit Broken Hill in 1892, to suggest ways to treat the sulphide ore.[45] He recommended relying on roasting and leaching processes to remove the zinc. This meant abandoning considerable investments and expertise in concentrating, a loss the Broken Hill companies were not prepared to accept. The advice was ignored and the companies turned instead to other authorities, closer to home. The South Australian copper mines at Wallaroo and Moonta had developed relatively sophisticated ore concentrating techniques, and their expertise and technology were adopted at Broken Hill. Captain Warren, an old hand from South Australia, pioneered new concentrating methods at Broken Hill's Block 10 mine, introducing his own patented vanner.[46] The Hancock jig, a South Australian invention, was another crucial component in this new technology. By 1894 the experiments were yielding favourable results and gradually Broken Hill companies began to produce concentrate that was high in lead and low in zinc from sulphide ore.

The solution to Broken Hill's sulphide problem brought new challenges: vast quantities of slime and tailings began to accumulate, as concentrators separated out the galena from the sulphide ore. By 1904, 6.5 million tons lay outside the mills, including a large amount of zinc whose recovery seemed virtually impossible.[47] The difficulty was that the sulphide ore contained appreciable amounts of rhodonite and garnet, in addition to the galena (lead sulphide) and the blende (zinc sulphide). Successful concentration required some difference in the specific gravity of the materials being treated but unfortunately for the mining companies, the zinc blende, the rhodonite and the garnet shared a very similar specific gravity. Even with the considerable advances in concentrating technique during the 1890s, mechanical concentration could not separate them.[48]

As the concentrating mills generated tailings rich in zinc, the mining companies of Broken Hill came to realize that by solving one problem they had created another. A way had to be found to recover the unrealized profit that lay in the tailings. 'The zinc product is dumped on the mines where it remains a vast asset, locked up', declared the Inspector of Mines in 1901, 'The zinc problem at Broken Hill is therefore the question of the hour'.[49] The flotation process, a novel method of ore treatment, was the solution to the zinc problem. Although its 'invention' soon became mired in controversy and litigation – and would remain so for some 20 years – Broken Hill metallurgists played a critical role in its development.[50]

Instead of utilizing differences in specific gravity, flotation separated the two materials by taking advantage of 'the surface tension of liquids and the adhesion of liquid films to the surfaces of minerals'. The result, noted a writer with the American Bureau of Mines, was 'concentration turned upside down'. He went on to define flotation: 'the process or processes by which the valuable minerals in a mass of finely ground ore are caused to float on a liquid into which the finely ground ore is fed'.[51] Although a revolutionary way to separate ore, what made the process so critical to the mining industry in the early years of this century was not its novelty *per se* but its efficiency. First at Broken Hill, and subsequently at most base and precious metals mining camps throughout the world, complex ores virtually impossible to treat by earlier methods were now treatable.

The environmental turn

This ability to recover the mineral content of virtually any ore came at a cost. Low-grade ores could be treated but only by adopting economies of scale. As a consequence, the capacity of mine mills expanded dramatically and the tailings or waste generated by the mills assumed massive proportions. Mining and metallurgical engineers tended to pay little attention, however, until society at large became increasingly concerned about environmental degradation. This came in the wake of Rachel Carson's *Silent Spring* (1962), the ritual celebration of Earth Day beginning in April 1970 (the same year that Greenpeace was launched), and the publication of *The Limits to Growth* (prepared for the Club of Rome in 1972) and E. F. Schumacher's *Small Is Beautiful: Economics as if People Mattered* (1973). Herbert Kellogg, a distinguished metallurgist at Columbia University, acknowledged the need for greater environmental awareness in his Sir Julius Wernher Memorial lecture in 1977. He drew attention to 'the urgent need for conservation of natural resources [which] obliges the metal production industry to adopt process designs that use less energy, require less capital investment and protect the environment and the health of workers'.[52] He returned to this theme the following year, in his Distinguished Lecture in Materials and Society entitled 'Toward a Materials-Conservation Ethic':

> Clearly, we are in environmental trouble today. We face real, identifiable risks to life and well being, both inside and outside our factories, in the air of our cities and in the water we drink. The problems may not yet be at a crisis stage, but they are serious enough to justify the sometimes costly efforts for environmental protection. Within recent years we have seen enactment of state and federal laws to control environmental pollution. I see this trend as inevitable, necessary, long-term, and involving an increasing number of industrial and individual actions subject to control. Gone are the simpler days of 1900 when the earth, the air and the water served as convenient sewers, with danger only to a minor fraction of the population.[53]

Kellogg included a graph to illustrate the relationship between increasingly low-grade ores and the energy required to recover their metal.[54] His point was that the growing reliance on lower grade ores came at a cost 'in labor, capital, energy requirement and environment damage', and that sooner or later these costs could well become prohibitive.

The need to confront the consequences of mining is even more pressing today. The end product – a bar of gold or silver or some other metal product – is not often

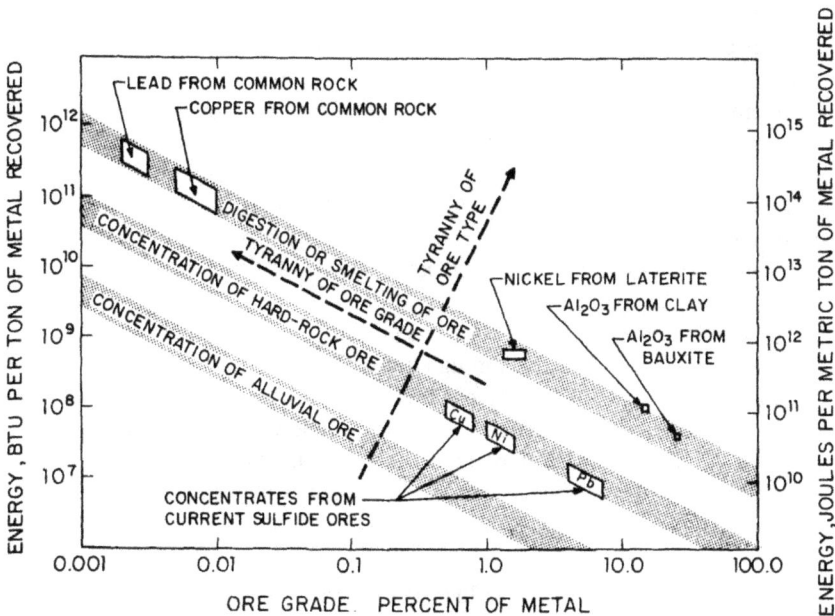

Figure 3.3 The tyranny of ore grade and type (reproduced by permission of Springer Nature)

associated with the environmental toll that its extraction entailed but it is past time that we began – in Bill McKibben's phrase – to connect the dots.[55] In an essay on 'The price of gold, the value of water', the writer Rebecca Solnit recalled another's comment that 'everyone who buys gold jewellery should get to deal with the consequences, too – such as the tailings it took to produce that much gold'. Solnit explained that ever since hearing that comment, she has imagined:

> a truck driver ringing the doorbell of a home to say something like, 'Ma'am, about that new wristwatch: would you like your seventy-nine tons on the front lawn or the back? You'll want to keep the kids and the dog off them 'cause of the acid and the arsenic'.[56]

As this pointed remark suggests, mining continues to produce a vast amount of waste, an industrial footprint that we can no longer ignore. And this footprint is a direct consequence of the engineering changes that began in such places as Butte, Johannesburg and Broken Hill.

The burgeoning field of environmental history has begun to engage the conventional mining historiography, revising (and complicating) what formerly seemed a straightforward linear narrative.[57] Although historians continue to examine the impact of mining, one area that has attracted relatively little attention – at least in North America – is its decline.[58] The topic did garner significant publicity in late 1984 when *Business Week* published a cover story on 'The Death of Mining'.[59] One might argue that this obituary came too soon – that rumours of mining's passing were greatly exaggerated – but the American industry is certainly a good deal less significant than in the past. The numbers employed as well as the industry's contribution to local and state economies continues to decline.[60]

The gradual eclipse of North American mining is not a straightforward declensionist narrative but a complex story involving the impact of various economic and political factors, as a government study concluded in 1986.[61] The major North American mining corporations increasingly sought new fields to locate and work mineral deposits around the globe. This shift in mining activity offshore highlighted fundamental changes in the world's markets. The US, as Gavin Wright argues,

> has not become 'resource poor' relative to others, but the unification of world commodity markets (through transport cost reductions and elimination of trade barriers) has largely cut the link between domestic resources and domestic industries. American corporations and engineers have been in the forefront of the globalization of the mineral economy. In essence, the process by which the United States became a unified 'economy' in the nineteenth century has been extended to the world as a whole.[62]

This reflected specific developments dating from the later nineteenth century rather than unique or exceptional features inherent in the US.

The American mining industry that emerged after the gold rushes stimulated the construction of transportation networks, enabled capital formation and encouraged an industrial model that was both remarkably efficient and relatively portable. In response to the conditions in such places as Butte, American mining engineers devised methods that were essentially generic and grounded in economies of scale, unlike European methods which were largely custom processes tailored to specific ores.[63] It follows that the changes that occurred in Butte, Johannesburg and Broken Hill described in these pages could be – and were – replicated elsewhere, a process that has gained force with globalization. We might also note that large and influential finance capital was associated with these three communities. The role of British capital in South African mining is well known and need not be described here. Broken Hill companies enjoyed close links with Melbourne capital and assumed an important role in Australia, both through the premier company, BHP, as well as through Collins House, a later financial group that profited especially from flotation's efficacy.[64] Butte's leading company, the Anaconda Copper Mining Company, was 'one of America's great corporations, by far the mightiest of world copper companies'.[65] In the years since mining's heyday in Butte, Johannesburg and Broken Hill, multinational mining corporations have replaced those earlier companies that were more grounded in specific regions of the world. This is a process that has yet to attract the scholarly attention it deserves, despite widespread and ongoing media coverage.[66] Mining historians need to move beyond the region or nation as their unit of analysis, to become as transnational as the major mining corporations.

Notes

1 The quotation is from an 1856 address that Abram Hewitt gave to the American Geographical and Statistical Society, 'Statistics and Geography of the Production of Iron', quoted in Allan Nevins, *Abram S. Hewitt* (New York: Harper & Brothers, 1935), p. 95. Harvey S. Perloff and Lowdon Wingo, Jr. describe 'the minerals–dominant economy' in their essay, 'Natural Resource Endowment and Regional Economic Growth', in: Joseph J. Spengler (ed.), *Natural Resources and Economic Growth* (Washington: Resources for the Future, Inc., 1961), pp. 193–97, and E. A. Wrigley relies on the concept of a mineral-based economy in his book, *Continuity, Chance and Change: The Character of the Industrial Revolution in England* (Cambridge: Cambridge University Press, 1988).
2 As Paul McGann noted, 'Typically for the 50 years before 1910, new discovery and technical progress combined caused decisive and continual shifts outward (and downward) of supply curves (showing increasing supply at constant prices) over the entire period'. ('Technological Progress and Minerals', in: Spengler (ed.), *Natural Resources and Economic Growth*, p. 78.) For a convenient source of data, see Christopher Schmitz's *World Non-ferrous Metal Production and Prices, 1700–1976* (London: Cass, 1979).
3 See for example Susan R. Martin, *Wonderful Power: The Story of Ancient Copper Working in the Lake Superior Basin* (Detroit, IL: Wayne State University Press, 1999). One can discern similar pattern in other parts of the world. Describing the European 'discovery' of copper deposits in sub-Saharan Africa, for example, mining historian A. J. Wilson noted that 'most of the big finds of the early years of the present century in both of the Rhodesias (Zambia and Zimbabwe) were more in the nature of a rediscovery of ancient mineral wealth than a sudden stumble on potentially rich resources'. (A. J. Wilson, *The Living Rock: The Story of Metals since Earliest Times and Their Impact on Developing Civilization* (Cambridge: Woodhead Publishing, 1994), pp. 259–260).

4 Cornwall played a critical role in these developments; for a useful overview, see Allen Buckley, *The Story of Mining in Cornwall* (Fowey, Cornwall: Cornwall Editions Limited, 2005), esp. pp. 94–97, 104–08, 121–22.

5 For evidence of this process of diffusion, see James E. Fell, Jr., *Ores to Metals: The Rocky Mountain Smelting Industry* (Lincoln, NE: University Nebraska Press, 1979); James E. Fell, Jr., 'Copper Smelting in North America: International Transfer and Technological Change', paper presented at the Fourth International Mining History Congress, Guanajuato, Mexico, 14 November 1998; Fred Quivik, 'Smoke and Tailings: An Environmental History of Copper Smelting Technologies in Montana, 1880–1930' (PhD thesis) (University of Pennsylvania, 1998). Lynn R. Bailey, *Shaft Furnaces and Beehive Kilns: A History of Smelting in the Far West, 1863–1900* (Tucson: Westernlore Press, 2002); Joël Kapusta and Tony Warner (eds), *Peirce-Smith Converting Centennial Symposium* (Warrendale, PA: John Wiley & Sons, 2009); and Warren Alexander Dym, 'Freiberg and the Frontier: Louis Janin, German Engineering, and "Civilisation" in the American West', *Annals of Science 68*(3) (2011), pp. 295–323.

6 See for example the discussion in Dym, 'Freiberg and the Frontier'. As he points out, 'European-trained engineers . . . met widespread nativism on the frontier, including hostility to chemistry and its "professors"', pp. 322–323.

7 Thos. H. Leggett, 'The Requirements of Modern Mining', *Mining and Scientific Press 95*(1) (6 July 1907), pp. 16–17 (this was a letter to the editor, in response to an earlier column with the same title).

8 H. W. Gepp, 16 August 1928, *Proceedings of the Australasian Institute of Mining Engineers (New Series) 71* (1928), pp. lxiii–lxiv. (Gepp was closely associated with the development of the flotation process at Broken Hill.) For a similar description, see the chapter 'The Origins of Modern Mining Methods,' in Harold Barger and Sam H. Schurr (eds), *The Mining Industries, 1899–1939: A Study of Output, Employment and Productivity* (New York: National Bureau of Economic Research, 1944), pp. 97–116.

9 As Barger and Schurr put it in the chapter cited in the previous note, 'No longer does the success of the mining enterprise depend upon the expertness with which the miner breaks the mineral and separates it from the waste; it is now a question of how well the engineer has designed mining and beneficiating operations on the basis of his geological data, and how carefully he has determined the geological structure and chemical nature of the ore deposits prior to the working out of suitable techniques' (*The Mining Industries, 1899–1939*, p. 115).

10 For a contemporary account, see Arthur C. Claudet, 'The Relation of Chemical Industry to Metallurgy', *Journal of the Society of Chemical Industry 29*(24) (1910), pp. 1421–1428, esp. 1424–1428 (the discussion that followed Claudet's presentation).

11 See Donald MacMillan, 'A History of the Struggle to Abate Air Pollution from Copper Smelters of the Far West' (PhD thesis) (University of Montana, 1973), subsequently published as Donald MacMillan, *Smoke Wars: Anaconda Copper, Montana Air Pollution, and the Courts, 1890–1920* (Helena, MT: Montana Historical Society Press, 2000); Quivik, *Smoke and Tailings*; and Timothy J. LeCain, *Mass Destruction: The Men and Giant Mines that Wired America and Scarred the Planet* (New Brunswick, NJ: Rutgers University Press, 2009).

12 Thomas A. Rickard (ed.), *The Economics of Mining* (New York: The Engineering and Mining Journal, 1905); *The Economics of Mining*, second edition, (New York: Hill Pub. Co., 1907); and James Ralph Finlay, *The Cost of Mining: An Exhibit of the Results of Important Mines Throughout the World* (New York: McGraw-Hill Book Co., 1909).

13 For example, 'By far the most important side of [the mining engineer's] work is the consideration of cost – that is, not that he should get the absolutely lowest cost per ton or per yard, but that he should get the best result at the lowest cost, and get his work of all kinds done at a minimum of cost. His commercial success is really the supreme test of his ability'. (James B. Lewis, 'The Training of a Mining Engineer', *Australian Institute of Mining Engineering 12* (1907), p. 34).

14 Roger Burt questions the 'American-ness' of much of this new technology and argues that it was often derivative (Roger Burt, 'Innovation or Imitation? Technological Dependency in the American Nonferrous Mining Industry', *Technology & Culture 41*(2) (2000), pp. 321–347). However, as Dianne Menghetti suggests in 'Invention and Innovation in the Australian Non-Ferrous Mining Industry: Whose Technology?' (*Australian Economic History Review 45*(2) (2005), pp. 204–19), the whole issue of ascribing national identity to technology is fraught with problems. See also Eugene S. Ferguson, 'The American-ness of American technology', *Technology & Culture 20*(1) (1979), pp. 3–24; Dianne Newell, 'All in a Day's Work: Local Invention on the Ontario Mining Frontier', *Technology & Culture 26*(4) (1985), pp. 799–814; and David A. Hounshell, 'Rethinking the History of "American Technology"', in Stephen H. Cutcliffe and Robert C. Post (eds), *In Context: History and the History of Technology – Essays in Honor of Melvin Kranzberg* (Bethlehem, PA, and London: Lehigh University Press, 1989), pp. 216–229.

15 A series of successful American contract bids underscored their ability to compete with their British counterparts. These included the award of a New Zealand locomotive order (1885), the Hawkesbury Bridge contract in New South Wales (1886), the Gokteik Viaduct in Burma (1899) and the Atbara Bridge contract in Sudan (1900). Contemporary news-papers, magazines and government publications indicate just how controversial these American successes were in the UK.

16 James Douglas, 'The Characteristics and Conditions of the Technical Progress of the Nineteenth Century', *Transactions of the American Institute of Mining Engineers 29* (1899), p. 665. Ironically, Douglas himself was a Canadian.

17 Waldemar Lindgren, 'The Development of the Metal Mining Industry in the Western States', *Mining and Scientific Press 93*(22) (1906), pp. 659–661, at p. 660.

18 I am relying here on Logan Hovis and Mouat, 'Miners, Engineers, and the Transformation of Work in the Western Mining Industry, 1880–1930', *Technology & Culture 37*(3) (1996), pp. 429–456.

19 Robert Bunnell Pettengill, 'The United States Foreign Trade in Copper by Classes and Countries, 1790–1932: An Explanation of the Trends and Fluctuations' (PhD thesis) (Stanford University, 1934), p. 123. For a useful statistical summary, see Christopher J. Schmitz, 'The Changing Structure of the World Copper Market, 1870–1939', *Journal of European Economic History 26* (1997), pp. 295–330; also his earlier arti-cle, Schmitz, 'The Rise of Big Business in the World Copper Industry 1870–1930', *Economic History Review 39*(3) (1986), pp. 392–410.

20 E. D. Peters, 'The Mines and Reduction Works of Butte City, Montana', in Albert Williams, ed., *Mineral Resources of the United States, Calendar Years 1883–1884* (Washington: Government Printing Office, 1885), p. 384, emphasis added.

21 For the development of the Manhès converter, see James Douglas, 'Treatment of Copper Mattes in the Bessemer Converter', *Transactions of the Institution of Mining and Metallurgy 8* (1899–1900), pp. 2–48 (the quotation in the text is from p. 8); Dr. James Douglas, 'Principles More Important Than Practice in Technical Education: Commencement Address, May 23, 1913', *Quarterly of the Colorado School of Mines 8*(2) (1913), p. 5; Larry Southwick, 'William Peirce and E.A. Cappelen Smith and their Amazing Copper Converting Machine', in: Kapusta and Warner (eds), *Peirce-Smith Converting Centennial Symposium* (2009), pp. 3–27; and Albert Pelletier, Phillip J. Mackey, Larry M. Southwick, and A. E. (Bert) Wraith, 'Before Peirce and Smith – The Manhès Converter: Its Development and Some Reflections for Today', in Kapusta and Warner (eds), *Peirce-Smith Converting Centennial Symposium* (2009), pp. 29–49.

22 Peters' series on Modern American Copper Smelting began in the *Engineering and Mining Journal 39*(14) (1885), pp. 228–229, and concluded 11 months later, in *Engineering and Mining Journal 41*(10) (1886), pp. 172–173.

23 Edward D. Peters, *Modern American Methods of Copper Smelting* (New York: Scientific Publishing, 1887), pp. 172–73.

24 William Gowland, 'Presidential Address', *Transactions of the Institution of Mining and Metallurgy 16* (1906–07), pp. 265–295, at p. 266. Gowland was at this point Professor of Metallurgy at the Royal School of Mines. Donald Levy echoed Gowland's comments five years later, in *Modern Copper Smelting*, being lectures delivered at Birmingham University greatly extended and adapted, and with an introduction on the history, uses and properties of copper (London: C. Griffin & Company, 1912), p. 11: 'To meet the enormous present-day demand for [copper] metal with the older methods and furnaces would have been impossible. The greatest stimulus to the adoption of these new or modified processes was the shifting of the chief producing centres from the older and more conservative influences to districts like the then newly awakening [American] West, where, with ever-increasing – almost limitless – supplies of ore available, and free from the necessity of considering the capital invested in old plants, the men in charge of the work, untrammelled by old smelting customs which might stand in the way of rapid progress, were in a position to develop their ideas with originality and vigour'.

25 For data on gold production by year and by country, as well as the world total, see Schmitz, *World Non-ferrous Metal Production and Prices, 1700–1976* (1979), pp. 79–91, esp. pp. 84–85. For much of the twentieth century, South Africa was the world's leading gold producer, reaching a peak in 1970, when it produced nearly 80 per cent of the world's gold. Since then, production has fallen dramatically; at the time of writing, four other countries each produce more gold than South Africa (China, Australia, the United States and Russia).

26 There is a large historiography devoted to gold mining in South Africa. For popularly written overviews, see John J. Stephens, *Fuelling the Empire: South Africa's Gold and the Road to War* (New York: Wiley, 2003) and Jade Davenport, *Digging Deep: A History of Mining in South Africa* (Johannesburg and Cape Town: Jonathan Ball Publishers, 2013). Since Hobson's work on South Africa, much has been written about its role within the late Victorian and Edwardian Empire, the context of which is explored in Andrew S. Thompson, 'The Language of Imperialism and the Meanings of Empire: Imperial Discourse in British Politics, 1895–1914', *Journal of British Studies 36* (1997), pp. 147–177, and earlier by Robert Vicat Turrell, 'Review Article: "Finance . . . The Governor of the Imperial Engine": Hobson and the Case of Rothschild and Rhodes', *Journal of Southern African Studies 13*(3) (1987), pp. 417–432. Cf. Peter Cain, *Hobson and Imperialism: Radicalism, New Liberalism, and Finance 1887–1938* (Oxford: Oxford University Press, 2002). Earlier work that covers some of the same territory but with an exclusive focus on mining includes Robert V. Kubicek, *Economic Imperialism in Theory and Practice: The Case of South African Gold Mining Finance 1886–1914* (Durham, NC: Duke University Press, 1979) and Jean Jacques Van-Helten, 'British and European Economic Investment in the Transvaal: With Specific Reference to the Witwatersrand Gold Fields and District, 1886–1910' (PhD thesis) (University of London, 1981). Note also the critical reviews of Kubicek's book by Van Helten, 'Mining and Imperialism', *Journal of Southern African Studies 6*(2) (1980), pp. 230–235, and Shula Marks, 'Scrambling for South Africa', *Journal of African History 23*(1) (1982), pp. 97–113. The following is useful for its reexamination of the Blainey thesis (concerning the significance of the deep-level mining interests): Peter Richardson and Jean-Jacques Van Helten, 'The Development of the South African Gold-Mining Industry, 1895–1918', *Economic History Review 37*(3) (1984), pp. 319–340.

27 Editorial, *Engineering and Mining Journal 42*(10) (1886), p. 163; cf. Hamilton Smith, Jr., 'Costs of Mining and Milling Free Gold Ores', *Engineering and Mining Journal 42*(10) (1886), pp. 168–169. The editorial also noted that 'Mr. Hamilton Smith, Jr's [article] on the Costs of Mining and Milling Free Gold Ores is one of the most instructive and as well as interesting articles on the subject that has yet appeared in print'.

28 For detail of Hennen Jennings' role in the adoption and success of the cyanide process in South Africa, see A. P. Cartwright, *The Corner House: The Early History of Johannesburg* (London: MacDonald & Co, 1965), pp. 95–98; Kubicek, *Economic*

Imperialism in Theory and Practice (1979), p. 63; and Clark C. Spence, *Mining Engineers & the American West: The Lace-Boot Brigade, 1849–1933* (New Haven: Caxton Press, 1970), pp. 303–317. See also the interview, 'Hennen Jennings, and Mining on a Big Scale', *Mining and Scientific Press 111* (1915), pp. 954–971, reprinted in T. A. Rickard (ed.), *Interviews with Mining Engineers* (San Francisco: Mining and Scientific Press, 1922), pp. 223–253. W. R. Ingalls' obituary provides a summary overview of Jennings' career: 'Hennen Jennings', *Transactions of the American Institute of Mining and Metallurgical Engineers* 66 (1922), pp. 819–824.

29 For comment on the importance of American mining engineers in South Africa, see Alpheus F. Williams, *Some Dreams Come True, being a sheaf of stories leading up to the discovery of copper, diamonds, and gold in southern Africa, and of the pioneers who took part in the excitement of those early days* (Cape Town: Howard B. Timmins, 1949), pp. 548–553. When the American mining engineer, Thomas Leggett, gave a paper on 'Present Mining Conditions on the Rand' (printed in the *Transactions of the American Institute of Mining Engineers 39* (1908), pp. 211–223), the then president of the (English) Institution of Mining and Metallurgy, Alfred James, acknowledged the primacy of American engineers on the Rand, referring to Leggett as 'one of those very engineers to whom South Africa is so much indebted' (in the 'Discussion of Leggett's paper', op. cit., pp. 856–859, at p. 857). More generally, see Shula Marks and Stanley Trapido, 'Lord Milner and the South African State', *History Workshop Journal 8*(1) (1979), pp. 50–80 at pp. 61–62; Robert Vicat Turrell, with Jean-Jacques van Helten, 'The Rothschilds, the Exploration Company and Mining Finance', *Business History 28*(2) (1986), pp. 181–205; M. Z. Nkosi, 'American Mining Engineers and the Labour Structure in the South African Gold Mines', *African Journal of Political Economy 2* (1987), pp. 63–80; Elaine Katz, 'The Role of American Mining Technology and American Mining Engineers in the Witwatersrand Gold Mining Industry, 1890–1910', *Economic History of Developing Regions 20*(2) (2005), pp. 48–82; Elaine Katz, 'The Contributions of American Mining Engineers and Technologies to the Witwatersrand Gold Mining Industry, 1890–1910', *Mining History Journal 13* (2006), pp. 12–30; and John Higginson, 'Privileging the Machines: American Engineers, Indentured Chinese and White Workers in South Africa's Deep-Level Gold Mines, 1902–1907', *International Review of Social History 52*(1) (2007), pp. 1–34.

30 See Davenport, *Digging Deep* (2013), pp. 176–177; also J. B. Taylor, 'Recollections of the Discovery of Gold on the Witwatersrand and the Early Development of the Gold Mines' (Cape Town: unpublished (1936), pp. 8–9). Taylor stressed that 'The success of the mines, both in the early days and later, in deep level mining, is largely due to the efficiency of that [engineering] staff' (p. 9).

31 For the changes to the labour regime, see Elaine N. Katz, 'The Underground Route to Mining: Afrikaners and the Witwatersrand Gold Mining Industry from 1902 to the 1907 Miners' Strike', *Journal of African History 36*(3) (1995), pp. 467–489, and Higginson, 'Privileging the Machines', as well as the overview by Peter Richardson and Jean Jacques Van-Helten, 'Labour in the South African Gold Mining Industry, 1886–1914', in: Shula Marks and Richard Rathbone (eds), *Industrialisation and Social Change in South Africa: African Class Formation, Culture and Consciousness, 1870–1930* (London: Longman, 1982), pp. 77–98.

32 The best account of the development and diffusion of the cyanide process is Alan L. Lougheed's self-published *Cyanide and Gold: The Cassel Cyanide Company and Gold Extraction, 1884–1927* (Buderim, Queensland, 2001). See also the sources cited in Note 28, above.

33 Ralph Stokes, Jas. E. Thomas, G. O. Smart, W. R. Dowling, H. A. White, E. H. Johnson, W. A. Caldecott, A. McA. Johnston, and C. O. Schmitt, *A Text-book of Rand Metallurgical Practice. Designed as a 'Working Tool' and Practical Guide for Metallurgists upon the Witwatersrand and Other Similar Fields* (London: Charles Griffin and Company, 1911), p. v.

34 Van-Helten, 'British and European Economic Investment in the Transvaal' (1981), p. 327.

35 R. G. Bevington, 'Inaugural Address', *The Journal of the Chemical, Metallurgical and Mining Society of South Africa 9*(1) (July 1908), pp. 5–6.

36 Note for example A. Atmore and S. Marks, 'The Imperial Factor in South Africa in the Nineteenth Century: Towards a Reassessment', *Journal of Imperial and Commonwealth History 3*(1) (1974), pp. 105–139; Marks and Trapido, 'Lord Milner and the South African State' (1899); Jean Jacques Van-Helten, 'Empire and High Finance: South Africa and the International Gold Standard 1890–1914', *Journal of African History 23* (1982), pp. 529–548; and Russell Ally, *Gold & Empire: The Bank of England and South Africa's Gold Producers 1886–1926* (Johannesburg: Wits University Press, 1994), esp. pp. 11–28.

37 In terms of transportation costs, it is telling that the short rail link connecting Broken Hill with South Australia became very profitable indeed: 'Next to the "Proprietary," it [the Silverton Tramway Co.] is the greatest dividend-paying "mine" on the Barrier, and probably there is no other Railway Company in the world, with only 35 miles of lines, that can boast of paying away in dividends over £1,180,877 in eighteen years [1889–1906]' (Leonard Samuel Curtis, *The History of Broken Hill: Its Rise and Progress* (Adelaide: Libraries Board of South Australia, 1908), p. 119). Blainey notes that the South Australian section of this railway 'became the most profitable ever built in South Australia. Four years after it was opened it was generating 44 per cent of the revenue for the entire railway system of South Australia'. (Geoffrey Blainey, *The Rise of Broken Hill* (Melbourne: Macmillan, 1968), p. 26). A direct rail connection east to Sydney from Broken Hill did not come until 1927. Brian Kennedy's *Silver, Sin, and Sixpenny Ale: A Social History of Broken Hill 1883–1921* (Parramatta, NSW: Melbourne University Press, 1978) provides a telling description of the community's lack of amenities. He has also written an intriguing comparative history of Broken Hill and Johannesburg: Kennedy, *A Tale of Two Mining Cities: Johannesburg and Broken Hill 1885–1925* (Melbourne: Melbourne University Press, 1984).

38 The following is based on J. Mouat, 'The Development of the Flotation Process: Technological Change and the Genesis of Modern Mining, 1898–1914', *Australian Economic History Review 36*(1) (1996), pp. 3–31.

39 In May 1888, the general manager informed shareholders that 'The new ore-dressing plant from Messrs. Fraser and Chalmers, of Chicago, U.S.A., is now on the ground . . . This plant is a duplicate of ore dressing machinery used at the Lake Superior and Annaconda [sic] copper mines in America' (BHP Directors' Report for 31 May, 1888, pp. 12–13). 'Mr. Holly, an American expert in ore concentration', supervised its construction and the plant was in operation by July 1889. (The quotation is from Leonard Samuel Curtis, *The History of Broken Hill: Its Rise and Progress* (Adelaide: Libraries Board of South Australia, 1908), p. 45.) For an overview of American mining engineering in Australia during this period, see J. Mouat, '"Just Now the 'Merican expert is the Prominent Man": American Mining Engineers and the Australian Mining Industry, 1880s–1910s', *Australasian Mining History Journal 6* (2008), pp. 136–149.

40 Note, for example, the remarks in E. T. Henderson, 'The History of Ore Treatment Processes in Broken Hill', *Proceedings of the Australasian Institute of Mining and Metallurgy, New Series 72* (1928), pp. 99–112 at p. 101, and Roy Bridges, *From Silver to Steel: The Romance of the Broken Hill Proprietary* (Melbourne: George Robertson & Co. Ltd, 1920), pp. 191–192.

41 Donald Clark, *Australian Mining & Metallurgy* (Melbourne: Chritchley Parker, 1904), pp. 371–372 and 388–389. The chapters of the book originally appeared as articles in the *Australian Mining Standard*. Clark was also critical of Broken Hill mining methods; see for example p. 356. For more general criticisms of Australian mining practice at this time, see Karl Schmeisser (a German mining expert who visited Australia at the request of a British-based mining company), *The Gold-fields of Australasia*, translated by Henry Louis (London: Macmillan, 1898), pp. 185–187.

42 Woodward dates this shift as 'Towards the end of 1893' (O. H. Woodward, *A Review of the Broken Hill Lead-Silver-Zinc Industry* (Sydney: West Publishing Corporation, 1965), 2nd ed., p. 78).

43 E. C. Andrews, *The Geology of the Broken Hill District* (Sydney: New South Wales Department of Mines, 1922), pp. 155–156; see also the table of values in Clark, *Australian Mining & Metallurgy* (1904), p. 388.

44 See for example Clark, *Australian Mining & Metallurgy* (1904), p. 371.

45 Bridges, *From Silver to Steel* (1920), pp. 184–187. The problem was not confined to Broken Hill. For example, in the same year that the German metallurgist visited Broken Hill, a New York mining annual declared that 'Among the most important problems of the mineral industry of today is the question of how to profitably utilize the enormous quantities (amounting to millions of tons) of mixed blende and galena that have been opened in Colorado, New South Wales, and other mining countries' (Stephen H. Emmens, 'The Treatment of Zinc-Lead Sulphides', in: Richard P. Rothwell (ed.), *The Mineral Industry: Its Statistics, Technology & Trade* (New York: McGraw-Hill Book Co., 1892), Vol. 1, p. 316).

46 For Warren's role, see Curtis, *The History of Broken Hill* (1908), pp. 59–60; James Hebbard, 'Evolution of Minerals Separation process on Central Mine', *Proceedings of the Australasian Institute of Mining and Metallurgy, New Series 10* (1913), p. 76; Geoffrey Blainey, *The Rise of Broken Hill* (Melbourne: Macmillan, 1968), p. 54; and especially Clark, *Australian Mining & Metallurgy* (1904), p. 371 and pp. 380–386.

47 Woodward, *A Review of the Broken Hill Lead-Silver-Zinc Industry* (1965), p. 79.

48 'Both [rhodonite and garnet] have a specific gravity closely approximating to that of blende, and it was realized very early that the separation of these minerals by ordinary water concentration was not to be expected' (Hebbard, 'Evolution of Minerals Separation process' (1913), p. 78).

49 Quoted in Woodward, *A Review of the Broken Hill Lead-Silver-Zinc Industry* (1965), p. 79.

50 In response to various authors who discounted the role that Broken Hill's metallurgists played in developing the flotation process, 'Members of the Broken Hill Branch of the Australasian Institute of Mining and Metallurgy' collectively wrote an article pointing out that 'In published literature of recent date, where much publicity has been given to flotation processes, credit for certain developments has inadvertently been given to others, where it truly belonged to Broken Hill metallurgists'. (Members of the Broken Hill Branch of the Australasian Institute of Mining and Metallurgy, 'The development of processes for the treatment of crude ore, accumulated dumps of tailing and slime at Broken Hill, New South Wales', *Proceedings, Australasian Institute of Mining and Metallurgy, New Series 80*(1930), p. 382). A more impartial authority noted as early as 1913 that Broken Hill was 'generally recognized as the home of flotation processes', and referred to 'the acknowledged leadership of Broken Hill engineers in this branch of metallurgy'. (H. N. Spicer, 'Evolution of Methods of Handling Slime - III, Australian Practice', *Metallurgical and Chemical Engineering 11*(1913), p. 315).

51 Oliver C. Ralston, *Answers to Questions on the Flotation of Ores* (Washington: US Bureau of Mines, 1917), p. 5; the earlier quotations in the text are also from this source.

52 Herbert H. Kellogg, 'Conservation and metallurgical process design', *Transactions. Section C, Mineral Processing & Extractive Metallurgy 86*(1977), C47–C57.

53 Herbert H. Kellogg, 'Toward a Materials-Conservation Ethic', *Metallurgical and Materials Transactions A, 9A*(12) (1978), p. 1700. Kellogg quotes King Hubbert's key point that 'the epoch of the fossil fuels can only be a transitory and ephemeral event – an event, nonetheless, which has exercised the most drastic influence experienced by the human species during its entire biological history' (p. 1698).

54 The graph appears on p. 1697.

55 See <http://billmckibben.com/index.html> (last accessed 2 June 2019).

56 The essay first appeared in *Sierra* magazine in 2000 and is reprinted in Solnit's collection, *Storming the Gates of Paradise: Landscapes for Politics* (Berkeley, CA: University of California Press, 2007), pp. 115–127. The quotation is from p. 127.

57 Gavin Bridge, for example, refers to 'the "heroic" version of US copper mining history, a quintessentially modern narrative about the triumph of man's ingenuity (gender implied) over nature . . . [T]his is a story of exploring a fickle yet ultimately bountiful nature, organizing labor power, energy, and finance to win riches from the earth's crust, and developing technologies of exploration and production able to beat nature's odds'. (Gavin Bridge, 'The social regulation of resource access and environmental impact: production, nature and contradiction in the US copper industry', *Geoforum 31*(2) (2000), p. 247.) More generally, see Katherine G. Morrissey, 'Rich Crevices of Inquiry: Mining and Environmental History', in: Douglas Cazaux Sackman (ed.), *A Companion to American Environmental History* (Chichester, West Sussex and Malden, MA: Wiley-Blackwell, 2010), pp. 394–409, as well as Andrew Isenberg, *Mining California: An Ecological History* (New York: Hill & Wang, 2005); LeCain, *Mass Destruction* (2009); and Liza Piper, *The Industrial Transformation of Subarctic Canada* (Vancouver: UBC Press, 2009).

58 Studies of the eclipse of mining in the west include Ralph Mann's pioneering study, *After the Gold Rush: Society in Grass Valley and Nevada City, California, 1849–1870* (Stanford, CA: Stanford University Press, 1982), as well as the more recent book by Eric L. Clements, *After the Boom in Tombstone and Jerome, Arizona: Decline in Western Resource Towns* (Reno, NV: University of Nevada Press, 2003). For a surprisingly persuasive argument that the American west is better off without its once-famed resource extractive industries, see Thomas Michael Power, *Lost Landscapes and Failed Economies: The Search for a Value of Place* (Washington: Island Press, 1996).

59 Patrick Houston, Zachary Schiller, Sandra D. Atchison, Mark Crawford, James R. Norman, and Jeffrey Ryser, 'The Death of Mining', *Business Week 2873* (1984), pp. 64–70. See also Michael P. Malone, 'The close of the copper century', *Montana: The Magazine of Western History 35* (1985), pp. 69–72; Malone, 'The Collapse of Western Metal Mining: An Historical Epitaph', *Pacific Historical Review 55*(3) (1986), pp. 455–464; and *The Competitiveness of American Metal Mining and Processing: A Report prepared by the Congressional Research Service for the Use of the Subcommittee on Oversight and Investigations of the Committee on Energy and Commerce, U.S. House of Representatives* (Washington, 1986).

60 See the comments in *The Competitiveness of American Metal Mining and Processing* (1986), p. 178, as well as the other sources in the preceding footnote. A recent report commissioned by the mining industry itself put mining's direct contribution to state employment at 1 per cent, to state labour income at 1.2 per cent, and its contribution to states' gross domestic product at 1.3 per cent *(The Economic Contributions of U.S. Mining in 2008*, A report prepared by PricewaterhouseCoopers for the National Mining Association, October 2010, E–2).

61 See *The Competitiveness of American Metal Mining and Processing* (1986), esp. p. 25 and pp. 177–186.

62 Gavin Wright, 'The Origins of American Industrial Success, 1879–1940', *American Economic Review 80*(4) (1990), p. 665.

63 The relative inadequacy of European methods when transplanted to the North American west during the nineteenth century has been noted by other scholars; see esp. Rodman Wilson Paul, 'Colorado as a Pioneer of Science in the Mining West', *Mississippi Valley Historical Review 47* (1960–1961), pp. 34–50; Spence, *Mining Engineers and the American West* (1970); Fell, *Ores to Metals* (1979); and Dym, 'Freiberg and the Frontier' (2011).

64 BHP had quickly become one of the most profitable of Australian companies, prompting Geoffrey Blainey to suggest that had the government in Sydney appreciated the fact, it would probably have rewritten the colony's mining law to prevent such wealth

being controlled by one company (*The Rise of Broken Hill* (1968), pp. 33–34). On Collins House, see John Kennett, 'The Collins House Group' (PhD thesis) (Monash University, 1980); Peter Cochrane, *Industrialization and Dependence: Australia's Road to Economic Development, 1870–1939* (St. Lucia, Queensland: University of Queensland Press, 1980), pp. 76–89; Peter Richardson, 'The origins and development of the Collins House group, 1915–1951', *Australian Economic History Review 27*(1) (1987), pp. 3–29; and Richardson, 'Collins House financiers', in: R. T. Appleyard and C. B. Schedvin (eds), *Australian Financiers: Biographical Essays* (South Melbourne: Macmillan, 1988), pp. 226–253.

65 Malone, *The Battle for Butte: Mining and Politics on the Northern Frontier*, 1864–1906 (Seattle and London: University of Washington Press, 2006), p. 206. The corporate history of Butte mining companies is convoluted, to say the least, but for a useful summary see Malone, *The Battle for Butte* (2006), pp. 200–207. One can gain a sense of Anaconda's importance from a list of industrial companies of the world in 1912 compiled by Christopher Schmitz: Anaconda ranks fifth in the world on this list (Schmitz, 'The World's Largest Industrial Companies of 1912', *Business History 37*(4) (1995), p. 87, 'Table 1, Leading World Industrial Companies. 1912. Ranked by Median Market Capitalisation of Issued Equity Stock ($ million)').

66 See Roger Moody's *Rocks & Hard Places: The Globalization of Mining* (London & New York: Zed Books, 2007), and Alain Deneault and William Sacher, *Imperial Canada Inc.: Legal Haven of Choice for the World's Mining Industries* (Vancouver: Talonbooks, 2012). Greg Palast provides a more polemical treatment in *The Best Democracy Money Can Buy: An Investigative Reporter Exposes the Truth about Globalization, Corporate Cons, and High Finance Fraudsters* (London: Pluto Press, 2002).

4 A comparative account of deep-level gold mining in India and South Africa

Implications for workers' lives

T. Dunbar Moodie

Introduction

The focus of this chapter is the history of deep-level hard rock gold mining. While gold has been mined for centuries, historically gold had been extracted from open-cast or relatively shallow pit mines or by using alluvial techniques. Industrial deep-level mining was initiated for gold in Nevada County in eastern California, for silver on the Comstock lode in Nevada and for copper in northern Michigan. Most such early deep-level mines seldom went deeper than 1,500 feet. The US seems to have been the first country to provide university-level mining engineering training for deep-level hard rock mining. As a result, American mining engineers spread across the globe, especially to Latin America and eventually South Africa (henceforth SA). British mining engineers were initially more typically trained through an apprentice system.

By the turn of the century, the world's deepest mines were gold mines, located in only two places, the Witwatersrand[1] gold mines of SA and the Kolar gold fields in India. For the first half of the twentieth century, these two regions vied with one another for having the deepest mines in the world. This chapter will compare deep-level mining in these two locations, with special reference to labour issues.[2]

The ore from the Indian mines contained much greater quantities of gold than from even the highest-grade mines in SA but compared to SA gold mining, Indian gold reserves were relatively tiny. To give a sense of differences in scale, one might note that, between 1900 and 1940, total mining employment at the Kolar gold fields hovered around on average 25,000 workers.[3] SA's largest single gold mine, Crown Mines, during this period, employed about the same number. Total employment on the SA gold mines in the twentieth century before 1931, when SA left the gold standard and the numbers jumped higher, averaged around 225,000.[4]

Despite this difference in volume, and notable geological differences, production processes on deep-level gold mines during the first half of the twentieth century in both countries were remarkably similar. While all the Kolar gold shafts are now closed, mined out to depths around 3.5 kilometres, SA continues to mine gold at depths between 3 and 4 kilometres. In fact, currently, eight of the ten deepest mines in the world are to be found in SA.[5] SA, however, no longer dominates international gold markets. With an open market for gold and virtual abandonment of a gold standard, prices have taken off and opencast gold mines now again take centre stage.

What makes twentieth century gold mining before the 1960s so interesting is not only that it operated at such great depths, but also for the first two-thirds of the century the price of gold was fixed. As a guarantee for currencies worldwide, national reserve banks bought all the gold that could be mined at a fixed price. Throughout most of a century characterised by two world wars – and the Cold War to boot – gold mines operated at very deep levels, mining a commodity whose price had changed – with a brief blip after the First World War – only once, during the Great Depression of the 1930s.

Geological conditions, with narrow reefs in deep-level hard rock gold mines in both SA and India, have always made mechanisation virtually impossible. Gold mining in both areas has hence been labour intensive. Even now, South Deep, the seventh deepest mine in the world, is the only contemporary deep-level gold mine attempting mechanised operations, so far with limited profitability. The total labour force at South Deep is roughly 2,500 workers. Because of certain unique geological factors,[6] South Deep may make it as the first mechanised deep-level mine, but, meanwhile, Sibanye's neighbouring KDC mine, which is being mined the old fashioned way, is booming – with about 40,000 workers. Given the fixed price of gold and need for huge work forces, the price of labour on deep-level gold mines was historically kept as low as possible.

These facts are well known.[7] The standard argument is that black mine wages have traditionally been kept down in SA by employing oscillating migrants, paying them as single men and housing them in barrack-like compounds.[8] My own contribution has been to argue that such a system worked only for migrants who resisted full proletarianisation[9] and were willing and able to return to sustainable subsistence rural households.[10] Mining companies were thus obliged to look deeper and deeper into Africa for labour as most traditional SA sending areas became overpopulated and underdeveloped and SA migrant workers turned to higher wages in secondary industry. The proportion of SA migrant labour on the gold mines dropped substantially to less than 30% by the early 1970s.[11] When the gold standard finally collapsed in the early 1970s, mine wages almost at once were doubled[12] and SA workers returned to deep-level mines (their proportion of the labour force reached roughly 70% by 1980), creating a largely wage-dependent migrant labour force that continues to the present day in an environment of massive retrenchments as many SA gold mines close down.

Attention to the Kolar gold fields, however, obliges us to broaden our perspective. Indian gold mining there competed with the deepest SA mines (admittedly on a much smaller scale), but with a low wage, stable, permanent (and hence proletarianised by my definition) labour force that reproduced itself for several generations before the lodes were mined out. My intention in this chapter is to look at the lessons to be drawn from the history of the Kolar fields for understanding the conditions for cheap migrant labour on the SA gold mines. I start at the beginning of modern industrial deep-level mining in both places.

Origins

Most gold in SA was laid down in the silt along the edge of a prehistoric lake millions of years ago. It now lies buried in narrow bands (called reefs) of black rock like a massive saucer seldom more than 35 centimetres thick, buried several kilometres beneath the surface of the highveld. That is why the SA gold mines spread out from the Witwatersrand in a semi-circle to the west and east. They're tapping the ragged edge of the gold ore "saucer" deep underground.[13]

We have noted that the Kolar gold fields were never as extensive as the SA gold reefs. They also differed geologically. The Kolar gold-bearing lodes are separated into almost vertical sheets of gold ore, more or less parallel with each other. Thought to be extrusions of magnum from ancient volcanic activity, their ore had a considerably higher gold content (initially at least twice as much as 'low-grade' SA mines and one-and-a-half times as much as SA's highest grade mines). They also tended to be wider than SA gold-bearing seams.[14]

The Indian fields had been worked during earlier periods down to the maximum feasible level achievable (about 100 metres) using means available to pre-industrial miners. In 1884, the British firm, John Taylor and Sons, struck gold beneath one of the ancient diggings and a rush of joint-stock companies was formed. Five mines survived the first rush, all under management by John Taylor and Sons, who struck a profitable deal with the newly formed princely state of Mysore (5% profit-sharing plus 2.5% of dividends).[15]

Since the area of the strike was agriculturally prosperous and relatively unpopulated, locals eschewed mine work and workers had to be imported. Most of them were Adi Dravida Tamils (*dalits* – 'untouchables') from the neighbouring state of Madras, which was going through a crisis in agricultural production at the time. They were wage dependent from the outset, grateful to be earning wages, however low, given their demeaned status scraping seasonal earnings at their places of origin, where land ownership was denied to them. They came with their wives and families and were provided with tiny shacks (8 feet square) by the company. There were eventually 40 to 60 shack villages along the gold lines.[16] These were thus 'immigrant' as opposed to oscillating migrant workers. At least half of them (those who worked in the hardest places) were not employed directly by the mine, but fell under sub-contractors, themselves Adi Dravida, who recruited them, supervised their work and paid them on behalf of the mine, collecting a cut of their wages.[17]

Indian underground mine workers worked three eight-hour shifts every day in the gold mines. Like the contract workers, permanent labour was also supervised by *maistries* (eventually all foremen were required by law to be trained and the company set up a school for the purpose[18]), but permanent crews were paid directly by the company and showed up for work much more reliably. Contracted workers, who received no attendance bonus, however, had an absentee rate of around 20%. It went up to 30% in some years. Contractors coped with this by inviting men coming off work to work double (and sometimes triple) shifts, thus earning the right to be absent for a day or two in their turn – and probably increasing the likelihood of accidents, given the rigours of underground work in deep mines with inadequate ventilation, let alone refrigerated air.

Before the turn of the century, Indian gold mining companies simply followed the reefs of ore as they found them down and across the lodes. Each of the five mines thus constituted a veritable warren of tunnels as well as sheer drops accessible only by ladders built into the walls. Workers were issued with candles but many early accidents occurred because workers slipped off ladders or simply stepped into unmarked drops in the dark.[19] Clad only in loin cloths with bamboo hats, they carried the broken ore by hand to the shafts, whence it was hauled to the surface by mules and draught oxen.

In 1897, a serious accident, apparently caused by panic about overcrowding on the ladders at the end of a shift, led to more than 50 men falling to their deaths. Those near the top seem to have knocked others off the steps on the way down. This was the point at which John Taylor and Sons decided that proper shafts needed to be constructed with cages to hoist the men to and from the depths. A coal-fired power station was constructed for back-up electricity but by 1902 hydroelectric power from the Sivasamudrum waterfall at Cauvery dam, 92 miles distant, brought reliable electricity to the mines. The mines were thus enabled to go deeper, despite mounting evidence of the dangers of heat exhaustion from mining, especially hand-hammer mining at depth. Air conditioning was not introduced until the 1930s. Underground mining techniques, however, came to resemble more and more closely the situation on the SA deep levels, also under development at this time.

SA deep-level gold mines were constructed by boring vertical shafts thousands of metres deep and then digging out horizontal tunnels at different levels to intersect the ore at different points. Narrow 'stopes' (blasting channels) led off wider winzes at different angles into the ore at the 'face'. To remove the ore, holes for explosives were drilled into solid rock. After blasting, gold bearing ore was 'lashed' (shoveled) down to the haulage and into trucks to be hauled to the vertical shaft ('tramming') and pulled to the surface for cyanide treatment. The largest proportion of underground labour on the mines was devoted to drilling and lashing in the stopes and development sections. For a century after gold was discovered on the Rand, through the 1970s, this work was done by teams of black workers under the supervision of white miners and their black 'boss boys' (later called 'team leaders').

Originally, drilling was done by hand – a labourious process in which a series of sharpened 'jumpers' (steel rods) of different length were sledge-hammered into the rock to a depth of 3 or 4 feet. Gradually, as with copper mining in northern Michigan, hand drillers (known as 'hammer boys') were replaced by men with machines – at first large cumbersome reciprocating drills but eventually smaller and lighter jackhammers. The shift from 'hammer boys' to 'machine boys' was complex, however, and not entirely to the liking of many mine managers.[20]

On the Kolar mines this shift was never fully implemented. When the Indian mines finally closed at the end of the twentieth century, 'hammer jumper coolies' still continued to work in some shafts. In SA, drilling (whether by hand or machine), lashing and tramming was invariably the work of black migrants. In India, it was usually done by the contracted workers who, of course, lived permanently around the mines.

White miners in SA, historically, supervised black work teams and were entrusted with blasting. Since 1989, although the technology has changed very little, white miners have been replaced at the head of the underground teams by better-educated black African miners with blasting certificates. In India, of course, *maistries* had always been native Indians, most of them Adi Dravida and illiterate, like their work teams.

Deep-level mining in both countries required heavy initial capital investment. The 'development' process (drilling shafts and non-gold-bearing rock to get to the ore) usually took several years in SA (less time in India because deep-level mines never expanded beyond the Kolar gold fields, although development sub-shafts had to be sunk as the mines got deeper) without showing any return at all. Moreover, in both countries, non-productive development work had to continue throughout the life of the mine.

Logistical exigencies

By the turn of the nineteenth century it was clear that deep-level gold mining would be capital intensive and logistically very complex. Given the fixed price, a premium had to be placed on costs and organisational efficiency. As a result, in both countries, the initial panoply of joint-stock mines was consolidated and administered by large holding companies. In the case of India, as we have seen, the industry was localised and small enough for one holding company to take complete control. In SA, the holding companies themselves were organised into a Chamber of Mines, which exercised tight control over the industry until at least the early 1970s.

The original design of SA deep-level mines was set up largely by American mining engineers, university graduates from some of the best universities in that country. Many of them had Californian experience, but most had also mined in other parts of the world.[21] A number of them came via El Calleo mine in Venezuela, but others had hard rock mining experience from California and Nevada.

Recent unpublished work by Keith Breckenridge,[22] however, has argued that the most important contribution of the Americans was their capacity for organisational and logistic planning. Foremost among them in this regard was Hennen Jennings, who orchestrated evidence to the 1897 Mining Industry Commission laying out a model for the organisation of the industry into holding companies – and indeed the government's enforcement of migrant labour by the administration of the pass laws after the Anglo-Boer War. One must be careful not to overemphasise the innovative organisational impact of the Americans, however, since John Taylor and Sons at the Kolar gold fields was essentially a British operation, managing five separate joint-stock operations on the fields and relying heavily on expertise from hard rock mining in Cornwall.

The model for engineering design and management of the SA gold mines after the Anglo-Boer War, however, was indeed the work of a young American mining engineer. The consolidation of Crown Mines was designed by Reuel Chafee Warriner, a youthful graduate of Lehigh University. He retrofitted several very rich but virtually mined-out early outcrop mines into a massive mining enterprise under the control of the Corner House holding company, using profits from the outcrops to develop deep-level mining south of Johannesburg. Such schemes were eventually introduced all along the reef.

In 1918, however, Reuel Warriner was fired by the Corner House because initial profit predictions were not being met. He was replaced by an Englishman named Walton. Cartwright, the historian of the Corner House companies,[23] thinks Warriner was made a scapegoat to assuage company shareholders, although his management style was indeed authoritarian and aloof. By 1922, Reuel Warriner's plan had produced the largest (and certainly the most profitable) mine on the Reef, but, by 1980, many were much larger. Historically, such mine managers took very seriously their responsibility to 'run a tight ship'. They exercised sovereign power over enormously demanding operations.[24]

Moreover, mines of this size required massive amounts of electricity to sustain their operations. As a result, SA gold mine holding companies, without the benefit of the hydroelectric power provided for John Taylor and Sons in India, also invested in coal mining, which in turn fueled an array of large power stations. Coal in the Transvaal thus also fell under the aegis of the Chamber of Mines, despite the Chamber's preoccupation with gold. The outcome, by the 1920s, was a minerals/energy/finance complex,[25] which continues even today as a major factor in the SA political economy.

Dealing with cost constraints on the Witwatersrand

In both India and SA, the fixed price of gold obliged deep-level mining to rely on cheap labour. In SA, the existence of higher- and lower-grade mines created an additional danger of competition that inevitably favoured the higher-grade mines. All except one of the original holding companies after the Anglo-Boer War, however, controlled a mixture of low-grade and higher-grade gold mines. The mining houses in Chamber of Mines thus had an incentive to protect investment in the low-grade mines. After the Anglo-Boer War, however, some higher-grade mines had actually already encouraged the formation of a contract labour system (similar to that on the Kolar mines), especially for tramming teams with chiefly sanction from Lesotho,[26] because productivity from those teams was markedly improved. Rapidly rising recruitment costs for such teams, however, raised the spectre of wage competition with its inevitable effects on returns to capital for low-grade and newly developing mines. By 1912, labour costs had become the Chamber's primary concern.

Immediately after the Anglo-Boer War, the Chamber of Mines had created a recruiting organisation, the Witwatersrand Native Labour Association (WNLA) which was able to monopolise recruitment from Mozambique and tropical Africa to the north, but failed to incorporate established recruiters from the SA 'native reserves' or the British Protectorates. Since 1907, the screws had been tightened on white labour, but increasing the black labour supply and keeping down black wages became an obsession after an investment spurt in 1908–1909 fizzled out and migrant labour from China began to be phased out. In September 1912, the mining houses established the Native Recruiting Corporation (NRC) under Chamber aegis to extend the monopsony on Mozambican and 'tropical' labour to SA and the British protectorates.

By the end of December, maximum wage schedules were set (by the Chamber's Committee of Consulting Engineers) for all member mines. Whatever the work done by a mine's black workers, the mine was obliged to pay no more than an average of 2/1d a worker per shift for its total black labour force (with the exception of hammer drillers, who were expected to lash their own stopes and drill one hole a day). Thus, almost from the moment it was established, the NRC imposed the infamous maximum average system on all member mines. In the first six months of 1913, average black wages dropped about a penny a shift.[27] Black workers were enraged and the black labour supply dropped across the board, especially for those who had been on piecework, whether on machines or in lashing and tramming.

Many managers simply ignored the rules and exceeded the maximum average in the early teething stages, but several of them testified to the Buckle Commission, established to investigate 'native grievances' after a wage strike of black workers in 1913,[28] that they had had to abandon piece work even though they were 'perfectly satisfied they were getting better value' out of it.[29] They had been forced by the imposition of monopsony to abandon the most effective (incentive-based) organisation of lashing and tramming. Piecework for machine drillers was reintroduced almost immediately, however, consequently this ensured they were the best-paid black workers on the SA gold mines.

The literature on the Kolar mines makes no mention of production incentives for workers there. Management was obsessed with the question of gold theft but productivity seems to have been taken for granted, at least in what I have read. Perhaps contractors provided incentives for their workers? They certainly tolerated absenteeism. The literature is silent on the matter. Piecework incentives would certainly help explain why half the labour force at Kolar (who did the most dangerous and strenuous work) was made up of contracted workers – also apparently slightly better paid on average than permanent labour.

Labour control problems at the point of production on Witwatersrand mines

Monopsony in recruiting never freed SA mine managers from anxiety about the production process itself. Since workers are more than merely mechanical 'means of production', problems of 'control' are endemic to capitalist productivity. Consulting engineers and mine managers were obsessed with 'scientific management' (efficient social organisation for greater profitability). It was lower-level supervisors, however, who were obliged to put scientific management into practice by organising underground labour so that black miners worked hard as well as putting in their time. Production targets for white miners (who, of course, were never subject to a maximum average system) were little help unless they could effectively organise the work of their black subordinates. Having precious few carrots (although sometimes they overmarked holes to reward assiduous drillers[30]), white miners literally had recourse to the stick to ensure worker performance in lashing and tramming.[31]

Assault in the push for production was taken for granted in the mines. Pohl, the union representative on the Low-grade Mines Commission in 1919, put the white supervisor's dilemma neatly:

> He is forced to get a job and he goes underground and he is told that [this and] that is required of him and of the natives. His boss – the mine captain or the shift boss or the underground manager, tells him at the same time 'you must not hit the boys, you will get fired if you do'. Well . . . if he does not get the work out of the boys, he stands a chance of being fired [as well]. What position do you put him in?[32]

Thus, since the introduction of the maximum average system in the SA mines, despite the logistical complexities of moving men, ore and materiel to and from the stopes, the 'rhythm of work' was maintained by violence – violence and the very movement of the ore line itself.[33] No doubt, on the whole, the experience of supervisory violence tended to be restricted to new workers, or to the very young. People learned by being beaten and, once skilled or promoted, avoided the worst of it. In fact, black team leaders often themselves became the ones doing the hitting. They were not 'boss' boys in name alone!

Black workers resigned themselves to the system, clambering to a precarious seniority based on tacit skill and strength. Managers resigned themselves to inefficiency, offset by abundant cheap labour, or sometimes struggled unavailingly against the current, seeking to establish competent management in a world without incentives. Invariably they ended up turning a blind eye to structural tendencies to violence as underground supervisors pushed for results at the point of production.

In the longer run, of course, production suffered. After things changed in the middle 1970s, workers also came to see this. When the maximum average system was replaced by the Paterson Scale, which sought to remunerate black mine workers in accordance with their skills and trained them more thoroughly, productivity leapt ahead. 'There were no criteria for choosing [team leaders] in those days – except how well a man could fight', a group of Vaal Reefs black mine workers, speaking of the days of the maximum average system, told Mark Ntshangase in 1984:

> If he was afraid of his men, he was fired, and an aggressive man chosen in his place. There was no training centre or ability training in those days. This was why people worked so badly. Nowadays men know their jobs and are not pushed around. No one has to be taught his job underground.[34]

When things changed it became clear that what Keith Breckenridge called 'the endless violence of mine work' had depended on structural consequences of the maximum average system. Methods to control wages had given rise to a system of production noted for the casual violence of its marginally efficient systems of supervision and labour control. Whereas reorganisation of production might eliminate some structural aspects of violence underground, useful alternatives to institutionalised violence after 1913 had to wait until the eventual abandonment of the maximum average system in the 1970s.

Ultra-exploitation, migrant labour and the compound system

In SA, wages barely suitable for single men were paid to migrant workers, thus exporting household reproduction costs to the sending areas. For future rural subsistence homestead proprietors, the fact that the mines provided food and housing (of however minimal quality) and a substantial cash proportion of their wages at the end of the contract, was a major incentive since accumulated cash could be invested in rural agricultural resources to fund retirement. Moreover the lobola system meant that bride wealth payments from each migrant younger generation also funded the retirements of their parents.

The system was certainly profoundly exploitative.[35] There was an element of reciprocity nonetheless. The system served capitalist class interests but it also enabled rural homestead proprietors to resist complete dependence on wage income. Resistance to proletarianisation was a battle which set up rural Africans for a series of rolling defeats over several generations as rural underdevelopment increased dependency on wages, but as long as the industry was able to spread its recruiting net more widely in order to recruit non-proletarianised workers deeper in southern and central Africa, a substantial measure of systemic and institutional stability was assured.

Workers from the rural areas living in the mine compounds had a high degree of associational power, however. They used it to establish what I have called a 'moral economy' (rooted in informal networks) that involved often very successful confrontations with mine managements, especially around late hoisting and the quality of compound life, including food, alcohol, sexuality and administrative arrangements. Between 1913 and 1920, there *were* serious confrontations about wages, largely initiated by Mozambican hammer drillers who resented white miners issuing them with 'loafer tickets'.[36] These conflicts abated in the 1920s with the abolition of the loafer ticket system, a 15% raise in the maximum average at the time of a rolling strike in 1920 and the post-war devaluation of the Portuguese escudo.

It was only in the 1940s, with the rise of the African Mine Workers Union (AMWU), that moral economy confrontations again turned to wages. I argue that the associative power of the most successful stay-aways during the 1946 strike was based on strike repertoires developed in moral economy demonstrations. When the strike was put down brutally by policing agents of the SA state, the limits of moral economy action were restored and migrant mine workers accepted low wages yet again. 'What could we do', old men in the countryside told me, 'we were beaten and we went back to work'. These were, however, the same old men who could all remember participating in demonstrations of several thousand workers confronting managements on specific late hoisting and compound issues.[37]

The fact that men were obliged to live in compounds reinforced powerful informal networks. These were decidedly not men without associational power. Except for the two rare and extreme post-war situations, however, they used such power to ensure acceptable living conditions and a social life that permitted adherence to rural values and support for rural subsistence, rather than wage raises per se. Access to the compounds (and the establishment of shaft-level worker organisation) was clearly essential for unionisation of the SA mines. Once access was granted by Anglo-American in the early 1980s, unionisation followed rapidly.[38] By then, however, the composition of the work force had also changed as subsistence agriculture collapsed and migrant workers and their families became more wage-dependent.

Worker discipline at the Kolar gold fields

At the Kolar gold fields, as we have seen, the resident, fully proletarianised, work force, lived in shack settlements in lines along the hills. Sirens rang out across the settlements to ensure attendance at each of the three shifts. Although there were high levels of absenteeism from contracted workers, work continued at rates quite comparable to the SA mines, and perhaps more effectively, because workers were experienced. Although there was occasional hitting by white mine officials,[39] violence was not necessary to induce inexperienced workers to learn their tasks and to work hard. Most workers had at least five years of mining experience. Even contracted workers took pride in their work.

Wages, however, were at least as low as for migrant workers in SA. The average wage seems to have been about 1/6d per day.[40] How was this achieved? For one thing, mine wages, in relative terms, were not very much lower than textile workers in nearby Bangalore. They were certainly higher than the earnings of Adi Dravidas' seasonal pickings in agriculture in Madras from whence they had come. In Madras, paraiya seasonal earnings, often paid in kind, varied between 2/8d and 6/6d *a month*. Moreover, mine pay ensured a regular income at all times of the year.

Nonetheless, an inquiry into the 1930 strike at Kolar concluded that gold mine worker's families spent less on food 'than a prisoner's jail ration'.[41] Levels of indebtedness, easily obtainable from, often Marwari, loan sharks, often exceeded six month's pay. Bankrupt mine workers apparently would regularly abscond, flee town and then return to work under a different name with someone elses's identity disc. There was a regular illegal trade in miner's identity discs.

Mysore Mines Regulations permitted management policing capacities that would have been quite unacceptable even in the rest of colonial India. Overt resistance to management control was criminalised. This was ostensibly aimed at the dangers of gold theft but also ensured a draconian regime of control.[42] Theft was indeed a problem. Often the ore from the Kolar shafts was streaked with visible raw gold. Hundreds of goldsmiths ran thriving operations in the townships around the shafts. But then white underground workers were never searched for smuggled gold so we don't know to what extent it was they who kept the goldsmiths in business. Workers on the edge of destitution and heavily in debt themselves, certainly never turned in their fellows. That was left to Punjabi security men who, like Zulu policemen on the early SA mines, were heartily disliked by Tamil-speaking underground workers. If worker resistance was criminalised (and at least some whites probably ran gold rackets on the side), then Tamil theft took on a certain air of legitimacy, Nair argues. Indeed, according to Nair:[43]

> despite their insertion in relations that were exploitative in the extreme, to most workers the money wage spelt liberty of a limited kind. Moreover, the capitalist workplace was one where new forms of collectivity could be forged, opening up possibilities for action that had long been denied to the agricultural labourer.

Nair argues that for Adi Dravida workers who came to the mines from destitution on the bottom rung of agricultural communities in Madras, a regular income opened up 'visions of plenitude on the margins of subsistence'.

Nair's evidence is derived from analysis of songs by musicians from among the miners' own number.[44] Unlike African migrants to the SA gold mines, there was no subsistence agriculture to which these 'paraiya' workers and their families could return. They were dependent for their livelihoods on regular, if meagre, wages and on ubiquitous moneylenders – and drinking places and 'fancy women' for an occasional night on the town. Their ballads celebrated opportunities for advancement, limited as they were, and at the same time poked fun at pretentious fellows who eschewed the eating of meat and sought to enter Brahmanic Hinduism by the back door or became Buddhists (some became Christians too).

The Kolar gold fields, argues Nair eloquently,[45]

> represented much more than just a wage to potential immigrants. The route to the world famous liquor shop . . . mapped out a transformed social geography of the region, passing by rich men, charlatans, generous men, prostitutes, teachers, reformers, moneylenders and thieves, while including European overseers, police officials, workers and managers, all of whom constituted the contours of a new social universe.

She captures here the widespread ambience of the occupational culture of proletarianised mine workers, whether Adi Dravida Indians, disreputable Welsh coal miners singing Methodist hymns in their pubs, or Thomas Klubock's Peruvian copper miners at El Teniente in the years before 1951.[46]

So it is perhaps the rural integrities of migrants to the SA gold mines before the 1970s, with their 'homeboy' drinking networks, their 'mine marriages' to younger men, their fear of town women, even their faction fights, all collective patterns of resistance to proletarianisation, that were unusual. We were 'on business' at the mines, old men in Pondoland told me, referring to investment in subsistence agriculture. Solidarity with their fellows, often ethnic, would enable them pick up deferred pay and return home to preside over a homestead, helping their neighbours to plough, enjoying their retirement at neighbourhood beer drinks,[47] recalling, as also did Basotho songsters, the privations and victories they had achieved by hard work and male solidarity at the rock face underground.

In the 1920s, management officials from the Kolar gold fields visiting the SA gold mines were quite envious of the compound system. They were obliged to adopt different tactics, however. Colin Simmons[48] reveals a more mundane side to Kolar worker quiescence than Nair:

[Mine controllers resisted paying higher wages by buying] the commitment of their labour by investing in a wide range of social infrastructure – including low-rental housing, day and night schools, canteens, hospitals, parks, pit-head baths, shops, clubs, *pucca* roads, perennial pure water supply, home electricity, regular refuse collection, the employment of teams of sweepers to keep the city clean, child welfare programmes, maternity provision, rehabilitation units, a dairy, the occasional issue of subsidized rice rations, and the organization and active encouragement given to cultural and sporting events.

Sometimes, however, such efforts at what Foucault would have called 'governmentality' also backfired in the face of the occupational culture described by Janaki Nair.

Labour insurgency

Did Indian gold mine workers finally manage to achieve representation and what sort of representation did they seek? Nair's argument on the matter is subtle and requires careful exposition.[49] There were two major strikes at the Kolar mines. One was in 1930, the second in 1940. 'Lightening strikes' on local issues occurred in the later 1930s between the two large events and the 1940 strike was followed by another massive strike in 1946, organised by communists with Congress support. Neither communists nor nationalists, however, understood the importance of caste issues. Their hostility to the Scheduled Caste Federation rapidly lost them support from Kolar mine workers.

The 1930 strike was not explicitly about wages. It had to do with an apparently 'governmental' decision by John Taylor and Sons to require, from January 1930, all mine workers to undergo medical examinations and provide thumbprints, ostensibly to meet requirements for a new Mysore Workmen's Compensation Act. On April 1, after 7,000 workers had already been fingerprinted, a note, written in Tamil, appeared near the housing lines at one of the mines. It called on all workers to refuse both signatures and fingerprints. A strike began at once and became a general strike across all the mines on April 7.

This general strike lasted 21 days. It was total and essentially peaceful, although police fired on pickets at several mines, wounding about 50 mine workers and killing two. Workers held mass meetings but refused to elect leaders and made their demands by putting up written notices. Their demands included full pay for the strike period and a provident fund, both of which were flatly refused by management. Perhaps most significant was a demand that contracted labour receive an attendance bonus – as company workers always had done. In this regard, the general strike meant that company workers were standing by their sub-contracted comrades. Indeed, in the end, management conceded this demand.

It is interesting that there were no demands for higher wages, although workers were certainly feeling financial pressures at the time. Debts to Marwari moneylenders had increased substantially. Fingerprints mattered in that regard too. The company had started to permit Marwaris to 'attach' worker debts from their pay cheques. Moreover, Marwaris had started to use thumbprints as collateral for debts. Thumbprints were also used when people went to prison.

Nair suggests that perhaps the most important single factor in the strike demand for abandonment of fingerprinting was that prints would have enabled the company to abandon miner's discs, issued by management and necessary to qualify for work on the mines. As we have seen, contract workers with heavy debt on the company books might be fired (or they might simply jump town to avoid debt collection), but they could always return to work under another name by purchasing a miner's disc, readily available for sale on the black market.[50]

On April 27, the Dewan (labour representative of the Maharaja) of Mysore was called in to negotiate a settlement. Mine management used the excuse to climb down gracefully. They agreed to abolish the entire system of fingerprint registration. Next day, all the mines were back at work. Within a year, contract workers were granted attendance bonuses and wages were paid fortnightly rather than each month. Nair notes, 'Most important, all unskilled and some skilled labourers were exempted from attachment of wages in case of debt judgements, and mine panchayats [works committees] won the right to hold elections'.[51] Nair suggests that these workers, given their 'paraiya' origins, had vehemently rejected any suggestion that fingerprints would enable them to be 'bonded' to the company. This was then, in her opinion, a strike on behalf of the 'freedoms' their proletarian status had brought for them.

Nair seems to find the 'anonymous' and 'leaderless' aspect of this strike by 18,000 workers puzzling. On the basis of the SA experience, however, the associative power of informal worker networks seems to me to be obvious. The sort of 'plebian' integrity discovered by E.P. Thompson for England before 1790, surely applies as much to the Kolar strike of 1930.[52]

Nair indeed cites Thompson, referring to 'a robust anonymous tradition'. She adds:[53]

> This invisibility proved no inconvenience to workers themselves; they successfully held meetings, restrained workers from rejoining work, formulated demands, and maintained peace in the area . . . Every move of KGF strikers

suggested a high degree of organisation and it was the 'invisibility' that proved difficult for company management.

Having read more than a hundred reports on mine disturbances in the SA compounds, I am not at all surprised by her findings (although in SA workers would shout their demands in unison rather than writing them).

Although she directly denies this, citing my work, it seems to me there *was* in fact a sort of 'implicit contact' at work in this strike and those that followed. Honour and dignity, Nair says, mattered as much for these workers as wages during this 1930 strike and the lightening strikes that followed during the 1930s. But surely that is precisely the meaning of 'moral economy'? Mine managements gave way in response to such actions, complaining about workers' lack of willingness to consult with them, but at the same time firmly rejecting any suggestion of trade union representation.

Whatever power was conceded to mine management, any suggestion that they return workers from the dignity of wage labour to erstwhile pariah boundedness was flatly denied them by massive (and anonymous) collective worker action. The Indian Congress movement that claimed the right to represent workers also condemned their strikes as not in 'the national interest'.

When, after another major strike in 1940, John Taylor and Sons was forced to recognise trade unions, it continued to ban union leaders, especially local leaders, from the area (as it was entitled to do under the indirect rule of the Maharaja of Mysore). K.C. Phillips, the most competent local union leader, was banned from Mysore and forced to operate from Kuppam over the provincial border in Madras. Independent labour leaders laid claim to represent the miners but rapidly lost support because they had no conception of the occupational culture of these Tamil mine workers.

Eventually, when the Congress was banned during the war, union officials from the Communist Party of India did manage to achieve some sort of representative status, but they lost it very soon over refusal to seriously address questions of caste raised by the scheduled caste movement because, communists claimed, the movement was opposed to the interests of 'the Indian nation'. These long-proletarianised Indian mine workers refused to ignore questions of caste indignity in the name of class or nationalist unity.

In the 1950s, the Kolar goldfields were nationalised by the Mysore state. In the 1970s, they were passed into the hands of the government of India. Most of the mines were nearly 4 kilometres deep at that point. The industry limped along with a rapidly reducing workforce until its closure in 2001. The resilient worker community that had been established over the past century dissipated as local wage work collapsed. Left behind is a proletarian community without work.[54]

'New workers' in SA in the 1970s and the rise of the NUM

Events in SA took a somewhat different turn, although the final outcome looks ominously similar. Largely more than 2.5 kilometres down by the 2000s, gold mining there has also been hit by retrenchments. By the sixtieth anniversary of the maximum average system in 1972, the Chamber was deeply divided between the reform-mindedness of Anglo-American and the belligerently reactionary labour control policies of Goldfields and before 1986 Gencor, the other really wealthy mining houses. During that year, the Chamber announced a substantial 25% increase in the maximum average payable by its mines. In November 1972, JCI (a small mining house under the control of Anglo) defied the Chamber and granted a further 30% increase, at the same time (along with Anglo) explicitly challenging the maximum average system. Other mining houses strongly objected and a compromise was worked out, but the skids were clearly under the system and within another year, it was gone. The Chamber scrambled to maintain an appearance of unity.

In March 1973, the Chamber announced that June raises would amount to 30%. In December, it pushed the annual increase up 44%. 'Minimum wage rates were raised by just over 122 per cent in two installments in 1974 and by 38 per cent the following year', according to Vic Allen's assessment.[55] Mine wages had taken off in real terms. Young men growing up in the sprawling rural slums created by Bantustan removals began to consider mine work. Anglo-American in particular began investing in skills training to improve productivity.

SA workers who 'joined' the mines in the 1970s were better educated and distinctly less tolerant than their predecessors of what union leader, Cyril Ramaphosa, was to call 'the industry's ancient industrial relations practices, its mindlessness, and its violence'. Pay for team leaders had risen faster than for any other underground occupational category. Resentments ran deep. Computerisation of mine records, fingerprinting and a return engagement bonus to ensure seniority, also meant that mine workers were expected to become 'career miners', returning to sending areas only for relatively brief and clearly demarcated periods.

Proletarianisation on the SA gold mines took place almost overnight – certainly within a decade. The new SA migrant workers were essentially wage dependent. Families of Basotho workers also increasingly settled along the border with SA. Collective action by mine workers became increasingly volatile. Progressive mining houses like Anglo-American began seriously to consider permitting trade unions. When the Wiehahn commission gave state imprimatur to such ideas, Anglo granted relatively easy compound access to Cyril Ramaphosa and the National Union of Mineworkers (NUM).[56]

While in no way wishing to downplay the organisational skills of Cyril Ramaphosa and his able lieutenants, nor the audacity of Anglo Gold Division in granting generous access to the compounds, I do want to emphasise the importance of the 'new' mine workers recruited in the 1970s. Better educated, more skilled, trammeled by short leaves and weekend commutes and much more personally humiliated by racism, these workers developed a proletarian structure of feeling that gave racial meaning and militant intensity to local confrontations, not only with management, but with white supervisors as well.

Conclusion

The lifetime of extractive industries like mining is relatively short. As I reflect on the comparative material presented here, I am reminded yet again of the relevance of context. On the one hand, Adi Dravida migrants to the Kolar gold fields escaped rural bondage to become deep-level underground wage workers in a traumatically dangerous environment. For them proletarianisation was a sort of liberation – and it enabled many of them to lead more meaningful lives until deep-level mining drew out to its inevitable end.

For early SA oscillating migrants, on the other hand, mine work was a fierce challenge which enabled them to resist proletarianisation in pursuit of an eventual good life as household heads at home. In both cases, however, as I read the evidence, they insisted on local standards of dignity, whether, in SA, being true to their rural autonomy (or, eventually, as migrant proletarians to their union[57]) or in India, affirming proletarian solidarity over against 'paraiya' outcast degradation. Inevitably, however, in both cases they represented life styles that are no longer possible.

In the Indian case, I have no idea what the closing of the mines has meant to those who worked in the Kolar gold fields and their families. Perhaps this is a research project worth undertaking. In the SA case, Thabang Sefalafala[58] is at the moment engaging with retrenched black migrant gold mine workers in informal settlements around Welkom, trying to understand their unavailing struggles to find wage-work. One factor emerges time and again in his findings, loss of meaning. Retrenched mine workers in SA dream of the collective solidarity of underground mine work, with all its dangers, but awaken, despairing, to a world of worklessness.

Notes

1 For descriptive purposes I expand the 'Witwatersrand' to include the crescent of mines that stretches from the Evander mines in the far east to the Free State mines in the south west.
2 For a comparison between Indian and South African coal mining (where the situation was very different) see Alexander, Peter, 'Women and Coal Mining in India and South Africa c1900–1940', *African Studies* 66(2–3) (2007), pp. 201–202.
3 For employment figures on the Kolar gold fields, see Colin Simmons, 'The Creation and Organization of a Proletarian Mining Labour Force in India: The Case of the Kolar Gold Fields, 1883–1955', in: Mark Holmstrom (ed.), *Work for Wages in South Asia* (New Delhi: Manohar, 1990).
4 For a convenient source for figures on employment on the SA gold mines, see the classic work of Francis Wilson, *Labour in the South African Gold Mines 1911–1969* (Cambridge: Cambridge University Press, 1972).
5 The other two deep mines on the list are copper mines in northern Ontario in Canada. Because of the different manner in which copper deposits were laid down geologically, these mines are fully mechanised.
6 It is claimed that several gold reefs lie close enough to one another at South Deep to make it possible to do trackless mining without diluting the ore too much. A similar project failed at much shallower levels at Western Areas, however.
7 The classic formulation of this issue remains F.A. Johnstone, *Class, Race and Gold* (London: Routledge and Kegan Paul, 1976).
8 The initial form of this argument was put forward by Harold Wolpe in his classic paper, 'Capitalism and Cheap-Labour Power in South Africa', *Economy and Society 1* (1972), pp. 425–456. Peter Alexander ('Challenging Cheap-Labour Theory', *Labour History 49*(1) 2008, pp. 47–70) makes use of evidence from the SA coal mines to challenge some of the basic assumptions of this theory, positing that SA state concern about massive urbanisation

was as important as economic factors in the creation of compounded labour on the gold mines. I make similar claims in my own work, T. Dunbar Moodie and Vivienne Ndatshe, *Going for Gold* (Berkeley: University of California Press, 1994), pp. 76–77, but would not want to argue that this obviated cost constraints imposed by international price-fixing.

9 I define proletarianisation as dependence of workers on wage labour, whether such workers were migrants or not.

10 For a careful account of the rural end of this phenomenon in Lesotho, see Colin Murray, *Divided Families* (Johannesburg: Ravan, 1977).

11 See Jonathan Crush, Alan Jeeves and David Yudelman, *South Africa's Labour Empire* (Boulder: Westview, 1991).

12 For the inside story of the collapse of the gold standard, see Lang, John, *Bullion Johannesburg* (Johannesburg: Jonathan Ball, 1986), pp. 425–433. V.L. Allen, *The History of Black Mineworkers in South Africa* (Keighly: Moor Press, 2003), pp. 309–334, provides an accessible account of 'the mine wage explosion' of the early 1970s.

13 A prehistoric meteor strike pushed the edge of ore to the surface on the Witwatersrand where it outcropped – awaiting discovery in 1886.

14 <Shodhganga.inflibnet.ac.in/bitstream/10603/14618/8/08> (last accessed May 31, 2019) provides a useful account of the geology of the Kolar Schist Belt.

15 For my account of the Kolar gold fields, I have relied on secondary sources. Most important are Janaki Nair, *Miners and Millhands* (Walnut Creek: AltaMira, 1998) and Colin Simmons, 'The Creation and Organization of a Proletarian Mining Labour Force in India: The Case of the Kolar Gold Fields, 1883–1955' (1990). I also accessed chapters from an unpublished Indian dissertation at <Shodhganga.inflibnet.ac.in: 8080/jspui/bitstream/10603/27092/13/13> (last accessed 31 May 2019), but have not been able to discover either the name of the author or the title. Shodhganga has not responded to an email request for more details.

16 Contractors also provided housing clusters, even more inadequate, but at higher rents.

17 Simmons, 'The Creation and Organization of a Proletarian Mining Labour Force in India: The Case of the Kolar Gold Fields, 1883–1955' (1990), pp. 87–90. Workers and their families at the Kolar gold fields were thus wage dependent from the outset. Unlike the Indian coal mines, no women worked underground on the gold mines. During the strikes of the 1930s and 1940s, however, there are frequent accounts of 'their wives' pulling out scab miners from work places. For a carefully developed example of similar solidarity from contemporary women at the SA platinum mines, see Asanda Benya, 'The Invisible Hands: Women in Marikana', *Review of African Political Economy* 42(146) (2015), pp. 545–560.

18 Simmons, 'The Creation and Organization of a Proletarian Mining Labour Force in India: The Case of the Kolar Gold Fields, 1883–1955' (1990), p. 87

19 There was a brisk trade in unused candles, which served to supplement pathetically low wages and led to workers often finding their way underground in the dark. There was also a thriving trade in stolen gold, leading to an oppressive penal system of management control from which white workers were excluded. Since there were several hundred goldsmiths in the Kolar mining town, rackets for stealing gold must have been manifold, certainly including white employees (Janaki Nair, *Miners and Millhands* (1998), pp. 41–86).

20 For a brief account of the complexity of the shift from hammer to machine 'boys', see Moodie and Ndatshe, *Going for Gold* (1994), pp. 50–54.

21 See Clark C. Spence, *Mining Engineers and the American West: The Lace-Boot Brigade* (New Haven: Yale University Press, 1993), pp. 304–317. He asserts that before the Anglo-Boer War 'half the mines on the Witwatersrand were being managed by California engineers'.

22 Keith Breckenridge, 'Made in America: Progressive American Mining Engineers and the Legal Foundations of the South African State', see www.kznhasshistory.net/seminars/breckenridge/2006 (last accessed 31 May, 2019).

23 Cartwright, A.P., *Golden Age* (Cape Town: Purnell, 1968), pp. 211–224.
24 For more detail, see T.D. Moodie, 'Getting the Gold Out of the Ground: Social Constraints on Technical Capacity in South African Deep Level Mining', *Journal of Historical Sociology 28*(4) (2015), pp. 572–594.
25 The classic account of South Africa's 'minerals-energy complex' is B. Fine and Z. Rustomjee, *The Political Economy of South Africa from Minerals-Energy Complex to Industrialisation* (London: Hurst, 1996).
26 Alan Jeeves, *Migrant Labour in South Africa's Mining Economy* (Kingston & Montreal: McGill-Queens University Press, 1985), pp. 158–160.
27 Transvaal Archives Department (henceforth TAD), K358, Evidence of C.W.Villiers, 3/3/14.
28 For a discussion of management indignation at the elimination of piecework brought about by introduction of the maximum average system, especially on Nourse Mines, see Moodie, 'Maximum Average Violence', *JSAS 31*(3) (2005), pp. 561–563. There were two commissions of inquiry established immediately after the introduction of the maximum average system. One, at Crown Mines, was conducted by Cooke on behalf of the Native Affairs Department. Black worker evidence to this inquiry (TAD, NLB 111, 1376/13/154) is redolent with complaints about wage cuts. The other by Buckle (TAD, K358) was occasioned by a strike of black workers immediately after the 1913 white miners' strike demanding that black migrant wages be doubled.
29 TAD, K358, Buckle's remarks to C.W. Villiers, March 3, 1914, p. 14.
30 See, for example, Cooke's handwritten notes of his interview with P.J. Rount, Mine Captain at #2 Shaft, Crown Deep, nd. (c. July 17, 1913). (TAD, NLB 111, 1376/13/154).
31 Moodie, 'Maximum Average Violence' (2005) is focused on this issue. See also Moodie and Ndatshe, *Going for Gold* (1994, pp. 61–67) for failures to reform inefficiencies of black supervisors underground because of the maximum average system.
32 Moodie, 'Maximum Average Violence', pp. 564–565.
33 See Keith Breckenridge, 'The Allure of Violence', *JSAS 24*(4) (1998), pp. 669–693.
34 Moodie, 'Maximum Average Violence', p. 566.
35 The system of exploitation was also global in its impact. To the extent that the international economy depended on a fixed price for gold, it was maintained on the backs of low-wage southern African migrant mine-workers.
36 'Loafer tickets' derived from an early system whereby white miners in SA were sub-contracted but 'rented' black workers provided by the company. If a hammer-driller did not produce a deep enough hole during a shift, he was served by his white miner with a 'loafer ticket' and was not paid for his day's work; nor was the white miner charged by the mine for that man's labour. Since it was the white miner's responsibility to measure the depth of holes there was a tendency to issue loafer tickets to workers who fell only marginally short. In that case the black miner would not receive his day's pay but the white miner could still blast the hole and ensure his rate of production at a discount. To resolve ongoing disputes about loafer tickets, white sub-contracting was abolished in the late 'teens.
37 On moral economy see Moodie and Ndatshe, *Going for Gold* (1994), pp. 81–86); also T.D. Moodie (1986), 'The Moral Economy of the Black Miners' Strike of 1946', *JSAS 13*(1) (1986), pp. 1–35.
38 For an account of the early years of unionisation on the SA gold mines, see T.D. Moodie (2010), 'Becoming a Social Movement Union: Cyril Ramaphosa and the National Union of Mineworkers', *Transformation* (2010), pp. 72–73.
39 For 'lightening strikes' about underground assaults at the Kolar gold fields, see Janaki Nair, *Miners and Millhands* (1998), pp. 155–156.
40 Figures in this paragraph are drawn from Janaki Nair, *Miners and Millhands* (1998), p. 88. For comparative purposes I have converted her figures in rupees to sterling at exchange rates for the period.
41 Janaki Nair, *Miners and Millhands* (1998), p. 131.

42 Indeed, one source I consulted reveals from the minutes of management meetings with John Taylor and Sons that mine officials were obsessed with the view that unionisation would lead to an increase in labour costs which, in their opinion, the mines could not afford. They were also particularly concerned that miners' phthisis might be included for coverage under the workman's compensation laws. Keeping costs down was indeed high on their agenda. See <Shodhganga.inflibnet.ac.in/8080/jspui/bitstream/10603/27092/13/13>

43 Janaki Nair, *Miners and Millhands* (1998), p.296.

44 David Coplan, *In the Time of Cannibals* (Chicago: University of Chicago Press, 1994), provides similar evidence from the songs of Basotho migrants to the SA mines.

45 Janaki Nair, *Miners and Millhands* (1998), p. 119.

46 Thomas Klubock, *Contested Communities* (Durham: Duke University Press, 1998).

47 For a moving account of rural Xhosa beer drinks, see Patrick McAllister, *Xhosa Beer Drinking Rituals* (Durham: Carolina Academic Press, 2006).

48 Simmons, 'The Creation and Organization of a Proletarian Mining Labour Force in India: The Case of the Kolar Gold Fields, 1883–1955' (1990), p. 94.

49 Nair's chapters in *Miners and Millhands* (1998) are the most illuminating accounts of Kolar miners' 'consciousness' available anywhere.

50 Fingerprinting on the SA mines did not commence until the 1970s, after the collapse of the maximum average system. It was introduced without worker resistance.

51 Nair, *Miners and Millhands* (1998), p. 136.

52 Nair, *Miners and Millhands* (1998, p. 139) briefly refers to Gramsci's conception of 'contradictory consciousness'. To my mind, Gramsci's notion, provocative though it is, ultimately implies some sort of 'false consciousness' that seems condescending to me.

53 Nair, *Miners and Millhands* (1998), p. 143.

54 See Parvathi Menon, 'Death of a Mine', *Frontline 19*(11) (2002).

55 Allen, *The History of Black Mineworkers in South Africa*, Volume II (2003), p. 328.

56 For an account of the rise of the union, see Steven Friedman, *Building Tomorrow Today* (Johannesburg: Ravan, 1987), pp. 355–392. See also Moodie, 'Becoming a Social Movement Union', (2010).

57 I omit in this chapter discussion of the decline of the NUM.

58 Thabang Sefalafala, 'Experience of Unemployment, the Meaning of Wage Work: The Dilemma of Waged Work Among Ex-gold Mineworkers in the Orange Free State Goldfields', PhD Thesis, University of the Witwatersrand, 2018.

5 Local moments in mining history

Some ideas on the relationship between foreign and native in Mexican silver mining

Alma Parra Campos

Introduction

It is a strange and difficult challenge to think about what a 'local moment' means in mining history. This is an exercise that I decided to tackle by way of contrast and comparison, and by resisting the temptation of looking at the many specific Mexican cases to which the term 'local' may apply. This was possible since the example of Mexico *vis-à-vis* other silver producers holds many local moments due to the fact that mining development in this country has been largely defined by the relationship with the 'foreign'.

The very first local–foreign encounter was the Spanish conquest itself, in the early sixteenth century, bringing a revolution into metal workings. Despite the artistic achievements of the pre-Hispanic native cultures in this area, techniques for the extraction of metals and metallurgy were rudimentary, compared to those practiced in Europe.[1] The introduction of underground larger-scale mining and the adoption of European intensive mining methods, which the Spanish brought to America using the experience of German experts in the early sixteenth century,[2] defined areas of development characterised by the abundance of gold and silver. This geological and geographical peculiarity determined the complete organisation and administration of the future country.

The Spanish conquest of the American territories also meant that their government apparatus had to mirror that in Spain, which was committed to the search for deposits of precious metals upon which the growing commercial interests of the expanding markets in Europe based their monetary transactions.

This gave the Spanish conquest a decisive orientation towards the encouragement of the mining sector in New Spain (Mexico), which also implied the dismantling of any previous order of governance and organisation among the native population. In turn, this translated into the submission of large sectors of the population to forced labour through a system called the *Repartimiento* in every economic activity which provided mining with a large number of labourers.[3]

This initial transmission of mining organisation resembling the Spanish organisation set the pace and forms in which the 'local' would respond to the 'foreign', even after achieving independence from Spain. One, and perhaps the best, way of tracing these responses and the interaction between 'local' and 'foreign' is looking at the major technological changes registered in Mexican mining and the forces behind it in the long run.

Behind the changes that permeated the improvement in Mexican mining there were three fundamental forces encouraging technological advance.

The first was the central role that governments assigned to mining by the Spanish crown during the three colonial centuries between the sixteenth and the eighteenth centuries,[4] and from 1821, by the republican governments of Mexico as an independent country.

The second was the changing role of the methods and resources used by the state, both colonial and independent, in promoting the mineral industry, basically precious metals.[5]

And the third was the role of the private sector in promoting technological change and development. Native, but very importantly also, foreign investors and inventors played key roles in technological advance.

Within these larger themes and constructs, technological advances in Mexican mining fell largely into two spheres – efforts to improve the traditional extractive and metallurgical methods such as the *patio* process,[6] which was prevalent during the three colonial centuries and the first two-thirds of the nineteenth century,[7] and efforts to introduce vastly different technologies, particularly in the area of power sources for machines, which came largely in the last third of the century.

During three centuries prior to the achievement of Mexican independence from Spain in 1821, both government entities and private companies imported older, proven technologies developed in other countries. For obvious reasons, these included those used in Spain, but also some from other European countries like Germany, and invented new processes and techniques, as well. There were different avenues for this, however. Innovations sponsored by the government under the rule of the Spanish crown went into specific fields defined by the Mining Guild.

The Mining Guild comprised institutions like the Mining Tribunal, the Mining College and the Mining Bank for the promotion of the industry. These institutions were instruments of the Spanish government created in the late sixteenth century, following the Bourbon Reforms that were implemented both in Spain and Spanish America throughout the eighteenth century.[8]

The innovations made in the private sector, by contrast, generally came in response to specific problems at individual mines or *haciendas* (the large plants where owners used the *patio* process to recover gold and silver). In the case of private enterprises, once an invention had shown its utility, the inventors sent their work to the mining institutions in order to obtain official recognition and to ensure that the inventors received the economic benefits coming from potential royalties. The government was committed to the control of innovations, and there was an explicit dedication to this principle because mining was the main generator of income for the state.

By way of comparison, other silver producers in Latin America, like Peru, Bolivia and Chile in South America, with a very similar background in terms of colonial Spanish domination, confronted similar processes regarding their mining experiences. These countries in many ways adopted the legislation and institutions created by the Spanish in New Spain.[9] However, their mining industies reacted locally and adapted the importation of technologies according to their own traditions, even pre-Hispanic, and also according to the quality of the ores on the different sites. Some grinding and milling artefacts like the *Maray* used to crush the ores, or the *Guairas*, clay furnaces used for smelting ore, were adopted as alternatives to new methods by the Spanish in the early sixteenth century and used until amalgamation processes became widespread in the later part of that century in Potosí.[10]

Colonial experience with foreign technology

In the early eighteenth century, attempts by the mining institutions created in America through the Spanish crown at promoting the use of technological advances other than those practiced in New Spain proved fruitless. The first attempts were modelled on Britain, as new advances where incorporated in British mines and industry in general as a result of the Industrial Revolution. References to the intention to import to New Spain the Newcomen steam engine started the 1730s, when problems of the flooding of mines were particularly acute, and the alternatives afforded by the use of 'modern machines' were expected to substitute the traditional winches or whims.[11] Other alternatives, however, involving more specialised engineering and considerable investments were the '*socavones*', or drift mining, which consisted in horizontal tunnels built along the veins of the mines, which used gravity to drain the mines or to connect with other tunnels for the same purpose.[12] While a serious investigation was carried out involving miners from New Spain, royal officials and British engine builders in several parts of the growing industrial areas, the project did not go ahead. Some of the issues at stake where the obstacles connected to getting continuous supplies or iron and metal parts indispensable for the workings of the machines, items that were not available in New Spain, but also scarce in terms of imports from Spain. The causes of failure were attributed to local conditions related to the disagreements between local miners and financiers, in addition to the lack of government intervention. Government officials in turn also blamed the feeble government commitment and decisions to grant sufficient funds to embark on such an enterprise. In fact, it required a complete transformation and more open flow of trade between Spain and its colonies, which during this period was still under the absolute control and supervision of the crown.

In the later eighteenth century, a clear and probably more organised effort to innovate and to raise awareness of the advances in the scientific methods of other European countries promoted a state-sponsored evaluation of the mining sector. A severe crisis in the supply of mercury, upon which the most prevalent system of refining in New Spain relied, initiated a search for technological improvements. This time, attention shifted towards Germany by organising a group of 'experts' that would visit and evaluate the most important mining centres.[13] It was a large-scale scheme considering experiments by the major silver producers of Spanish America, namely Mexico and Peru, to introduce the Börn barrel system of refining metals and eventually substitute the *patio* method.

Frederick Sonneschmidt and Fausto de Elhuyar, the latter being the representative of the Mining Tribunal, guided the German group. From the beginning, the Freiberg School that trained experts faced 'adverse' conditions. After visiting the major mining centres, they only managed to put into practice one experiment, that is, the Börn barrel system of refining with mercury, in which miners and government officials had placed their hopes. The great expectations which guided those who had invited the Germans soon were greatly disappointment. The promise of producing 135 tons of ore in 20 hours, something that used to take six weeks under the *patio* process, was unrealistic under Mexican conditions, and figures reveal that there was no great advance in using the process when compared either with foundries or the *patio* system. The experiment carried out in the mines in Guanajuato, Mexico by another member of the German team, Fischer, was even challenged by local miners and observers who expressed many complaints to the Mining Tribunal in a scornful manner.[14]

Local conditions proved to be, if not better, more suitable, stable and reliable in terms of inputs and profits. In comparison with other regions like Zacatecas, it was also found that the cost of the experiments with barrels was almost twice as high as the cost of refining by traditional methods of local smelting. During the experiment in Zacatecas, it was found that investing 284 pesos in smelting would render 2072 pesos in value of ore, while investment in barrels was 529 pesos and would yield 1922.70 pesos in value of ore.[15] Although Sonneschmidt did not admit to the failure of his experiment,

> The results were then computed and they added up to a dismal failure for the new process. Less silver had been freed than was customary by the *patio* process, the loss of mercury had been enormous, the silver had come off impure – even when it was refined by smelting the yield was of low value – and of course, the costs had been very great.[16]

A similar effort was simultaneously carried out in Peru in 1788. A German team headed by Thaddeus Von Nordenflicht also tried to adopt the Börn barrel system instead of the *patio* system.[17] Having started the construction of the barrels machine for their experiment, they managed to conduct trials throughout 1789 and 1790. Furthermore, in 1793, they carried out more trials with local miners *in situ* to compare the feasibility and superiority of the different methods.[18] The results showed that although the quantities of silver obtained by different methods were similar, the loss of mercury was greater in the barrel experiments. The Viceroy of Peru decided to cancel the trials. He was convinced that Nordenflicht had failed in proving that the new method by barrels had brought substantial improvements in the process of refining ores.[19] The meagre results were proof that the Spanish experience with the foreign technology in the most important silver producing countries in Spanish America, Mexico and Peru, were not better than the most common practices performed locally. Sonneschmidt himself admitted in his major work on amalgamation that after being committed to the introduction of the Börn barrel system of refining ores in Mexico, he had found no real alternative to *patio*.[20]

By contrast, the mining regions of what is now Peru, Bolivia and Chile, in the mid-sixteenth century started to adopt different amalgamation processes including that of the *patio*, developed in Pachuca, Mexico as a natural consequence of transmission of knowledge within the Spanish empire. However, this was adopted in combination with extensive use of foundries, which had already been outnumbered in Mexico by the growth of *patio* establishments.[21] Moreover, in South America rather than in New Spain, other systems like the *Barba* method of amalgamation, which allowed for more domestic refining of minerals in smaller pots and pans using fire to accelerate the process of refining, were also widespread. This method combined well with foundries, which naturally depended on fire.

But it can be argued that there was an additional local element influencing the differences in metallurgy. In South America, pre-Hispanic cultures are known to have developed sophisticated systems of smelting ores which merged with more industrial smelting methods brought by the Spaniards.[22] As mentioned above, the *Guairas*, small clay furnaces, continued to be used in parallel to the method introduced by the Spanish.[23] Peter Bakewell has suggested that blended technologies should be analysed also through cultural evidence like language, by examining the large quantity of terms adopted from the conquered cultures. This, however, applies more to the Andes than to Mexico. In New Spain, particularly in the northern regions, mining responded to the pattern of conquest where the Spanish 'had the upper hand culturally',[24] since the mining centres were founded and developed as the conquerors occupied those territories. By comparison, in the Andes, 'nearly all the main mining centres developed within the area of pre-contact high culture'[25] where, for a considerable time, natives continued independently, based on their own technology, to extract and refine ores. 'In Mexico, [the] Spanish learned, it seems, little about mining from native people'.[26]

The experiences of New Spain and South America, although embedded in the same framework of the Spanish empire, show that local conditions constantly shape the technological inventions that disseminate in a more global context. Geography, pre-existent populations and their technological knowledge, the availability of natural resources such as water, timber and very importantly, the quality of the particular minerals of each region, made for the difference in the ways in which technology was either advanced or impeded.

Revolt and foreign incursions

The next event worth mentioning came during a period of greatest upheaval both in Europe and in New Spain. During the late eighteenth century and the early nineteenth, closer contact with Britain derived from a consistent penetration of British interests into the region, fostered by the opening of Spanish American colonies to 'controlled' free trade, prior to Mexican independence. War in Europe and wars of independence in Spanish America meant that an important number of British merchants with close links to the trade of some Spanish ports like Cadiz strengthened contacts with revolutionary countries in America in general and, in the case of New Spain, managed to seize the opportunities in the mining areas. Through direct contacts with local miners who were aware of the widespread diffusion of the steam engine in England, but also by using their influence on the government that focused on keeping mining as a major provider of income, the merchants fostered many initiatives to import major technological innovations to the Mexican mines.

After the wars, which started in New Spain in 1810, Mexico finally achieved independence from Spain in 1821. Soon after, the interest in promoting and developing mining technology was renewed, although the methods to achieve these goals became substantially different.

The construction of a new nation under republican governments was a large departure from colonial rule. The country emerged from a prolonged ten-year war and in need of financial resources to put the government apparatus to work. Aside from the foreign aid that poured into the country in the form of loans, mainly from Great Britain, the government took other measures to avoid insolvency and bankruptcy.

The transition from the colonial period to the national era consisted of key changes in government policy that affected mining. Unable to maintain the bureaucratic structure that had supported some of the colonial administrative tasks and with the idea of obtaining additional revenues, the newly elected governments liberated economic activities previously administered by the Spanish as government monopolies. During Spanish rule, many commodities, such as mercury, salt and gunpowder, all of them basic supplies, were subject to government control as a source of revenue and to monitor their distribution. Their liberation from these constrains consisted in granting concessions to private entrepreneurs or companies for the administration and commercialisation of such commodities, mainly quicksilver,[27] the basic ingredient of the amalgamation process and other productive activities like coinage.[28] Though the government retained control of gun powder, quicksilver and salt were given into private hands during most of the nineteenth century. Many people in Mexico perceived the relaxation of the rigid policies of colonial times as favourable to both the expansion of mining and the greater utilisation of scientific advances, as this was conceived as positive liberal practice and as an incentive to compete and therefore increase productivity. In practice, however, the weakening of government control did not always lead to the greater use of new technologies. There were many obstacles to this.

The political transformations resulting from the progressive rupture with Spain caught the mining industry by surprise and, in fact, slowed down the implementation of new technology. A case in point was the introduction of steam power. In 1805, the successful introduction of the steam engine at the mercury mines in Almadén in Spain encouraged the Spanish crown to investigate importing this English technology into Mexico. Although the steam engine had been tested in many places in England with a view to increasing power and cutting costs and had in fact created a revolution in production there, the intentions for introducing the technology into Mexico reflected a more limited, but specific goal. The steam engine would be used to pump water out of mines. The demonstration of this utility in some areas was intended to serve as a model to promote its use everywhere.

In 1820, just a year before independence from Spain was achieved, Thomas Murphy, a prominent Irish merchant who had taken advantage of the opening of markets resulting from the wars, became one of the most important beneficiaries of the Mexican wars of independence, as he received the concession to make the necessary contracts to bring in the new technology. By organising a company, Murphy would join the majority of the most important local miners. This initiative, however, soon crumbled to nothing after the war. Although linked to private enterprise, Murphy's project still depended strongly on state support, and that support evaporated in the 1820s.[29]

Murphy's initiative was hindered by the drawbacks that government officials had constantly been highlighting in the documents drawn up by both the Mining Tribunal and the Royal Mining College in the latter part of the colonial era. These institutions, which were acquainted with the local conditions of mining production, foresaw problems related to insufficient coal supply, but moreover debated the benefits that Murphy would obtain, like tax exemptions in detriment of the local miners who were decimated by the wars. The costs and supplements of the contracts to bring the steam engine into Mexico rose to GB£5900, reaching a total of more than £20,000 after charges for the agent, Murphy, transport, insurance and other services.[30]

Foreign technology in the flesh

Consequently, the use of steam power in Mexican mining developed some years later than other initiatives. The key impetus came from Britain and the technological strategies that British companies, which arrived in Mexico from 1824, used in developing mines in Mexico during the 1820s.[31] The British believed that many mines in Mexico were underdeveloped and that European technology was superior to that in use in Mexico. So great was British confidence in using their own technology that during the 1820s there was a great increase in the formation of companies both in Britain and in the recently independent countries in Latin America. Entrepreneurs founded 35 companies in England and Wales, and eventually, some 26 of them operated mines in Latin America. As Disraeli remarked in his writings on American mines referring to Mexico:

> the mineral treasures of America are, as yet, most imperfectly developed[32] . . . the mining works of America, we shall find them the productions and operations of an unscientific people, who possessed a great inducement for labour, and paid little for it.[33]

For the British investors in Mexico, as had been the case with the Spanish, the most attractive use of the steam engine was the steam pump, but its introduction in Mexico quickly led to its use in crushing, milling and amalgamating in *patio*, which, said one commentator, was a 'rude, expensive and tedious' method.[34]

Experiments in using steam may have been carried out here and there in Mexico, but still little is known about them. They received little publicity. The first known experiments with steam power originated at two of the most important Mexican mining companies: The Adventurers of Real del Monte, with properties in Bolaños, and the Anglo Mexican Mining Company enterprise that operated widely in the state of Guanajuato. These efforts attracted much attention. In 1826, one report said:

> five big steam engines (five Wolf and two Taylor) [have] been sent from England, together with a steam engine for mortar and another two smaller sawing machines that together weighed 1,500 tones . . . of which [have] been transported from the coast to Real del Monte without damage.[35]

But this optimism had to be tempered with reality. There were other, more depressing, reports about the difficulties of transporting the equipment and installing the machinery, coupled with the frustrations with specialised technicians and the very long time needed to train workers. There was little modern infrastructure in these regions, and the need to import parts and other equipment from England greatly increased costs. Although companies like Real del Monte discovered the big advantages of using steam-powered machines for reducing costs and draining water from the mines, there were long delays before the engines could be put to work. In 1830, for example, the enterprise received a steam pump from Falmouth, England via Vera Cruz, but it still took another three months before it could be installed and put into operation.[36]

Further evidence from the other side of the Atlantic reinforced this view, as Roger Burt has written in his research on John Taylor's records when comparing the success of British companies formed during the same period. In the same way that distance created administrative difficulties, it also caused serious supply problems. One of the cornerstones of Taylor's success in British mining was economy in the purchase of mine supplies, largely from local dealers. In Real del Monte, with no established industrial infrastructure, this was virtually impossible. Almost everything, from pen nibs to steam engines, had to be brought from England. This entailed not only a long and arduous sea crossing – with a constant prospect during the early years of the only convenient ports of disembarkation being blockaded by the Spanish – but also extensive overland haulage along more than 200 miles of poor roads from the coast to the high sierras.[37]

And even once installed, the steam technology could sometimes not be made to work. For example at Tlalpujahua, another of the British mining companies, the steam pumps failed to drain the water from the mines because of the lack of an industrial base that could modify machinery and guarantee replacement parts. And in the mines themselves, as one author on this subject has stated, the division of labour which the British instituted at Real del Monte for instance, prevented specialised technicians from teaching others how to use the devices, a practice that continued for decades.[38]

Meanwhile, as the British were developing the use of steam power in Mexican mines, they also gave their attention to using power in metal recovery. As the German experts had been in the latter part of the eighteenth century, the British were very keen on modifying the *patio* system, which had consistently been retained as the main method for the refining of precious metals. As mentioned above, since the mid-sixteenth century, the experiments of Bartolomé de Medina had seen a widespread expansion of this method which consisted basically of the use of quicksilver, sulphates, salts and water for the separation of gold and silver. However, by this date a lot more attention had been given to the fact that the *patio* system had been particularly suitable for a specific kind of metal. Since colonial times, mining experts had identified a type of ore named '*galena*', that is, silver-bearing lead ores that were considered as being 'rebellious' metals since they yielded poor results if subjected to amalgamation in *patio*. Amalgamation in *patio* was particularly suitable for free-lead ores, which were common in many mining regions. Therefore, when '*galena*' was found, it was preferably smelted in order to obtain the best results.[39] Most probably this was one explanation of the difficulties faced when other methods were in trial.

Despite the fact that the process carried out in *patio* establishments was considered a completely 'unscientific' and empirical method, this practice supported the unprecedented expansion of silver production of New Spain in the eighteenth century, superseding other methods like smelting.

The majority of technicians, engineers and miners that arrived from Britain with the mining companies came mainly from Cornwall, an area traditionally known as a large producer of copper and tin. However, due to the extensive European knowledge of metallurgy in silver at that time, they relied on this to commit their efforts to Mexican mines.

John Taylor, a Cornish mining engineer and contractor, developed a large scheme for the formation of two of the most important British companies in Mexico, The Real del Monte Company and the Bolaños Company, although he kept close contact with the rest of the other companies like the Anglo Mexican and the United Mexican Mining Association. In 1829, he published his *Records of Mining*, which included the major topics on mining engines and technological novelties of the time. Notoriously one of the articles, written by Henry Vivian, also an important miner in Cornwall, praised the benefits of the Börn barrel method practiced in Freiberg.[40] When the British again brought attention to refining methods, they insisted on new trials of the barrel method of extraction. Despite the previous failure of the German experiments in the later eighteenth century in Mexico, the Börn barrel system was still considered as an alternative for 'rebellious' ores to *patio* treatment.

The unstable conditions of quicksilver supply, then carried out by independent suppliers to Mexico, stimulated the search for quicksilver-saving methods. The Real del Monte Company became the consistent leader of this changeover later in the mid-nineteenth century.

But as early as 1826, H. George Ward wrote that at the San Agustin Hacienda in Guanajuato the new steam power generated enough energy for grinding the ores and powering the 'rotating barrels of Freiberg'. British companies continued their efforts at introducing this new technology at many Mexican mines, and they invested large sums of money in doing so, but many of these efforts failed to work out. The investments did not pay off in greater production.[41]

Further experiments were carried out for the sake of quicksilver savings in the metallurgical process, and minor adaptations of tools and adjustments to machines were made. Taylor was aware of the need to innovate in such areas. In 1829, he reported that:

> At the suggestion of my excellent friend Mr. Roger Morgan, made during the arrangement of many articles of apparatus for the Mines of Real del Monte in Mexico, whither he afterwards proceeded to take charge of the processes of reduction of the ores, I included a set of Retorts for the separation of the mercury from the amalgam, as it appeared to us both that improvement might be expected in this operation.[42]

Though it is not certain to what extent of the diffusion these kinds of trials had an impact in the long run, they seemed to have rendered some results in the short term by reducing costs (of mercury) or time required in particular phases of the processes involved in refining silver.

Unlike John Taylor, who engineered the formation of the British mining companies without ever visiting Mexico, Richard Trevithick, a Cornish mining engineer who successfully adapted steam engines for mining in Cornwall, played a key role in the attempts at introducing steam power to mines in Pasco, Peru.

As in Mexico, the wealth of the Peruvian mines had attracted the attention of many British investors. Already under Spanish rule an initiative was spearheaded by Francois Uvillé, who found that the mines of Peru were in need of rehabilitation and 'hearing also that these mines were richer in silver ore than those of Mexico, he conceived the idea of introducing the steam engine'.[43]

As in Mexico, the use of the steam engine was intended, primarily, for the draining of mines. Uvillé travelled to London and bought a machine in Fitzroy Square, which was the Trevithick model, for a first trial in Peru.

Back in Peru, Uvillé proceeded to obtain contracts for the exploitation of the mines in 1812 and travelled to Britain again, where he met Richard Trevithick at Camborne, starting a thorough investigation about the feasibility of the adoption of the steam engine. With that intention, Uvillé entered into contact not only with Trevithick but also with Bolton and Watt, and other Cornish miners like Trevarthen, Henry Vivian and William Bull, all of them involved in the modernisation of the mines in Cornwall at the time.[44]

Uvillé's main concerns were the geography, the adverse climatic conditions and the altitude of the mines in Peru. But 'Capt. Trevithick was not deterred from the pursuit'.[45] It seems that it was Trevithick's determination that made it possible to overcome such problems, as he had experience in adapting the basic steam engines to mine workings. Many accounts of advances in the field of mining grant him all the credit. As John Taylor put it:

If the names of these celebrated engineers do not appear so often in the following account as might have expected, it must not, therefore, be concluded that they had not their share in the general improvement. Steam engines had undergone alterations; and to Captain Trevithick, is due the introduction of the form which is now generally adopted, and from which doubtless considerable advantage has arisen.[46]

Trevithick finally provided the apparatus for the construction of nine engines at a cost of $10,000.[47] The machinery arrived in 1815, leading Trevithick to travel to Peru in 1816,[48] much earlier than the machines that arrived in Mexico.

The expectations in Peru were high, as expressed in the *Lima Gazette* the following year:

Immense and incessant labour, and boundless expense, have conquered difficulties hitherto esteemed altogether insuperable; and we have, with unlimited admiration, witnessed the erection, and astonishing operation of the first steam-engine. It is established in the celebrated and royal mineral territory called the mountain of Yaiiricocha, in the province of Tarma; and we have the felicity of seeing the drain and the first shaft in the Santa Rosa mine, in the noble district of Pasco . . . we anticipate a torrent of silver, that shall fill surrounding nations with astonishment.[49]

Trevithick remained in Peru as an engineer in the House of Mint. He had also built machinery for that establishment.[50] According to Anthony Burton:

Thanks to his intervention, Peruvian mining was transformed from an industry that had scarcely changed for centuries into one that was modern, efficient and profitable. He may have been forced to abandon the silver mines himself, but others were eventually able to restart the work and they flourished once more.[51]

Indeed, others had to continue his work and the relaunching of the use of steam power. During the wars of independence in Peru, most of the area was devastated, had its mining workforce depleted and the machines destroyed.

Although Spanish American countries achieved independence during the 1820s, the interests in mining remained the same as they had been before independence. Peru was no exception. Various mining companies were formed to exploit the riches of the area. The Pasco Peruvian Mining Company was the one which took the Pasco mines, where Trevithick had been involved. Their expectations based on Alexander Von Humboldt's accounts and the revival of the interest in the steam engine gave a second impulse to experiment with it. 'It will appear the Mines in possession of this Company have hitherto been very productive, more particularly when the steam-engines were applied in draining them'.[52] The company created to exploit these mines raised 10,000 shares of £10 each. 'It . . . secured contracts for a long term of years for some valuable

Mines on the celebrated heights of Pasco, in the district of Yauricocha and Province of Tarma'.[53]

British investment in Mexican as well in South American mines also found itself hampered by the London Stock Exchange crisis that followed a speculative fever in American mines and investments in the 1820s. Due to the 1820s' boom and crisis, companies and entrepreneurs held back funds that might have supported the installation of more innovative technologies, and this in turn meant that British mine managers in Mexico reconsidered the view that extant mining and extractive methods there were more suitable for local economic conditions. Ward himself knew how to distinguish between the economic benefits derived from skilled workers and the costly problems attributed to the introduction of new technology. More than once he said that '*amalgamadores*' (a specialised branch of craft workers who by experience only could calculate the correct ingredients used in the *patio* system and the timing for metal separation) offered the best guarantee of success, not the large-scale spending of British companies on labour-saving technology.[54]

In the long run, the profits that the British extracted from their investments along with their insistence on new technology were beneficial to those who took over the mines,[55] but in the short run, new technology was a mixed blessing with mixed results. The legacy of British technological advances in Mexican mining in the early nineteenth century was a limited one. Some English mining engineers remained in Mexico where they used advancing technology adapted to older circumstances and traditional local mining methods, and they provided a link to continued technological change. But the British in general began to migrate their investments towards other alternative ventures, sometimes connected to mining. Most British capital concentrated in the administration of some regional houses of minting created in the 1820s in the provinces and devoted much attention to investment in Mexican government bonds. Mexico as a destination for British migrant miners lost its attractiveness and was overtaken by better opportunities in other English-speaking mining countries like the US or Australia, just at the time that steam power and other technologies were becoming important throughout Cornwall, the main source of Mexican imports and also through the mining world.[56]

The dressing up of traditions, the continuity of traditional local methods as a form of technological innovation and the obstacles confronted

Despite its difficult beginnings in the 1820s, by the latter part of the nineteenth century, the use of the steam engine was becoming widespread in Mexico. It was used to power a myriad of activities in mining. But even then, the technologies used for recovering metal from ores remained a mixture of different methods that ranged from the old *patio* process to the newer smelting techniques. By the later nineteenth century, this process demonstrated many new techniques slowly incorporated over the years. The perfection of the process reflected the determination of many people[57] who could be better described as alchemists rather than scientists. Despite their knowledge of the advances in eighteenth and nineteenth century science, these individuals relied more on acquired skills than on proven and reproducible scientific methods.[58]

Much the same can be said for the engineering and mechanical aspects of mining and for the problem of draining water from mines. The nineteenth century witnessed many efforts to improve techniques, but only rarely did mining people find acceptable solutions. Nonetheless, there were continuous initiatives from the final days of the Spanish era through the end of the nineteenth century. Both the state and the mining companies, along with the intelligentsia of New Spain, later independent Mexico, were all responsible for a series of initiatives that arose from the constant interplay of economic interests and aspirations and from the changing technological challenges in the mining industry. Ultimately, these visions gave way to a clear understanding that Mexico was and would continue to be a mining nation *par excellence*. The nation's income depended heavily on mineral exports, because outside of mining its industrial sector was too small and growing too slowly to provide an alternative.

A good approximation for assessing the technological initiatives in Mexican mining, as an expression of local factors and as an explanation of the impossibility of adopting foreign methods, along with the obstacles to development and implementation, are the official records for industrial patents. While these registers, however useful, do not show the extent to which technological advances were actually applied in the industry and in the different mining regions, the list of patents granted does give a complete picture, direction and magnitude of the changes proposed in each application.

After a brief interval in the introduction of innovations by foreign companies, the government of Mexico as well as the Mexican intelligentsia wanted to reaffirm and expand the role that mining played in the country. Independence from Spain certainly offered the possibility, at least in theory, of implementing a new project of development having Europe, and especially Great Britain, as a model. The British withdrawal from Mexican mining, at least in terms of new investment, seems to have renewed the interest of Mexican miners, who sought to revitalise the industry through the new institutions of the new republic.

Other countries like Peru, Bolivia and Chile tried similar experiments, in order to restore conditions for the recuperation and promotion of mining by mirroring colonial institutions in republican governments, while also adapting to the redirection of their mining enterprises into other products such as nitrates and guano production in Peru or copper in Chile.[59]

The Mining Tribunal in Mexico, which for better or worse had been a channel for development under the old regime, had then been shut down and substituted by a minor government office with diminished functions. After independence from Spain, 'while initial enthusiasm was shown and the main expectations were placed in foreign investors, the owners of big mines were not concerned with the provision of a new organism that supported the sector'.[60]

The dismantling of the Mining Tribunal and the other important colonial institutions after independence from Spain, like the Banco de Avío (Mining Bank) and the Royal College of Mines, required new institutions that could redirect technological innovation. But the new nation-building process sought greater diversification and a stronger internal economy, both of which seemed to require the weakening of colonial institutions and the links that had favoured the strengthening of mining. The old system practiced under Spanish rule of granting recognition for innovations gave way to the Patent Register, and the administration of the new system fell into the hands of the Secretaria de Fomento (or the Ministry of Promotion) of the new independent state governments. Despite different attitudes and styles in treating innovations, between 1832 and 1913, new governmental administrations did not undertake consistent and definite legal reforms. The patent laws, which were in effect during this prolonged period of time, required the presentation of an application and the accompanying documentation that explained the case, and the financial proof of the invention. Far from being effective, as the Mining Tribunal had been in previous times, the new law proved hard to enforce. There was little opportunity to experiment with innovations in different mining regions – this was instead left to independent inventors and promoters. Thus, the Ministry of Promotion had, like many other branches of the new republican government, very limited resources to finance and experiment with new technological innovations.

Revision of an important source at the National Archives in Mexico of more than 175 patent applications related to mining provides a general picture of how the promotion of innovation, the maintenance of tradition and the recognition of some elements that could constitute important changes in mining activity were carried out. The majority of patent applications were related to various amalgamation processes using quicksilver. During these years, there were 104 applications for innovations related to refining minerals. Only two of these excluded the use of mercury. Second were the improvements to crushing and grinding that preceded the process of amalgamation itself, and of those, eight focused on draining mines and one on smelting. The rest dealt with raw materials, drilling machinery and mechanical engineering. Overall, the tendency of these innovations was not the adoption of new methods, but rather improving well-known technologies.

The continued reliance on the amalgamation process tied Mexican mining to mercury throughout the nineteenth century. The only fundamental changes here came with the supply. It shifted from a monopoly controlled by the state to a monopoly controlled mainly by the Ezpeleta Spanish merchants at the turn of the eighteenth century, the Rothschild's interests in the first half of the nineteenth century[61] and the Barron and Forbes' interests in the following decades until the cyanide process was introduced by American companies in the later part of the same century, establishing a new revolution in mining practices in Mexico.[62]

The patents granted reflected the fact that the major problem in Mexican mining was the dependence on the use of quicksilver. Patent applications that promised to save or eliminate its use were more likely to be received favourably.

The goal, according to the Ministry of Industry, that was found repeatedly throughout the nineteenth century:

was to rid the Republic from the need to use one of the most important branches of wealth that is imported from abroad and today monopolized by a foreign speculator, and substituting it for another amalgamation method that brings together the advantages of time economy and can extract almost all the silver contained in the minerals by using ingredients that are all found in the country.[63]

One of the traditions inherited by the College of Mining was the importance of saving mercury and other supplies used in processing ore and metal and the need to maintain a true industrial orientation in the mining sector, and these aims were maintained by decrees for the granting of patents by the new Ministry of Promotion. This was not a smooth process, however. The Ministry was overloaded by complaints from those opposed to the granting of some patents, all of which were published in the *Diario Oficial*. In disputed cases, the Scientific Section of the Ministry gained the responsibility to develop independent scientific opinions on any patent application. The Scientific Section also responded to the grievances of miners complaining of the use of patents to establish monopolies that would allegedly limit mining development.

Numerous proposals came from well-known figures in Mexican scientific circles, such as John Bowring, Maumejean, Saint Claire Duport and Boivin. The treatment given to their patent applications reflects the treatment of all. Maumejean wrote in 1857 that:

> at my age, being sixty-one years old and without ambition now ... I am interested in the future and the prosperity of the Republic ... I will give the advantages that result in the exploitation of my privilege half to the Mexican College of Mining and another half to invalidated soldiers.[64]

The mining patents issued reflected a number of factors. Many inventions devolved from the geological conditions in whose context they originated. Others reflected politics and finance. Still others reflected the interrelated and sometimes conflicting investments of Mexicans and foreign nationals. In the 1840s, for example, the so-called 'rebellious' ores or 'black metals' of northeastern Mexico fostered inventions based on variations of the quantities of supplies used in the process of amalgamation.[65] 'Rebellious metals' were those which contained larger quantities of minerals like lead. Consequently, many essays on metallurgy centered on developing new processes designed to eliminate those residues in processing. Ignacio Retes, a miner in the state of Sinaloa, claimed to be able to produce large quantities of silver by reducing the quantity of copper sulphates in the mixture sent to the *patio*.[66] Saint Clair Duport proposed the use of electricity.[67] In contrast, the central mining regions around Guanajuato favoured improvements in crushing and milling.[68] Miners of other regions insisted on extra efforts to improve the process of mixing the ingredients in the *patio* in less time.[69]

The end of bonanzas also contributed to many new inventions in an effort to forestall decline, but these developments also led to calculations on how such work could be used.

In the later nineteenth century, mainly from the 1880s, patent applications grew in number for other reasons. Again, foreign investment, technological developments in other countries and the consolidation of supply centres providing machinery to the industry were key to this. Technological advances in the American southwest were particularly important, and inventors there, or their agents, sought to obtain patents in Mexico. It was, for example, not a coincidence that the Mexican consul in San Francisco, Francisco Godoy, became a key individual in introducing new patents for the technology of drilling machinery. San Francisco was, after all, a major supply centre for mining in the entire American west, and it certainly extended its influence into Mexico.[70]

Preliminary conclusions

Looking at the technological advances in Mexican mining during the nineteenth century leads us to the conclusion that the relationship between the 'local' and the 'foreign' shaped different epochs of technological development from the times of the Spanish conquest, by implanting Western methods of mining and interacting with uses and conditions fermented in the practices of local mining. Long after Mexican independence in 1821, the overall tendency was to introduce minor improvements to older methods, notably to the *patio* process. And this was reflected in the legal structure.

Traditional mining bodies like the *diputaciones mineras* (that is to say, the representatives of prominent local miners of different regions to national authorities), acted on a local basis, and technological advances came from local producers. Later, in the latter part of the nineteenth century, technological change came largely from outside Mexico. It was during those years, roughly from 1867 to 1900 and after, that processes like lixiviation and cyanide treatment were introduced by foreign investors who had acquired the patent rights to those technologies. And the introduction of those processes also came at a time when Mexico became a more economically and politically stable country.

Much of this chapter, apart from suggesting further research of local archives to offset the limitations of the Mexican patent records for the nineteenth century, also considers the investigation of similar or different processes in other silver producer countries, both within Latin America and the rest of the world, in order to be able to compare fully in a global context.

Unlike the colonial records of royalties and the College of Mining records in Mexico, the Patents Register does not follow up the decisions on inventions and their experimentation in different regions. The knowledge gained from the Patents Register, on which much of this work is based regarding the nineteenth century, has to be accompanied by more detailed research on their application and diffusion in different regions of the country. Very few files contain detailed calculations on the impact of new technologies that can yield comparative data in terms of productivity, time- or labour-saving outcomes. Some cases show, to a minor extent, the use of some of the proposed improvements registered in the patent documents. However, this is not common. More efforts should be made to investigate the ways in which innovations impacted mining and became generalised practices.

Nevertheless, it is fair to suggest that the impact of the major improvements in mining that came with the adoption of foreign new practices in the different phases of mining concerned either extraction or metallurgy.

In the short run, all innovations acted as a force of modernisation of mining which required not only the commitment of financial resources but also of initiatives of different sectors of society, from those in government to those in the private sector.

In previous centuries, the Spanish crown constantly pursued the renovation of the mining sector, through institutions and mining-oriented policies that nurtured and strengthened this industry, resulting in a surge of production between the 1770s and the 1800s. The same could be said for the renewed interest in innovation reflected in the introduction of new technologies that originated in Britain, after Mexico obtained independence from Spain, which was conducive to a great bonanza between the 1840s and 1850s.

Mexican and some South American mines were flooded or destroyed by the effects of the wars of independence, running from 1810 to the early 1820s. Their decay promoted renewal through large injections of capital, specialised labour and long-awaited technologies. As mentioned above, the companies confronted countless difficulties in importing, installing and putting into operation the steam engines intended for the drainage of mines. But at the same time, they extended their uses to other tasks, which in the long run must have had consequences for productivity. It is true that the companies themselves channelled their capital to other investments in the late 1830s, but the mines went back to the previous owners of the rights of exploitation and they were able to reinitiate operations in very comfortable conditions. They took advantage not only of the new infrastructure and rehabilitation carried out by the British in their mines in Mexico, but continued the use of the technology the British companies had introduced, as was the case with the steam engine.

Another inheritance may well be perceived in the ways the British approached the mining problems through different and up-to-date scientific methods. This possibly does not apply completely to the metallurgical processes, since the *patio* remained the most important system for the refinement of gold and silver. But again, it probably made the Mexican miners aware of the need to try newer methods to reduce the consumption of quicksilver, which was clearly shown in the inventions presented in the Patents Register during the whole of the nineteenth century. Coexistence of traditional methods with new imported ones may also have accounted for innovation on a smaller scale. Particularly in those cases where parts for, and replacements of, machines were not at hand.

Notes

1 Some of the techniques of metal workings carried out before the Spanish conquest are recorded in the early chronicles and Codex; see: *Códice Florentino*, ed. facsimil. México, AGN, 3 Vols.
2 Robert West, *The Mining Community in Northern New Spain: The Parral Pining District* (Berkeley and Los Angeles: University of California Press, 1949), p. 16. They introduced the water power stamp mill and amalgamating processes.

3   The organisation of the economy by the Spanish crown imposed a system of taxation which included the '*tributo*' among the indigenous population of the New Spain. Part of this system consisted of granting the main landowners or mine owners quotas of indigenous labourers from adjacent regions to work on their properties. In Mexico this was known as *Repartimiento*, while South America adopted the name '*mita*'. The vast literature on this subject regarding Mexico points out that by the end of the eighteenth century, despite the difference in regions and moments, compulsory mining labour was progressively substituted by free labour aided by the '*partido*', which meant that mining workers could obtain a small percentage of the returns of a specific mine, according to the minerals they extracted during a certain period of time. Roberto Moreno, 'El régimen de trabajo en la minería mexicana siglos XVI y XVII', in: *Minería Mexicana* (México: Comisión de Fomento Minero, 1984), pp. 86–90; Cuauhtémoc Velasco, 'Los trabajadores mineros en la Nueva España, 1750–1810', in: Pablo González Casanova (eds.), *La Clase Obrera en la Historia de México* (México: Siglo Veintiuno, 1980), pp. 239–299; Brígida Von Mentz, *Trabajo, Sujeción y Libertad en el Centro de la Nueva España* (México: Porrúa, 1999); Enrique Tandeter, 'Forced and free labour in late colonial Potosí', *Past and Present 93* (1981), pp. 98–136; Matthew Smith, 'Laboring to choose, choosing to labor: coercion and choice in the Potosi Mita', *Past Imperfect 10* (2004), pp. 21–44.

4   Peter Bakewell (ed.), *Mines of Silver and Gold in the Americas* (Variorum: Ashgate Publishing 1997), pp. xiii–xvi.

5   Both the colonial government or later the republican administrations devoted considerable efforts to creating institutions and organisations either in finance, or in education and science and promoted a legal framework in other fields for the advancement of mining. Some of the colonial institutions were transformed following independence, while others were dismantled. Cuauhtemoc Velasco, Eduardo Flores Clair, Alma Laura Parra Campos and Edgar Omar Gutiérrez, *Estado y Minería en México 1767–1919*, (México: Fondo de Cultura Económica, 1988).

6   The *patio* process was the most extended metallurgical method used in New Spain since 1555, and its use continued until the late nineteenth century.

7   Developed by Bartolomé de Medina in the sixteenth century in Pachuca, Mexico, it became widespread. It became the most important alternative to foundries or any other metallurgical process practiced in the whole of New Spain. Based on sources of the sixteenth, seventeenth and eighteenth centuries, twentieth-century authors like Bargalló and Robert West have ably described the amalgamation in *patio* as follows:

> it embodied the mixing of finely crushed ore with water and usually three reagents, mercury, salt and impure copper sulphate. Often other ingredients, such as tequizquite, a lime compound, and iron fillings were added. The mixing of the reagents with the water soaked ore (*masa*) was done in a spacious, rock-floored courtyard (*patio*), surrounded by adobe or stone walls. The mixing was usually accompanied by driving mules through the ore mud. The mixed material was then heaped into piles (*montones*) and permitted to stand for weeks (in some patios as long as six months) during which time the silver ion was chemically separated from its compounds and the metal amalgamated with mercury. The time necessary for amalgamation varied with air temperature; the process required a longer period in winter than in summer. After amalgamation the treated ore mud (*incorporado*) was washed in water tanks or troughs (*lavaderos*); there the worthless slime was carried away, leaving the heavier amalgam and free mercury. This residue was then placed in canvas bags and much of the free mercury was filtered out. The residual amalgam was pressed into bars and placed in a small furnace (*buitrón*), where the mercury volatilised and the metallic silver-gold alloy remained, retaining the shape of the original amalgam bar. The bars were taken to the official assayer (*ensayador*), who recast,

assayed, and stamped each bar with an identity number and a figure indicating fineness. (Robert West, *The Mining Community in Northern New Spain: The Parral Mining District* (1949), pp. 32–33.)

Also the classic work on metallurgy in Spanish America: Modesto Bargalló, *La Minería y la Metalurgia en la América Española Durante la Epoca Colonial* (México: Fondo de Cultura Económica, 1955).

8 The most important Spanish colonial institutions were the Mining Tribunal, in charge of the issue and enforcement of the Mining Laws and the solution of mining disputes amongst mining organisations, mining representatives and independent miners, it established the policies for the Royal Mining College, and the Mining Bank which intended to provide and organise financing to mining. Walter Howe, *The Mining Guild of New Spain* (Cambridge: Harvard University Press, 1949), gives a comprehensive overview of these colonial institutions, also David Brading, *Mineros y Comerciantes en el México Borbónico, 1763–1810* (México: Fondo de Cultura Económica, 2010), from the 1st English edition 1971.
 9 John Fisher, *Minas y Mineros en el Perú Colonial 1776–1824* (Lima: Instituto de Estudios Peruanos, 1977), pp. 71–113.
10 Modesto Bargalló, *La Minería y la Metalurgia en la América Española Durante la Epoca Colonial* (1955), p. 40 and pp. 95–96.
11 Carlos Sempat Assadourian, 'La bomba de fuego de Newcomen y otros artificios de desagüe: un intento de transferencia de tecnología inglesa en la minería novohispana, 1726–1731', *Historia Mexicana 50*(3) (2001), pp. 385–457, showed that the machines were intended for the Real del Monte area where 34 whims already served to drain the mines at a rate of 50,000 '*arrobas*' (Spanish volume measure equivalent to 11.502 litres) of water during 24 hours at a depth of 184 metres. The machines from England would almost double the rate to 90,000 in 24 hours.
12 Inés Herrera, 'Los socavones aventureros', *Historias 28* (April–September 1992), pp. 75–86; this author explains how this technique was known since medieval times and adopted in very few cases in New Spain, and later in Mexico, due mainly to the high costs and the availability of cheap labour.
13 In the early 1970s, several scholars from Britain and the US took interest in Mexican mining, some of their works, although key for future research, have been underestimated. Such is the case of Clement G. Motten, *Mexican Silver and the Enlightenment* (Philadelphia: University of Pennsylvania Press, 1950). This is a pioneer work, which reports the first 'foreign' incursions in Mexican mining before the independent era. New research is being carried out by scholars, whose conclusions on the impact of the German 'visitors' are divided. Omar Escamilla, 'Un metalurgista germano en Guanajuato y Michoacán, las cartas de Franz Fischer (ca.1757–ca.1814 a Ignaz Von Born 1789–1814', *Boletín Archivo General de la Nación 19* (January–March 2008), pp. 98–120.
14 *Archivo General de la Nación*, (México: Fondo Minería 073, Vol. 43) (January 1790); David Brading, *Mineros y Comerciantes en el México Borbónico* (México: Fondo de Cultura Económica, 1975), p. 444.
15 Eduardo Flores Clair, 'El fracaso del método de Börn en la minería novohispana' (forthcoming).
16 Clement G. Motten, *Mexican Silver and the Enlightenment* (New York: Octagon Books, 1972), p. 48.
17 Mervyn Lang, 'La tecnología minera alemana en la minería virreinal', in: *Actas VIII, Sociedad Española de las Ciencias y de las Técnicas*' (2004), pp. 55–62.
18 Fisher, *Minas y Mineros en el Perú Colonial 1776–1824 (*1977), pp. 138–139.
19 Fisher, *ibid.*
20 Federico Sonneschmidt, Prologue to *Tratado de la Amalgamación de Nueva España* (Paris: Mégico, 1825).

21 Mary Van Buren, 'Estudio etnoarqueológico de la tecnología de fundición en el sur de Potosí, Bolivia', *Textos Antropológicos 14*(2) (2003), pp. 133–148.

22 Tumialán de la Cruz and Pedro Hugo, *Compendio de Yacimientos Minerales del Perú* (Lima, INGEMMET, 2003).

23 Modesto Bargalló, *La Amalgamación de los Minerales de Plata, Fundidora Monterrey* (México: Compañía Fundidora de Fierro y Acero de Monterrey, 1969), pp. 161–165.

24 Peter Bakewell, Preface to Langue Frédérique and Carmen Salazar Soler, *Diccionaire des Termes Miniers en Usage en Amérique Espagnole (xvi–xix)* (Paris: Recherche sur les Civilisations, 1993), pp. xvi–xvii.

25 *Ibid.*

26 *Ibid.*

27 María Eugenia Romero Sotelo, 'El mercurio y la producción minera en la Nueva España 1810–1821', *Historia Mexicana* 65(4) (2000), pp. 349–377; Inés Herrera Canales, 'Mercurio para refinar la plata mexicana', *Historia Mexicana 60*(1) (1990), pp. 27–51; Alma Parra, 'Lionel Davidson and the Rothschilds in Mexico', *The Rothschild Archive Review* (April 2007–March 2008), pp. 27–34.

28 Alma Parra Campos, 'Control estatal vs. control privado: la casa de moneda de Guanajuato en el siglo XIX', in: Bátiz Vázquez, Jose Antonio and José Antonio Covarrubias, *La Moneda en México, 1750–1929* (México: Instituto Mora, 1998), pp. 155–168. Alma Parra Campos, 'La Anglo Mexican Mining Association y la Casa de Moneda de Guanajuato', *Mundo de Antes* 12(2) (2018), pp. 127–149.

29 *Archivo, General de la Nación*, (hereafter *AGN*) Fondo Minería, vol. 28, file 5 and fs. 147–300. See for example Bryan Earl, *Cornish Mining: The Techniques of Metal Mining in the West of England, Past and Present* (Truro: D. Bradford Barton Ltd, 1968) and Frank D. Woodall, *Steam Engines and Waterwheels: A Pictorial Study of Some Early Mining Machines* (Bath: The Bath Press, 1975); *AGN*, Minería, vol. 28, fs. 55–74, 78–96.

30 *Ibid.*, vol. 28, fs. 373–474.

31 Alma Parra and Paolo Riguzzi, 'Capitales, compañías y manías británicas en las minas mexicanas', *Historia 71* (September/December 2008). The formation of the companies implied the arrival of capital, engineers, technicians and a flow of migrants providing the workforce of these enterprises.

32 *An Inquiry into the Plans, Progress and Policy of the American Mining Companies, with Considerable Additions by Disraeli the Younger* (Prime Minister Benjamin) Third edition (London: John Murray, 1825), p. 131.

33 *Ibid.*, p. 14.

34 Based on reports of Mr Garby and Mr Moyle, British engineers contracted by the Anglo Mexican Company, *ibid.,* p. 34 and William Rawson, *The Present Operations and Future Prospects of the Mexican Mine Associations Analysed by the Evidence of Official Documents English and Mexican and the National Advantages Expected from Joint Stock Companies Considered in a Letter to the Right Hon. George Canning, by . . .* Second Edition (London: Hatchard and Son, 1825), pp. 34–35.

35 H. George Ward, *Mexico en 1827* (Mexico: Fondo de Cultura Económica, 1981), p. 364 (from the first English edition of 1828).

36 R. W. Randall, *Real del Monte: Una Empresa Minera Británica en México* (México: Fondo de Cultura Económica, 1977), p. 121; Ward, *Mexico en 1827* (1981), p. 367.

37 Roger Burt, *John Taylor: Mining Entrepreneur and Engineer 1779–1863* (Buxton, Derby: Moorland Publishing Co. Hartington, 1977), p. 42.

38 Eduardo Flores Clair, 'Conflictos de trabajo de una empresa minera, Real del Monte y Pachuca, 1872–1877' (BA Thesis) (UNAM, 1989), pp. 177–190.

39 Robert West, *The Mining Community in Northern New Spain* (1949), p. 25.

40 Henry Vivian, 'Description of the process of Amalgamation as practised in Freyberg', in: John Taylor (ed.), *Records of Mining* (London: John Murray, 1829), pp. 21–50.

41 H. George Ward, *Mexico en 1827* (Mexico: Fondo de Cultura Económica, 1981), p. 558.
42 John Taylor, 'Description of retorts for the distillation of the mercury from amalgamated methods', in: John Taylor (ed.), *Records of Mining* (1829), pp. 141–143.
43 Henry Boase, 'On the introduction of the steam engine to the Peruvian mines', in: *Transactions of the Royal Geological Society of Cornwall Vol. 1* (London: William Philips, 1818), pp. 212–223.
44 *Ibid.*
45 *Ibid.*
46 John Taylor, 'On the duty of steam-engines', in: John Taylor (ed.), *Records of Mining* (London: John Murray, 1829), pp. 149–166.
47 *Ibid.*
48 Francis Trevithick, *The Life of Richard Trevithick with an Account of his Inventions* (Cambridge: Cambridge University Press, 2011), pp. 226–228.
49 Charles F. Partington, *A Popular Descriptive Account of the Steam Engine, with an Appendix of Patents and Parliamentary Papers.* Third Edition (London: John Weale, 1836), p. 41.
50 Hyde Clark, *Life of Richard Trevithick*, (England?: Publisher not identified, 1848?), 30 pp.
51 Anthony Burton, *Richard, Trevithick, Giant of Steam* (London: Aurum Press, 2000), p. 193.
52 Henry English, *A General Guide to the Companies Formed for Working Foreign Mines* (London: Boosey and Sons, 1825), p. 94.
53 *Ibid.*, pp. 49–50.
54 *Ibid.*, p. 561.
55 Alma Parra, 'La familia Rul y Pérez Gálvez y el despegue de la minería post-independiente en Guanajuato' (PhD Thesis) (ongoing). In fact, the previous owners of the mines were the ones who took over the mines, benefitting from receiving the mines recovered from the damaged state resulting from the wars of independence.
56 Thomas Lean and Brother, *Historical Statement of the Improvements Made in the Duty Performed by the Steam Engines in Cornwall, from the Commencement of the Publication of the Monthly Reports, Compiled at the Request of the British Association for the Improvement of Science* (London: Simpkin, Marshall and Co., 1839).
57 Elías Trabulse, 'Aspectos de la tecnología minera en Nueva España', in: *Historia de la Ciencia y la Tecnología, Lecturas de Historia Mexicana* (México: Fondo de Cultura Económica, 1994), pp. 218–264; Alma Parra, 'The persistence of the patio system in Guanajuato during the XIX century', in: *IV International Mining History Congress* (Golden, Colorado, 1994).
58 Alma Parra, 'Experiencia destreza e innovaciones en la minería de Guanajuato en el siglo XIX', *Historias 58* (2004), pp. 69–81.
59 In 1821 for instance, the Peruvian government established the Dirección General de Minería, De la Cruz and Hugo, *Compendio de Yacimientos Minerales del Perú* (2003) p. 11.
60 Cuauhtémoc Velasco et al., *Estado y Minería en México, 1767–1910* (México: Fondo de Cultura Económica, 1988), p. 122.
61 Alma Parra, 'Mercury's agent: Lionel Davidson and the Rothschilds in Mexico', *Rothschilds Archive Review* (April 2007–March 2008).
62 Inés Herrera, 'Mercurio para refinar la plata mexicana en el siglo XIX', *Historia Mexicana 40*(1) (1990), pp. 27–52.
63 *Archivo General de la Nación*, Patentes, Box 1, file 60.
64 *Archivo General de la Nación*, Patentes, file 336.
65 Classic books on metallurgy make this distinction. Alvaro Barba, *Arte de los Metales en que se Enseña el Verdadero Beneficio de los de Oro y Plata por Azogue* (Lima: Real Tribunal de Minería, Imprenta de los Huérfanos, 1817), p. 75. Sonneschmidt, Prologue to *Tratado de la Amalgamación de Nueva España* (1825), p. 55.

66 *Archivo General de la Nación*, Patentes, file 67.
67 *Archivo General de la Nación*, Patentes, file 60.
68 *Archivo General de la Nación*, Patentes, file 76.
69 *Archivo General de la Nación*, Patentes, file 27.
70 See Lynn R. Bailey, *Supplying the Mining World: The Mining Equipment Manufacturers of San Francisco, 1850–1900* (Tucson: Westernlore Press, 1996) and Gray Brechin, *Imperial San Francisco, Urban Power, Earthly Ruin* (Berkeley, Los Angeles, London: University of California Press, 2001).

6 Coal mining, migration and ethnicity

A global history[1]

Ad Knotter

Introduction

Between Spitsbergen (Svalbard) Islands (Norway) in the far north, South Island (New Zealand) in the far south, Vancouver Island (Canada) in the far west, and Hokkaidō Island (Japan) in the far east, coal mining has been (and in fact still is) a truly global industry.[2] From the nineteenth century the development of industrial and transport technologies required the supply of coal-based energy in every part of the world, and to provide for this energy coal mining expanded globally.

Nineteenth and early twentieth century globalisation, including colonialism, to a great extent depended on a transport and industrial revolution, based on coal as a supplier of energy. The Suez Canal, for instance, both instrument and symbol of colonialism, is hardly imaginable without steamships, and there were no steamships without coal, of course. The same holds for railways all over the world, which, together with steamships, formed the infrastructure of colonialism. The exploitation of coal mines expanded with rising energy needs in transport and industry. Wherever in the world coal was found, even in the most desolate and remote areas, mines were opened and to enable exploitation mine operators had to find, mobilise and direct workers to these sites.

Being place-bound by geology, often originating in isolated places, and always labour intensive, coal mining was dependent on migrant labour in almost every district. At the start, experienced miners were recruited from other mining areas. Migration trajectories, return and circular migration, resulting in ethnic diasporas of skilled miners, can be traced in many coal-mining districts. British miners and engineers were the expert workers of the coal-based energy revolution in the nineteenth century, who transferred their knowledge to develop coal industries in the British Empire, in the United States and in other parts of the world.

Early migration of skilled groups of workers to introduce mining skills was supplemented by waves of inexperienced migrants from the surrounding countryside, and soon also from more distant places, regions and countries. Workers had to be found who could be coerced or motivated to move hundreds of kilometres from their place of birth, not only changing places, but their entire way of life. Cross-border migratory labour connected coalfields, regions and countries, and mobilised new groups of workers of a variety of national and ethnic descent.

Therefore, the history of labour in the coalfields is not only a global but often a transnational history as well.

These salient features of mining labour have generated a lot of research, especially in labour history. In this research the common saying that 'everybody was black down there' has been increasingly provided with a question mark.[3] In histories of coal-mining communities and mining labour, issues of ethnicity and culture have become major topics.[4] While in this kind of research many insights have been gained on issues like class (solidarity), race (discrimination) and ethnicity (ethnic identity) in the history of coalfields in different parts of the world, the range of topics related to migration and ethnicity in coalfield history is far from exhausted. Migration and ethnic mobilisation of labour for the coal mining industry were often closely linked to the transition from agriculture to industry, the creation of a wage labour market, and the formation of an ethnically stratified coal-mining proletariat. These processes were not easy or straightforward. In areas were labour markets were underdeveloped, mining labour was often combined with subsistence agriculture and could only be employed seasonally. Special recruitment mechanisms like subcontracting were employed to bridge the gap between agriculture and industry. In other cases, force was used to recruit ethnic groups with a perceived inferior status. The history of coal mining witnessed a whole array of force to press workers into the mines. In many cases, this was realised by a combination of physical and economic coercion through indenturing, debt and other bonds. Forced labour and ethnic or racial discrimination were often closely related, but also when miners worked for wages, mining labour markets were systematically structured and institutionalised by wage discrimination and ethnic stratification.

In this chapter I give an overview of the most relevant issues in the history of migration and ethnicity in coal mining from a global perspective. The chapter is roughly divided into four parts. The first part deals with the recruitment of migrants and ethnic minorities in upcoming coalfields in the nineteenth century and the first few decades of the twentieth century in areas where wage labour markets were underdeveloped. In these cases mining companies had to rely on temporary migrants, peasant miners, from the land; on special recruitment systems, mainly subcontracting; or on systems of coercion, such as convict labour, indentured or contract labour, or blunt force and violence.

The second part of the chapter deals with diasporas of specific groups, such as skilled British workers, who migrated all over the world to introduce mining skills and often took a position of privilege vis-à-vis unskilled workers from other ethnic groups. British migrants were particularly important in the coal mines in the white settler colonies of South Africa, Canada, Australia and New Zealand, and also in the emerging coal industry of the United States. Another mobile group were Polish miners, both in Europe and in the United States.

The third part deals with state-regulated migration to the coalfields of Western Europe (Britain excepted), which started after the First World War and became fully developed after the Second. This was the so-called guest worker system, which brought migrants at first from Eastern European and later from the

Mediterranean countries to Germany, France, Belgium and the Netherlands. I pay special attention to Turkish migrants in Germany, Italian migrants in France and Belgium, and Moroccan migrants in France, Belgium and the Netherlands.

The fourth part of the chapter looks at the social relationship of migrants and established workers in the mining communities. How did different ethnic groups live together? How did they cope with racial discrimination and ethnic segregation? To what extent could new groups of workers and their families integrate in the mining community?

Peasant miners and oscillating migration

In emerging coalfields, large parts of the labour force were recruited seasonally as peasant migrants from the land. In this way labour supply and the agrarian seasons were interconnected, coal extraction proceeding in reverse tandem with agrarian seasons. Seasonal peasant workers were recruited both locally and as temporary migrants. A striking example of the local recruiting of peasant miners is the labour system in the Zonguldak coalfield in Turkey, where an intricate system of rotational work was installed in the nineteenth century (since 1860), forcing peasants from villages in the region to work underground during several weeks of the year.[5] In 1965 rotational workers were still drawn from some 377 villages located throughout the province of Zonguldak. Force was no longer used, however, as working in the mines had become a family tradition, handed down from father to son.[6]

In the large Jahria coalfield in India (opened in the 1890s), 'recruited' or seasonal workers comprised 50 to 75 per cent of the workforce by the 1920s, compared with 'settled' migrants who made up 15 to 25 per cent, and 'local' workers amounting to only 5 to 10 per cent. Peasants and landless labourers, who were seasonally unemployed, and often indebted, came to work in the coalfields to keep the village households functioning. The mining workforce, male and female working together in family teams, was mainly 'low caste' and 'tribal' (so-called Adivasi or 'aboriginals'); 'upper caste' were to be found only in the supervisory grades.[7]

Seasonal peasant workers, mainly from Russia, were also recruited on a massive scale for the Donbass coal mines in Ukraine. In mid-1880s, 60 to 70 per cent of Donbass workers were migratory, and in 1904 31.7 per cent of the coal cutters were away from the mines in the summer.[8] The local population being persistently reluctant to enter the mines, Russian migrants and migratory workers formed the rank-and-file of the mining labour force. In 1889 only 5 per cent of miners were of local origin.[9] The seasonal migration of peasant miners in the Donbass persisted well into the early Soviet period (1920s). Again, most of them (about 75 per cent) were Russians. Seasonal migration greatly diminished after the forced collectivisation and so-called 'dekulakisation'.[10]

The importance of oscillating migrants working in the South African gold mines has been established in numerous studies;[11] in Witbank's coal mines, too, by 1925 about 75 per cent of the miners were migratory workers from rural areas of Mozambique. Mozambicans were recruited by the Witwatersrand Native Labour

Association (WNLA or 'Wenela'), which had established a kind of monopoly to obtain mine and colliery labour from the Portuguese territory of Mozambique. In 1920s over 80 per cent of African workers in the Witbank coal-mining district were recorded as 'Portuguese'. Although since 1907 there had been a tendency among 'detribalised' families to settle around the collieries, the proportion of 'permanent' miners, permitted to live with their wives, was officially restricted in 1926 to 15 per cent.[12]

Also, in the Appalachian coalfields in West Virginia (United States), the great wave of black migrants from the South before and during the First World War initially consisted of small-scale peasants or sharecroppers. When work became irregular or wages declined substantially, they returned to these homes until work in West Virginia improved, often in a seasonal pattern. It was not until the 1920s that they started to settle in coal towns, where semi-rural life was often maintained by miners' families, who cultivated gardens and raised livestock.[13]

Migration and labour contractors

Where the distance between mine owners, managers and migrant miners was large, systems of intermediation emerged, with a major role for labour contractors. They shaped both the recruitment and deployment of labour, and the systems of control inside and often outside the mines. Recruiting migrants using subcontractors seems to be a common device in countries with underdeveloped labour markets and a rural population reluctant to enter the mines. The contractors had closer ties to the rural population than owners and managers did, and this made it easier for them to recruit labour from their home villages.

Subcontracting had been known in the British coal industry since the eighteenth century.[14] Elsewhere in the world, the system was widely used to recruit and control both local and migrant labour. In Chinese coal mining, until the 1920s the largest part of the labour force, up to between 60 and 80 per cent, was recruited by contractors to work the face as well as to haul and tunnel underground. Apart from supplying labour, many contractors also had to provide most of the materials to work the mine.[15] The large Japanese-owned Manchurian mines, however, incorporated labour contractors in their system of management and met severe shortages of labour in the 1920s by sending their own agents to the Hebei and Shandong regions south of Manchuria, which formed an important reservoir of labour for these mines.[16]

In Indian coal mining, large groups of migrant labour were recruited by labour contractors, engaged for the entire labour process, from the hiring of the labour force to the supervision of the cutting and loading of coal. At its lowest end, the system relied on gang masters (so called *sardars*), who led groups of 15 to 40 miners, supervising work and receiving and distributing wages. The system was closely related to seasonal migration, as it enabled a stable but flexible connection between demand in the coalfields and supply in more or less remote villages. A contractor recruited relatives and personal friends from his home village or thereabout, and made every effort to ensure that his 'gang' would return to a particular mine next year. He advanced train fares, food and money to his co-villagers, later to be deducted from wages earned, obliging workers to stay with him and to work at a particular colliery.[17] In this way mining companies were able to get a hold over the migratory labour force.

A similar recruiting system existed for the Russian seasonal miners in the Ukrainian Donbass. Agents (*verbovshchiki*) went to the villages to persuade peasants to work in the mines, paying their travel and living expenses. These advances were later deducted from wages, keeping the worker in debt from the beginning. Mostly, these recruiters also acted as *artel'schick*, the leader of a team (or *artel'*) of up to 30 peasant miners, often friends and relatives from one village, who negotiated with the employers on behalf of the team, coordinated the work and organised living arrangements. Much like his Indian counterpart, the contractor took responsibility for arranging for sufficient numbers of miners, and might provide horses for transport and foremen and gang bosses to supervise daily work. He received the sum negotiated for piecework and in turn paid his workers by the shift.[18]

In Japan, too, a system of recruitment by labour contractors was generally used in coal and other mines. A contractor (known as *hamba-gashira* in the Hokkaido coalfields in the north and *naya-gashira* in Kyūshū coalfields in the south) hired several groups of between 10 and 20 mine workers from farming backgrounds, provided lodging and supervised labour underground. On behalf of the mine owners, the contractors had complete authority over the workforce, both at work and in daily life. They recruited the miners, supervised them at the production site and controlled their life at their lodges.[19] The system was a means to secure a regular supply of workers. In the southern Kyūshū coalfield the contracting system was also used to recruit families, to include females to work as haulers underground, but also at the surface. The system declined there only with the demise of female work in the underground teams after longwall mining had replaced pillar mining in the 1920s. This in turn was made possible by the massive recruitment of Koreans, who were initially also recruited by Korean *naya-gashira*.[20] Many of these Korean migrants returned to their home villages during the months of the summer harvest (July and August).[21]

Forced labour

The solutions employed by mine owners and the state to remedy the shortage of labour for the mines included various forms of compulsion or force. In this way the colliery owners were able to tie in a permanent supply of mining labour. In early modern Britain, systems of coercion were used to tie workers and their families to the mines, like the so-called 'colliery serfdom' in eighteenth-century Scotland, and the 'yearly bond' in the Durham mines.[22] Systems of coercion of this kind were not confined to early modern Europe however. In the coalfields of British India (Bengal), semi-feudal bonds were common in older collieries, in some cases even until the 1950s. Mine owners there had purchased large tracts of land near the pits and had developed a service tenancy arrangement, whereby peasants were granted a small piece of land in return for working a certain number of days in the company mine instead of paying rent, on pain of eviction.[23] In this way the colliery owners were able to tie in a permanent supply of mining labour. The system had been applied by early starters in the Indian coalfields; for more recently established enterprises, like in the Jahria coalfield, other means of obtaining the desired number of workers had to be used, mainly in the form of subcontracting migratory peasant miners (see above).

In this Indian example, and also in the case of the Zonguldak coalfield in Turkey, economic and extra-economic coercion were used to mobilise local labour for the mines. We find several examples, also in Chinese coal mining, well into the twentieth century, be it in the form of convict labour, debt servitude or servile labour.[24] In other cases, force was used to bring in migrant labour. In colonial Zimbabwe (Southern Rhodesia), the Wankie Colliery (opened in 1902) relied heavily in its early years on so-called *chibaro*, indentured labourers supplied by the Rhodesia Native Labour Board. In 1918, 40 per cent of the black labour force at the Wankie Colliery still belonged to this category, as against 60 per cent 'voluntary' labour, often migrants passing on their way from Northern Rhodesia to South Africa, who would work at Wankie for several months before resuming their journey. Thereafter, structural changes in the labour market freed the Wankie Colliery more or less from *chibaro* labour: by 1927 the percentage had dropped to five.[25]

In the Dutch East Indies (Indonesia), the labour shortage the Ombilin coal mines (West Sumatra) faced when they first began operating was 'solved' by the forced employment of convict labourers, both political and criminal prisoners, from other parts of the colony. Their number fluctuated up to 2,400 in 1898. Later, Chinese and Javanese 'contract labourers', too, were employed. They were not 'free', but bound to work for several years under the complete jurisdiction of the mine. Convict and contract labourers dominated the growing number of miners until the first half of the 1920s. After that, they gradually tended to disappear and were replaced by free labourers.[26]

Convict labour was, in fact, a fairly common recruitment device both in the start-up and the more advanced phases of coal-mine development. In the nineteenth-century southern United States, convict labour drawn predominantly from among African Americans was regularly used in the coal mines of Georgia, Tennessee and Alabama after the abolition of slavery (the employment of slaves had been common in the mines before).[27] Well into the twentieth century, convict labour played a fundamental role in setting the conditions under which free miners laboured, and provided a steady source of labour. With many black miners staying after their release, their experience as convict labourers in fact prepared large numbers of blacks for the slightly less harsh regime they would endure as free miners. So, in 1910 over 50 per cent of black coal miners in the Birmingham (Alabama) district had learned their trade as convicts. In this way the system offered both an instrument for disciplining the black labour force and for securing a steady flow of cheap labour for the mines.[28]

In Japan, labour scarcity at the start of the Hokkaido and Kyūshū coal mines (from the 1880s) was also solved in this way. Later on, convict labour was replaced by a system of recruitment by labour contractors (see above), but this system confined the freedom of the miners to such an extent that it could also be considered a form of forced labour. During the First World War coal mining grew strongly because of the economic boom and it became difficult to procure more labour from the agrarian villages in Japan itself. Labour shortages were now solved by transferring Korean migrants from rural areas in colonised Korea.[29] After 1939, the coercive mobilisation of Koreans in the mines and other industries became increasingly important in the Japanese war economy. Between 1939 and 1945 more than 300,000 Koreans were sent to Japanese mines, most against their will. Koreans were almost exclusively used as underground face workers. In Hokkaido, for example, Koreans comprised over 40 per cent of the coal-mining labour force, but they accounted for 60 to 70 per cent of underground workers.[30] Some 40,000 Chinese prisoners of war were employed in the Japanese mines as well.[31]

Convict labour and other forms of forced labour were also introduced in the newly built Kuzbass basin in Siberia to meet the demands of forced industrialisation in the Soviet Union in the 1930s.[32] In the German Ruhr, forced migrant labour, both civilian workers and prisoners of war, had already been used during the First World War. These were mainly Belgians and Poles from the occupied Russian territories (at that time Poland did not exist as an independent state).[33] The experience with this kind of *Arbeitseinsatz* during the First World War prepared the ground for the development of an extensive system of forced labour to support the war economy of Nazi Germany during the Second World War, both in Germany itself and in the European occupied territories.[34] In this system, ethnic discrimination and forced labour were closely interrelated as most of the deployed workers were so-called *Ostarbeiter* and prisoners of war from Poland, Ukraine and Russia, and were considered by the Nazis to be of an inferior 'race'.

British and other miners in white settler colonies and the United States

Much of the global expansion of coal mining in the nineteenth and twentieth centuries was possible only because of the migration of skilled groups of workers from Great Britain. They introduced mining skills and techniques, and often continued to hold privileged positions afterwards. The migration trajectories of British miners can be traced in almost every coalfield in the British Empire, but also in other parts of the world, for instance in the beginnings of coal mining in Brazil.[35] With the global expansion of coal mining from the nineteenth century onwards, British miners moved from coalfield to coalfield in British settler colonies such as South Africa, Canada, Australia and New Zealand to develop the industry there. In the South African coal mines of Natal and Transvaal a high proportion of senior staff was Scottish; others came from Wales, Northumberland, Cornwall and elsewhere in Britain.[36] In Australia, the coalfields of the Newcastle district in New South Wales were populated by English and Scottish miners, who also brought their tradition of trade unionism.[37] The West Coast mining district of New Zealand was an enclave of British mining practice as well.[38]

Also in the Canadian coalfields, both in the east (Nova Scotia) and the west (Vancouver Island), the British were the first to develop a mining industry and continued to arrive afterwards.[39] The British-born miners retained their prominence over the years.[40] Unskilled, casual labour on Vancouver Island was at first recruited from the native population, providing an auxiliary source of labour, but from the 1870s the aboriginals were increasingly displaced by Chinese migrants. Efforts were repeatedly made to exclude Asian labour, but the Chinese remained a critical part of the collieries' workforce into the twentieth century, both above and below ground. The Chinese were employed especially in longwall mines, where craft labour had been replaced by semi-skilled labour, under the supervision of a small number of whites.[41] White miners of British descent opposed the recruitment of Chinese labour, not only out of racial prejudice, but also in defence of craft positions.[42]

The South African sociologist John Hyslop has made a case for treating white workers of British origin in the settler colonies of the Empire as part of an 'imperial working class' for which 'whiteness' was a core component of identity.[43] This view is only partially convincing, as there were also British immigrants in South Africa, Canada and Australia, who brought radical socialist and later communist ideas to the colonial coalfields and propagated interracial solidarity.[44] More importantly, without suggesting that there was no racism involved, the opposition of white British miners to the entry of other ethnic and racial groups was inextricably linked to the defence of craft and skill in the mining industry.

The transfer of experience and technological skills, acquired at the coalface in the British mines, was essential for the development of the mining industry in the United States as well. The migration of British coal miners to the United States reached a provisional high in the 1860s and the early 1870s. In 1870 British immigrant miners (57,214) accounted for more than 60 per cent of all foreign-born miners (94,719) in the country. Once arrived, they moved from coalfield to coalfield in different US states.[45] Many of these immigrant miners were young single men who travelled from mine to mine on a seasonal basis. Arriving with cheap tickets for the summer season, they would return to Britain for winter work, or travel a miners' circuit through different mining states. Depressions, like that of 1873, drove recently arrived migrant colliers back to their former homes in Britain. In this way transtlantic immigrant networks became conduits of British influence in American mining practices, not least in trade unionism.[46] Up to 1900, British-born immigrants were still dominant in coal mining in Illinois, Pennsylvania, West Virginia and Kansas. Native-born miners were often also of British descent. Like everywhere else, ethnic networks were important in the migration patterns of British miners.[47] Welsh miners tended to cluster in communities around their own churches and to intermarry in their own group.[48] Welsh mine owners and managers often selected other Welshmen for their mining staff and workforce, thereby creating ethnic Welsh settlements.[49]

At the end of the nineteenth century, Italians, 'Slav' and other migrants from Southern and Eastern Europe increasingly started to work in the coal mines in the United States. The 'Slav' and Italian migrant miners were mostly of peasant origin and unskilled. Their working in the mines went hand in hand with the introduction of coal-cutting machines and the deskilling of mine work. The proportion of British miners diminished, but they kept a position as foremen and skilled workers.[50] The Irish, 'Slav' and Italian miners generally acted as labourers with a lower status. In these hierarchies, social and ethnic differences went together.[51] However, despite the condescending attitude that Anglo-Saxon miners displayed towards the 'new European' immigrants,[52] Eastern European and Italian migrants soon took an active part in the miners' struggles and had to be accepted as members in the miners' union branches.[53]

Relationships between 'white' and African-American miners in trade unionism were much more ambivalent.[54] While there were several coal-mining districts, in Alabama for instance, where British miners participated in interracial unions,[55] the racial policies of organised labour were far from uniform. A debate on black workers, race and organised labour in the United States, referred to as the 'Gutman-Hill debate',[56] started several decades ago and has continued in extended discussions about the importance of 'whiteness' in American working-class history, especially after publication of David Roediger's *The Wages of Whiteness* in 1991.[57]

Price Fishback explained the difference in the experiences of African-American miners in the American coalfields in relation to racial discrimination and assimilation by the tightness or looseness of labour markets there. African-Americans did better in the Alabama and West Virginia labour markets, because employers were constantly seeking new workers, and black migrants found ample employment. This contrasted with coalfields further north, where limits on employment growth constrained African-American immigration.[58] In West Virginia a large number of black miners worked side by side with other ethnic groups and were easily accepted into the trade union, United Mine Workers of America. Fishback relates the arrival of different migrant groups in the United States mines to different phases of exploitation: most British immigrants came with coal-mining experience and helped train American workers. They played a major role in the early development of the United States coal industry in the mid-1800s. Later, in the 1880s, and even more so between 1890 and 1910, in the Pennsylvania and Midwest mining regions inexperienced immigrants from Eastern and Southern Europe were employed on a massive scale to fill the need for unskilled labour. The coalfields in the low-wage Southern states (Kentucky, Virginia, West Virginia, Alabama) in turn attracted more African-American workers, migrating north from the Deep South.

The Polish diaspora in Europe and the United States

The Poles were the first and initially the most mobile among migratory mine workers in Europe. From the last few decades of the nineteenth century, they were mobilised on a massive scale to work in the coal mines of the Ruhr in Germany.[59]

At that time, almost all of them were Prussian citizens. By 1910, at least one-quarter of all Ruhr miners were Polish-speaking. Having arrived in the Ruhr, they formed ethnic communities, based on social organisations such as churches, trade unions, a Polish press and sports clubs. The confrontation with a foreign and often hostile German milieu helped to foster a common national identity among these migrants, who had hitherto a locally oriented peasant background.[60]

After the First World War, Poles from the Ruhr and from Poland itself moved to coalfields in northern France,[61] and to a lesser extent in Belgium and the Netherlands.[62] In France, the German-Polish migrants, called *Westphaliens*, brought a Polish press, social clubs, societies and other organisations, and in this way were able to hold on to a Polish ethnic, religious and national identity in a rather conservative fashion. This ethnicised segregation was consciously promoted by the mining companies, which sought both a fragmented workforce and ideological-cultural means to curb labour militancy.[63]

As both employers and the French state were interested in recruiting Polish workers, a bilateral treaty was signed in 1919 to regulate the arrival of Polish citizens to the mines. This *Convention entre la France et la Pologne relative à l'émigration et l'immigration* can be considered the first in a system of state-regulated migration of mine workers in Europe (see below). The employers cooperated in a Société Générale d'Immigration, which undertook a systematic programme of prospecting in Poland.[64] This resulted in a new wave of Polish outmigration of workers of peasant origin. Between 1920 and 1930 a total of 490,000 Polish migrants came to France, while about 60,000 left. A climax was reached in 1929 and 1930. The depression of the 1930s led to the expulsion of many of the Polish migrants who had arrived during the boom period in the 1920s. Almost all migrants (92 per cent in Nord-Pas-de-Calais) worked in underground positions, while supervisory personnel were mostly French.[65] After the Second World War, however, the Poles left the French mines *en masse*, be it to their home country or to other jobs: while there had been 46,000 miners of Polish descent in 1946, there were only 283 in 1981.[66]

In the early 1920s, Poles from the Ruhr and northern France also moved to the Belgian and to a lesser extent Dutch mines. In 1922 Belgian coal owners started to recruit in Poland itself. The number of immigrant Poles grew to several hundred each year. Individual migration developed alongside organised recruitment. In September 1930, there were 11,993 Polish mine workers in Belgium. In the depression years several thousand miners were dismissed again and had to return to Poland.[67] In the second half of the 1920s, Dutch mining companies started to recruit directly in Poland and in other countries in Central, Eastern and Southern Europe. As in Belgium, in the wake of official recruitment, Polish migrants travelled from mine to mine in Germany, Belgium and France.[68]

Initially, the origin of Polish migration to the Ruhr and the Pennsylvanian anthracite region in the United States was quite similar. In the 1870s, both migrant groups tended to come from traditional mining areas in Upper Silesia. By the 1890s, however, the regions of origin were diverging sharply. In the Ruhr, almost all Polish migrants came from the German rural provinces of East Prussia. The majority of Polish migrants to the United States now came from the Austrian (Galicia) and Russian empires.[69] In Pennsylvania 'old' migrants from Great Britain and Ireland and their American-born children constituted the core of the 'native' workforce, continuing to occupy the jobs with the highest status. In both regions, recently arrived Poles generally possessed the lowest social standing, at least until other immigrants began to arrive from the 1890s on; they suffered from significant discrimination because of their 'foreign' language, religion, habits and peasant background.[70] As a reaction, Poles developed an outspoken ethnic identity, supported by organised sociability.[71]

State-regulated labour migration to northwestern Europe

Shortages of labour in several continental Western Europe coalfields had already emerged in the expansion years before the First World War, and they re-emerged after that war and in the 1920s. While in that period Great Britain and Germany were able to build a mining labour force from their own internal labour supply, France, Belgium and the Netherlands witnessed severe labour shortages. To counteract these labour shortages in the interwar years, not only in the mining industry, Western European countries developed systems of regulated migration based on bilateral treaties, especially with newly formed states in Eastern Europe such as Poland, Yugoslavia and Czechoslovakia, but also with Italy. These agreements set up official migration channels alongside spontaneous individual migration. State involvement was triggered by protectionist labour-market policies, increasing state involvement in welfare arrangements, and concomitant costs of unregulated migration for both employers and the state.[72]

As mentioned above, the immigration agreement concluded between France and Poland on 7 September 1919 can be considered the first in a series of treaties of this kind. It was very soon followed, on 30 September, by a treaty with Italy, which in the early 1920s brought a first wave of Italian migrants to the coalfields in northern and central France, mainly from central and northeastern Italy.[73] Belgium concluded an agreement with Italy to regulate migration in 1923.[74] In general, the Belgian state cooperated closely with employers' organisations in the mines, and this was also the case in the recruitment of smaller numbers from Czechoslovakia, Yugoslavia, Hungary and Poland.[75] In the Netherlands, in the second half of the 1920s, recruitment campaigns by Dutch mining companies in Czechoslovakia, Yugoslavia, Poland and Italy brought new groups of inexperienced migrant workers to the coal mines. The campaigns were organised jointly on the basis of bilateral agreements. However, during the depression of the 1930s, most of the newly arrived migrants were dismissed and sent home.[76]

After the Second World War, the system of bilateral migration agreements became a general device to recruit 'guest workers' for the northwest European mining industry from Mediterranean countries. It all started with Italy. Just after the war, urgent energy needs, both in France and Belgium, prompted governments to call for a *bataille de charbon* (a 'battle for coal'), but labour supply in these countries fell short. Italy became the preferred country for the recruitment of migrant labour for the mines in France, Belgium and somewhat later also the Netherlands. France concluded an agreement with Italy on 26 February 1946 to arrange for the arrival of migrants in exchange for the delivery of a fixed amount of coal to Italy for each miner.[77] Some 200,000 were recruited, but most of them left again after the expiration of their contract. The employment of Italians in French coal mining rose to a highpoint of 11,023 in 1958; thereafter their number diminished to 1,687 in 1981.[78]

A few months later than France, on the 20 June 1946, Belgium concluded a comparable agreement with Italy. Between 1946 and 1958, 141,151 Italians were officially recruited to work in the mines.[79] The movement into and out of Belgian coal mines amounted to about one-third of the total number of underground miners in 1955–1956, twice the rate in France or Germany. Nevertheless, in Belgium the number of Italians working in the mines was much larger than in France: between 1948 and 1957 it fluctuated between 33,000 and 47,500 at the end of each year. After the Marcinelle disaster in 1956, which killed 269 miners, among them 136 Italians, the number of Italians employed in Belgian coal mines diminished from 44,000 in 1957 to 2,500 in 1975.[80] The Netherlands was a relative latecomer in the interstate quest for migrant workers from Italy: on 4 December 1948 an agreement was signed with the Italian government to recruit for the coal mines, although on a much smaller scale than in the French or Belgian cases.[81]

The proliferation of these kinds of agreement for the state-sponsored recruitment of migrant workers has to be considered a response to shortages of unskilled or semi-skilled labour in the mines.[82] Demand for unskilled labour had increased, relatively at least, because of the introduction of new mining methods and technologies. The Italian migrants were recruited mainly from the agrarian central and southern parts of the country, had no earlier experience with mine work, and had to learn the skills on the job. This is one explanation for the high turnover, other explanations being the miserable conditions of their lodgings (often camp-like dwellings), poor working conditions and their status as subordinate workers in general. As soon as new opportunities arose in their home country, the migration flow from Italy dried up. Western European countries started to negotiate with other countries in the Mediterranean periphery to find new supplies of mine labour. Instead of Italy, in the 1960s Spain, Yugoslavia, Greece, Tunisia, Turkey and Morocco became preferred countries of recruitment for the coal and other industries in continental Western Europe. This time, Germany, too, joined the group of recruiting countries. From each of these countries, 'guest workers', as they were called (from the German word *Gastarbeiter*), were again recruited on the basis of bilateral treaties to regulate migration according to the needs of the economy.[83]

In all coal-producing countries in Western Europe (excluding Britain), 'guest workers' were employed in coal mining on a relatively large scale. To provide for workers in Germany, for instance, a German–Turkish *Regelung der Vermittlung türkischer Arbeitnehmer nach der Bundesrepublik Deutschland* (Settlement to Procure Turkish Employees for the Federal Republic of German) was signed in 1961 to regulate the selection and mediation of Turkish contract workers by German agencies in Istanbul and Ankara. By 1963, 10,200 Turkish miners were already employed in German coal mining, the largest group among the 27,130 foreign workers in that industry.[84] After a relapse during the recession of 1966–1967, their number rose again, from about 5,200 in 1969 to 19,800 in 1973, accounting for 74 per cent of all foreign workers in coal mining.[85] The stark fluctuation of these numbers before, during and after the recession of 1966–1967 reflects the general position of foreign workers as a flexible reserve army, both in coal mining and other sectors. Although by 1973 several German mines were staffed almost exclusively by miners of Turkish descent, in most cases Turks were to be found only at the lower end of the job ladder. As with the Moroccan workers in other European coal-producing countries, they were often employed to ensure exhausted or unprofitable mines could be closed down smoothly.[86]

Moroccan migrants were the last group of miners recruited for the coal industry on the basis of bilateral treaties. Their main destinations were France, Belgium and the Netherlands.[87] In the late 1950s and early 1960s, migration to northern France, Belgium and the Netherlands started to grow, at first spontaneously, then in Morocco itself through recruitment by employers' organisations, but also through family networks. Treaties with Morocco to regulate migration officially were signed in 1963 with France, in 1964 with Belgium, and in 1969 with the Netherlands. The treaties enabled the entry of a growing number of Moroccan immigrants into the mines of these countries, in two waves: the first until the recession of 1966–1967, the second in the early 1970s. A total of 20,495 Moroccan immigrants arrived in France between 1960 and 1965; their number reached 78,000 in 1977.[88] Migration to Belgium and the Netherlands was less substantial in absolute terms (several thousands), but Moroccans became by far the largest group of foreigners in the Belgian and Dutch mines in the 1970s.[89] In the decaying coal industry in these countries, mining companies were unable to hold a local workforce. Miners and their sons took a bleak view of future opportunities in coal mining and started to look for employment outside the mines. Moroccan miners were brought in on short-term contracts to compensate for shortages of local workers and to help pit closures to proceed orderly.[90] In the 1970s, they were recruited for the Lorraine coal mines in France with this same goal in mind.[91]

Migrants in the mining community

While debates on ethnic and racial discrimination and segregation, as against solidarity and integration, in miners' unions have a certain tradition in mining labour history, the focus of historical research since the 1990s has shifted from the relationship of class, race and ethnic identity in workers' struggles to other aspects. New approaches have allowed scholars to shed light on issues such as racial discrimination, ethnic segregation, social integration and the intricate processes of identity formation among migrants in the context of mining communities. At the same time, the concept of 'mining community' itself, as a closed, homogeneous and often isolated settlement, dominant for a while in (especially British) sociology and mining history,[92] has come under scrutiny, precisely because of the diversity of its inhabitants.[93]

The shift towards the study of migration and ethnicity in mining communities is clearly visible in the landmark volume on comparative coalfield history edited by Stefan Berger *et al.*, published in 2005. Several chapters deal with 'identities', 'communities' and with the 'interlocking spheres of workplace, neighbourhood, family, and working-class organisations', including the one by Leen Beyers on 'ethnic, class and gender identities at street level' in the Belgian miners' colony (*cité*) of Zwartberg.[94] In this article she arrives at a fairly positive assessment of the interethnic interaction between Flemish, Polish, Czech and Italian neighbours. Elsewhere, she compares the construction and deconstruction of ethnic boundaries between second-generation migrants of Polish and Italian origin and Belgian nationals in this *cité*.[95] Both groups of migrants really succeeded in being accepted as 'Belgians' only after the arrival of new groups of Islamic migrants, predominantly from Turkey. The (perceived) distinctiveness of these new outsiders led to the view by the local population (many of them former migrants or descendants of migrants) that migrants from former migration waves had successfully integrated.

Comparable issues have been raised in the German Ruhr in discussions on the similarities and differences between Polish migration in the more distant past and Turkish immigration in the recent past.[96] While historical research has established a clear relationship between the segregation and discrimination of Polish miners before the First World War and the emergence of a strong feeling of national identity among them, the supposedly easy integration of Polish migrants in the past has repeatedly been invoked in public discourse as a counter-story pointing to the lack of integration of their Turkish counterparts today. In this discourse, the supposedly opposite behaviour of Polish and Italian migrants on the one hand and Turkish migrants on the other serves to disqualify the ability of Islamic migrants to adapt to 'Western' culture. From a historians' perspective this is much more ambiguous however. Opportunity structures and circumstances change considerably over time, which makes it difficult to compare (*ceteris paribus*) trajectories of migrants of different descent arriving in different periods of time. Some historians have argued that the difference might not be that salient, because, depending on the timeframe under consideration, it often takes several generations for migrant minorities to overcome segregation and discrimination and to integrate socially and culturally.[97]

The comparability of 'old' (nineteenth- and early twentieth-century) and 'new' (late twentieth-century) migrants has been questioned, however.[98] Both present-day society and the (ethnic and religious) composition of the 'new' immigrant groups in the United States and in Europe differ dramatically from earlier ones. What seems important from a historical point of view, also in the context of ethnic relations among migrant miners, is that the social construction of ethnic and racial differences changed over time, for instance in the case of Irish, Italian and 'Slav' migrants in the United States, who were initially seen as racially inferior and who only gradually became 'white'.

This 'process of whitening' has been explained partly by the mass migration of African-Americans to the north, which enabled other (European) ethnic groups to be redefined as 'whites' as opposed to 'blacks'. This is consistent with David Roediger's ideas about 'whiteness' as a constructed racial identity in opposition to 'blackness'.[99] Roediger's arguments have been criticised, however, as rather one-dimensional, because they ascribe a uniform racial identity to an abstract 'white' working class, which itself remained sharply divided along lines of ethnicity and other divisions.[100] Roediger himself cites the American writer Upton Sinclair, who in his novel *King Coal*, gave a rather bleak picture of interethnic relations in a Colorado coal town around 1917:[101]

> There were most rigid social lines in North Valley, it appeared. The Americans and English and Scotch looked down upon the Welsh and Irish; the Welsh and Irish looked down upon the Dagoes and Frenchies; the Dagoes and Frenchies looked down upon Polacks and Hunkies, these in turn upon Greeks, Bulgarians and 'Montynegroes', and so on through a score of races of Eastern Europe: Lithuanians, Slovaks, and Croatians, Armenians, Roumanians, Rumelians, Ruthenians – ending up with Greasers, niggers, and last and lowest, Japs.[102]

Conclusion

What becomes clear from this overview is that there was a huge difference between the migration and settlement of skilled miners, like those from Britain, but also from other countries, and the recruitment of groups of unskilled workers from rural areas. What stands out as common in the cases mentioned is that these recruitments were often targeted at ethnic groups of a lower social status, not only because they were rural or unskilled, but also because they were considered inferior from a cultural or ethnic viewpoint.

Ethnic descent differentiated not only groups of migrants, however, but also an 'established' workforce from 'outsiders': ethnic minorities who were considered and treated as people of a lower status than the dominant ethnic group. In many cases the recruitment of new ethnic groups proceeded in parallel with technological innovations in the mining industry, which deskilled large parts of the work and required an enlargement of the workforce by unskilled or semiskilled workers. Examples include the massive recruitment of Koreans in Japan in the late 1920s, when work gangs in longwall mining replaced family teams at post and pillar mining; the employment of Chinese labourers in longwall mines on the Canadian west coast, replacing craft labour by semi-skilled labour; the entry of 'Slav' and Italian migrant miners in the Pennsylvania coal mines, which went hand in hand with the introduction of coal-cutting machines and the deskilling of mine work; and the recruitment of inexperienced Italian migrants in the French and Belgium mines after the Second World War, whose deployment had been made possible by the introduction of new mining methods and technologies.

The ethnic divisions in the workforce were therefore blurred with divisions of skill and hierarchy. This explains the negative, or sometimes even racist, attitudes of early arrivers, mainly skilled white miners, and their descendants, towards newcomers, not only blacks, as in South Africa and the United States, but also newly arrived migrants of other complexities and looks, such as Eastern Europeans or Italians. Status and ethnic prejudice enforced each other also in the social relations between migrant families in the mining communities. The longer-term consequences of these divisions are, however, much less clear. Comparisons of 'old' (late nineteenth- and early twentieth-century) and 'new' (late twentieth-century) migration reveal great differences, which can perhaps be attributed to the shorter timeframe of the latter, but more likely to the persistent discrimination of the new migrant groups in Europe and the United States from the 1970s. The picture, however, is far from complete, as it does not include knowledge about the dynamics in and around coalfields in other world regions. The degree to which other places follow the European and North American pattern remains to be ascertained.

Notes

1 Reworked and somewhat shortened version of a survey article I wrote for the *International Review of Social History 60* (December 2015), Special Issue 23: 'Migration and Ethnicity in Global History: Global Perspectives'.

2 Cameron C. Hartnell, 'Arctic Network Builders: The Arctic Coal Company's Operations on Spitsbergen and Its Relationship with the Environment' (PhD thesis) (Michigan Technological University 2009), available at http://digitalcommons.mtu. edu/cgi/viewcontent.cgi?article=1288&context=etds; last accessed 6 August 2015; Louwrens Hacquebord (ed.), *Lashipa: History of Large Scale Resource Exploration in Polar Areas* (Groningen: University of Groningen, 2012); Len Richardson, *Coal, Class and Community: The United Mineworkers of New Zealand, 1880–1960* (Auckland: Auckland University Press, 1995); John Douglas Belshaw, *Colonization and Community: The Vancouver Island Coalfield and the Making of the British Columbian Working Class* (Montreal: McGill-Queen's University Press, 2002); Suzanne Culter, *Managing Decline: Japan's Coal Industry Restructuring and Community Response* (Honolulu: University of Hawaii Press, 1999); Ann B. Irish, *Hokkaido: A History of Ethnic Transition and Development on Japan's Northern Island* (North Carolina: McFarland, 2009). For an overview of current coal exploitation worldwide, see Ulrike Stottrop (ed.), *Kohle Global. Eine Reise in die Reviere der Anderen. Katalog zur Ausstellung im Ruhr Museum vom 15. April bis 24. November 2013* (Essen: Klartext Verlag, 2013).

3 Robert H. Woodrum, *Everybody Was Black Down There: Race And Industrial Change in the Alabama Coalfields* (Athens Georgia: University of Georgia Press, 2007); Leen Beyers, *Iedereen zwart: het samenleven van nieuwkomers en gevestigden in de mijncité Zwartberg, 1930–1990* (Amsterdam: Aksant, 2007); see also: *id.*, 'Everyone Black? Ethnic, Class and Gender Identities at Street Level in a Belgian Mining Town, 1930–50', in: S. Berger, A. Croll and N. LaPorte (eds), *Towards a Comparative History of Coalfield Societies* (Aldershot: Ashgate, 2005), pp. 146–163; Marcel Deprez *et al., 'Siamo tutti neri!': des hommes contre du charbon: études et témoignages sur l'immigration italienne en Wallonie* (Seraing: IHOES, 1998); M. Cegarra *et al., Tous gueules noires. Histoire de l'immigration dans le bassin minier du Nord-Pas-de-Calais* (Lewarde: Centre historique minier du Nord-Pas de Calais, 2004).

4 Ian Phimister, 'Global Labour History in the Twenty-First Century: Coal Mining and its Recent Past', in: Jan Lucassen (ed.), *Global Labour History. A State of the Art* (Bern etc.: Peter Lang, 2006), pp. 573–589; see also Berger *et al. Towards a Comparative History of Coalfield Societies*.

5 Erol Kahveci, 'Migration, Ethnicity, and Divisions of Labour in the Zonguldak Coalfield, Turkey', *International Review of Social History 60* (2015), pp. 207–226; Donald Quataert, *Miners and the state in the Ottoman Empire. The Zonguldak Coalfield, 1822–1920* (New York etc.: Berghahn Books, 2006).

6 Delwin A. Roy, 'Labour and Trade Unionism in Turkey: The Eregli Coalminers', *Middle Eastern Studies 12*(3) (1976), pp. 125–172, 126–134.

7 Dilip Simeon, *The Politics of Labour Under Late Colonialism: Workers, Unions and the State in Chota Nagpur 1928–1939* (New Delhi: Manohar, 1995), p. 28; *id.*, 'Coal and Colonialism: Production Relations in an Indian Coalfield, *c.* 1895–1947', in *International Review of Social History 41* (1996), pp. 83–108, 93–94; C.P. Simmons, 'Recruiting and Organising an Industrial Labour Force in Colonial India: The Case of the Coal Mining Industry, c. 1880–1939', *The Indian Economic and Social History Review 13* (1976), pp. 455–585, 458–460.

8 Theodore H. Friedgut, *Iuzovka and Revolution. Vol. I: Life and Work in Russia's Donbass, 1869–1924* (Princeton: Princeton University Press, 1989), pp. 209, 215, fn 80, 217 and 221.

9 *Ibid.*, pp. 211–212.

10 Tanja Penter, *Kohle für Stalin und Hitler. Arbeiten und Leben im Donbass 1929–1953* (Essen: Klartext Verlag, 2010), pp. 39–42.

11 T. Dunbar Moodie (with Vivienne Ndatshe), *Going for Gold: Men, Mines, and Migration* (Berkeley: University of California Press, 1994); Ruth First, *Black Gold: The Mozambican Miner, Proletarian and Peasant* (Manchester: Palgrave Macmillan, 1983); Alan Jeeves, *Migrant Labour in South Africa's Mining Economy: The Struggle for the Gold Mines' Labour Supply 1890–1920* (Kingston: McGill-Queen's University Press, 1985).

12 Peter Alexander, 'Oscillating Migrants, 'Detribalised Families' and Militancy: Mozambicans on Witbank Collieries, 1918–1927', *Journal of Southern African Studies 27*(3) (2001), pp. 505–525, 507, 509, 517; *id.*, 'Challenging Cheap-labour Theory: Natal and Transvaal Coal Miners, *ca* 1890–1950', *Labor History 49*(1) (2008), pp. 47–70, 53–54.

13 Ronald L. Lewis, 'From Peasant to Proletarian: The Migration of Southern Blacks to the Central Appalachian Coalfields', *The Journal of Southern History 55*(1) (1989), pp. 77–102, 87–88.

14 Arthur J. Taylor, 'The Sub-contract System in the British Coal Industry', in: Leslie S. Pressnell (ed.), *Studies in the Industrial Revolution Presented to T.S. Ashton* (London: University of London, 1960), pp. 215–235, 229 and 234; for empirical qualifications see James A. Jaffe, *The Struggle for Market Power: Industrial Relations in the British Coal Industry, 1800–1840* (Cambridge: Cambridge University Press, 1991), pp. 54–56.

15 Tim Wright, '"A Method of Evading Management": Contract Labor in Chinese Coal Mines before 1937', *Comparative Studies in Society and History 23* (1981), pp. 656–678, 659, 663–665, 669.

16 Limin Teh, 'Labor Control and Mobility in Japanese-Controlled Fushun Coalmine (China), 1907–1932', *International Review of Social History 60* (2015), pp. 95–119.

17 Simeon, *The Politics of Labour Under Late Colonialism*, pp. 27 and 149; Simmons, 'Recruiting and Organising an Industrial Labour Force', pp. 471–482.

18 Friedgut, *Iuzovka and Revolution. Vol. I*, pp. 234, 260–263 and 269–271.

19 On this 'lodge system', both at the iron mines and the coalmines, see Nimura Kazuo, *The Ashio Riot of 1907: A Social History of Mining in Japan* (Durham: Duke University Press Books, 1997); pp. 161–178.

20 Yukata Nishinarita, 'Technological Change and Female Labour in Japan', in: Masanori Nakamura (ed.), *Technological Change and Female Labour in Japan* (Tokyo: United Nations University Press, 1994), pp. 59–96; W. Donald Smith, 'The 1932 Asō Coal Strike: Korean-Japanese Solidarity and Conflict', *Korean Studies 20* (1996), pp. 94–122, 96–98.

21 Michael A. Weiner, *The Origins of the Korean Community in Japan, 1910–23* (Atlantic Highlands: Manchester University Press, 1989), p. 66.

22 Alan B. Campbell, *The Lanarkshire Miners: A Social History of their Trade Unions 1775–1874* (Edinburgh: John Donald Publishers, 1979), pp. 9–12; Sydney Webb, *The Story of the Durham Miners (1662–1921)* (London: The Labour Publishing Company Limited, 1921), pp. 7–15; Thomas S. Ashton and Joseph Sykes, *The Coal Industry of the Eighteenth Century* (Manchester: Manchester University Press, 1929), pp. 70–99.

23 Simmons, 'Recruiting and Organising an Industrial Labour Force', pp. 463–471; Simeon, *The Politics of Labour Under Late Colonialism*, p. 26.

24 Tim Wright, *Coal Mining in China's Economy and Society 1895–1937* (Cambridge: Cambridge University Press, 1984), p. 165.

25 Ian Phimister, *Wangi Kolia: Coal, Capital and Labour in Colonial Zimbabwe 1894–1954* (Harare/Johannesburg: Witwaterstrand University Press, 1994), pp. 11 and 76; see also (on the Rhodesian gold mines) Charles van Onselen, *Chibaro: African Mine Labour in Southern Rhodesia 1900–1933* (London: Pluto Press, 1980), pp. 99 and 104–114.

26 Erman Erwiza, 'Miners, Managers and the State: A Socio-political History of the Ombilin Coal-Mines, West Sumatra, 1892–1996' (PhD Thesis) (Amsterdam, 1999), pp. 36–41; Erwiza Erman, 'Generalized Violence: A Case Study of the Ombilin Coal Mines, 1892–1996', in: Freek Colombijn and Thomas J. Lindblad (eds), *Roots of Violence in Indonesia: Contemporary Violence in Historical Perspective* (Leiden: Brill, 2002), pp. 105–131.

27 Ronald L. Lewis, *Black Coal Miners in America: Race, Class, and Community Conflict 1780–1980* (Lexington: University Press of Kentucky, 1987), pp. 3–12; Alex Lichtenstein, *Twice the Work of Free Labor: The Political Economy of Convict Labor in the New South* (London: Verso, 1996).

28 Lewis, *Black Coal Miners in America*, pp. 33–34; Brian Kelly, *Race, Class, and Power in the Alabama Coalfields, 1908–21* (Urbana: University of Illinois Press, 2001), pp. 90–94.

29 Tom Arents and Noriiko Tsuneishi, 'The Uneven Recruitment of Korean Miners in Japan in the 1910s and 1920s: Employment Strategies of the Miike and Chikuhō Coal Mining Companies', *International Review of Social History 60* (2015), pp. 121–143. Also see: Yutaka Kusaga, *Transfer and Development of Coal-mine Technology in Hokkaido* (Tokyo: The United Nations University, 1982), pp. 24–26, 39–42, and 59–64; Weiner, *The Origins of the Korean Community* (Manchester: Manchester University Press, 1989), ch. 3; Ken C. Kawashima, *The Proletarian Gamble: Korean Workers in Interwar Japan* (Durham: Duke University Press Books, 2009), pp. 25–45; Regine Mathias, *Industrialisierung und Lohnarbeit. Der Kohlebergbau in Nord-Kyūshū und sein Einfluss auf die Herausbildung einer Lohnarbeiterschaft* (Vienna: Department of East Asian Studies, 1978), pp. 159–162; Michael Weiner, *Race and Migration in Imperial Japan* (London and New York: Psychology Press, 1994), pp. 112–113, 133–135 and 150.

30 Weiner, *Race and Migration in Imperial Japan*, p. 205; W. Donald Smith, 'Beyond *The Bridge on the River Kwai*: Labor Mobilization in the Greater East Asia Co-prosperity Sphere', *International Labor and Working-Class History 58* (2000), pp. 219–238, 223–226.

31 Laura E. Hein, *Fueling Growth: The Energy Revolution and Economic Policy in Postwar Japan* (Cambridge: Harvard University Asia Center, 1990), pp. 35–41.

32 Julia Landau, 'Specialists, Spies, "Special Settlers", and Prisoners of War: Social Frictions in the Kuzbas (USSR), 1920–1950', *International Review of Social History 60* (2015), pp. 185–205..

33 Kai Rawe, '. . . *wir werden sie schon zur Arbeit bringen!'. Ausländerbeschäftigung und Zwangsarbeit im Ruhrkohlenbergbau während des Ersten Weltkrieges* (Essen: Klartext Verlag, 2005).

34 Klaus Tenfelde and Hans-Christoph Seidel (eds), *Zwangsarbeit im Bergwerk. Der Arbeitseinsatz im Kohlenbergbau des Deutschen Reiches und der besetzten Gebiete*

im Ersten und Zweiten Weltkrieg, Band I: Forschungen (Essen: Klartext Verlag, 2005). For the Belgian and French cases see also Nathalie Piquet, *Charbon – Travail forcé – Collaboration. Der nordfranzösische und belgische Bergbau unter deutscher Besatzung, 1940 bis 1944* (Essen: Klartext Verlag, 2008).

35 Clarice Speranza, 'European Workers in Brazilian Coal Mining, Rio Grande do Sul, 1850–1950', *International Review of Social History 60* (2015), pp. 165–183.

36 Alexander, 'Challenging Cheap-labour Theory', p. 51; Peter Alexander, 'Race, Class, Loyalty and the Structure of Capitalism: Coal Miners in Alabama and the Transvaal, 1918–1922', *Journal of Southern African Studies 30*(1) (2004), pp. 115–132, 119, n. 20.

37 Ellen McEwen, 'Coalminers in Newcastle, New South Wales: A labour aristocracy?', in: Eric Fry (ed.), *Common Cause. Essays in Australian and New Zealand Labour History* (Sydney: HarperCollins Publishers Limited, 1986), pp. 77–92, 79–80; Robin Gollan, *The Coalminers of New South Wales. A History of the Union, 1860–1960* (Melbourne: Melbourne University Press, 1963), pp. 17–19; Andrew Reeves, '"Damned Scotsmen": British Migrants and the Australian Coal Industry, 1919–49', in Fry, *Common Cause*, pp. 93–106.

38 Len Richardson, 'British Colliers and Colonial Capitalists: The Origins of Coalmining Unionism in New Zealand', in Fry, *Common Cause*, pp. 59–75; Len Richardson, *Coal, Class & Community: The United Mineworkers of New Zealand, 1880–1960* (Auckland: Auckland University Press, 1995), pp. 3–28.

39 On Nova Scotia: Del Muise, 'The Making of an Industrial Community: Cape Breton Coal Towns, 1867–1900', in: Don Macgillivray and Brian Tennyson (eds), *Cape Breton Historical Essays* (Sydney: College of Cape Breton Press, 1981), pp. 76–94; Paul MacEwan, *Miners and Steelworkers. Labour in Cape Breton* (Toronto: S. Stevens, 1976).

40 John Belshaw, *Colonization and Community: The Vancouver Island Coalfield and the Making of the British Columbian Working Class* (Montreal: McGill-Queen's University Press, 2002), pp. 40 and 52–54, 59–60; *id.*, 'The British Collier in British Columbia: Another Archetype Reconsidered', *Labour/Le Travail 34* (1994), pp. 11–36. Allen Seager and Adele Perry, 'Mining the Connections: Class, Ethnicity, and Gender in Nanaimo, British Columbia, 1891', *Histoire Sociale/Social History 30* (1997), pp. 55–76, 67–69 and 73.

41 Belshaw, *Colonization and Community*, pp. 117–122.

42 *Id.*, 'The British Collier in British Columbia', p. 35.

43 Jonathan Hyslop, 'The Imperial Working Class Makes Itself "White": White Labourism in Britain, Australia, and South Africa before the First World War', *Journal of Historical Sociology 12* (1999), pp. 398–421.

44 Neville Kirk, 'The Rule of Class and the Power of Race: Socialist Attitudes to Class, Race and Empire', in *id.*, *Comrades and Cousins: Globalization, Workers and Labour Movements in Britain, the USA and Australia from the 1880s to 1914* (London: The Merlin Press, 2003), pp. 149–238; William Kenefick, 'Confronting White Labourism: Socialism, Syndicalism, and the Role of the Scottish Radical Left in South Africa before 1914', *International Review of Social History 55* (2010), pp. 29–62; Jonathan Hyslop, 'Scottish Labour, Race, and Southern African Empire c. 1880–1922: A Reply to Kenefick', *International Review of Social History 55* (2010), pp. 63–81; Lucien van der Walt, 'The First Globalisation and Transnational Labour Activism in Southern Africa: White Labourism, the IWW, and the ICU, 1904–1934', *African Studies 66* (2007), pp. 223–251.

45 Amy Zahl Gottlieb, 'Immigration of British Coal Miners in the Civil War Decade', *International Review of Social History 23* (1978), pp. 357–375.

46 John H.M. Laslett, 'British Immigrant Colliers, and the Origins and Early Development of the UMWA, 1870–1912', in *id.* (ed.), *The United Mine Workers of America: A Model of Industrial Solidarity?* (University Park Penn: Penn State University Press, 1996), pp. 29–50, 30–31.

47 *Id., Colliers Across the Sea: A Comparative Study of Class Formation in Scotland and the American Midwest, 1830–1924* (Urbana/Chicago: University of Illinois, 2000).

48 Ronald L. Lewis, *Welsh Americans: A History of Assimilation in the Coalfields* (Chapel Hill: The University of North Carolina Press, 2008), p. 8.

49 *Id.*, 'Networking Among Welsh Coal Miners in Nineteenth-century America', in: Stefan Berger, Andy Croll, and Norman LaPorte (eds), *Towards a Comparative History of Coalfield Societies* (Aldershot: Ashgate, 2005), pp. 191–203.

50 Quoted in Laslett, 'British Immigrant Colliers', pp. 46–47.

51 Lewis, *Welsh Americans*, pp. 189–249.

52 See also Michael A. Barendse, 'American Perceptions Concerning Slavic Immigrants in the Pennsylvania Anthracite Fields, 1880–1910: Some Comments on the Sociology of Knowledge', *Ethnicity 8* (1981), pp. 96–105.

53 Laslett, 'British Immigrant Colliers', p. 49; see also Mildred A. Beik, 'The UMWA and New Immigrant Miners in Pennsylvania Bituminous: The Case of Windber', in: Laslett, *The United Mine Workers of America*, pp. 320–344.

54 *Cf.* Joe Trotter, 'The Dynamics of Race and Ethnicity in the U.S. Coal Industry', *International Review of Social History 60* (2015), pp. 145–164.

55 Alexander, 'Race, Class, Loyalty and the Structure of Capitalism', pp. 118 and 126.

56 Alex Lichtenstein, 'Herbert Hill and the "Negro Question"', *Labor: Studies in Working-Class History of the Americas 3*(2) (2006), pp. 33–39. See also Joe William Trotter, *Coal, Class, and Color: Blacks in Southern West Virginia, 1915–32* (Urbana: University of Illinois, 1990); Daniel Letwin, *The Challenge of Interracial Unionism: Alabama Coal Miners, 1878–1921* (Chapel Hill: The University of North Carolina, 1998); Kelly, *Race, Class, and Power*, pp. 6–15 and 118–122.

57 David R. Roediger, *The Wages of Whiteness: Race and the Making of the American Working Class* (New York: Verso, 1991). For a review, see Eric Arnesen, 'Up from Exclusion: Black and White Workers, Race, and the State of Labor History', *Reviews in American History 26*(1) (1998), pp. 146–174; Bruce Nelson, 'Class, Race and Democracy in the CIO: The "New" Labor History Meets the "Wages of Whiteness"', *International Review of Social History 41* (1996), pp. 351–374.

58 Price V. Fishback, *Soft Coal, Hard Choices: The Economic Welfare of Bituminous Coal Miners, 1890–1930* (New York/Oxford: Oxford University Press, 1992), pp. 171–197.

59 Christoph Kleßmann, *Polnische Bergarbeiter im Ruhrgebiet 1870–1945. Soziale Integration und nationale Subkultur einer Minderheit in der deutschen Industriegesellschaft* (Göttingen: Vandenhoeck & Ruprecht, 1978); John J. Kulczycki, *The Foreign Worker and the German Labor Movement: Xenophobia and Solidarity in the Coal Fields of the Ruhr, 1871–1914* (Providence: Berg Publishers, 1994); *id.*, *The Polish Coal Miners' Union and the German Labor Movement in the Ruhr, 1902–1934: National and Social Solidarity* (Oxford: Berg Publishers, 1997); Richard C. Murphy, *Guestworkers in the German Reich: A Polish Community in Wilhelmian Germany* (New York: Columbia University Press, 1983).

60 *Cf.* Diethelm Blecking, 'Integration through Sports? Polish Migrants in the Ruhr Area, Germany', *International Review of Social History 60* (2015), pp. 275–293.

61 Christoph Klessmann, 'Comparative Immigrant History: Polish Workers in the Ruhr Area and the North of France', *Journal of Social History 20*(2) (1986), pp. 335–353; Janine Ponty, *Polonais méconnus. Histoire des travailleurs immigrés en France dans l'entre-deux guerres* (Paris: Publications de la Sorbonne, 1988); Philip H. Slaby, 'Industry, the State, and Immigrant Poles in Industrial France, 1919–1939' (PhD Thesis) (Ann Arbor, 2005); Donald Reid, 'The Limits of Paternalism: Immigrant Coal Miners' Communities in France, 1919–45', *European History Quarterly 15* (1985), pp. 99–118, 100; Gary S. Cross, *Immigrant Workers in Industrial France: The Making of a New Laboring Class* (Philadelphia: Temple University Press, 1983), pp. 81–84.

62 Pien Versteegh, *De onvermijdelijke afkomst? De opname van Polen in het Duits, Belgisch en Nederlands mijnbedrijf in de periode 1920–1930* (Hilversum: Verloren, 1994).

63 *Cf.* Philip Slaby, 'Dissimilarity Breeds Contempt: Ethnic Paternalism, Foreigners, and the State in Pas-de-Calais Coalmining, France, 1920s', *International Review of Social History 60* (2015), pp. 227–251.

64 Reid, 'The Limits of Paternalism', p. 102; Cross, *Immigrant Workers in Industrial France*, pp. 55–63.

65 Ponty, *Polonais méconnus*, pp. 69–72; Klessmann, 'Comparative Immigrant History', pp. 337–338.

66 Rolande Trempé, 'La politique de la main-d'œuvre de la Libération à nos jours en France', *Revue Belge d'Histoire Contemporaine 19* (1988), pp. 55–82, 70.

67 Frank Caestecker, *Alien Policy in Belgium, 1840–1940: The Creation of Guest Workers, Refugees and Illegal Aliens* (New York/Oxford: Berghahn Books, 2000), pp. 47, 60, 67–68, 92–94, 117–123, 176–182, 216–225 and 243–345.

68 Serge Langeweg, *Mijnbouw en arbeidsmarkt in Nederlands-Limburg. Herkomst, werving, mobiliteit en binding van mijnwerkers tussen 1900 en 1965* (Hilversum: Verloren, 2011), pp. 129–130, 140–148 and 153.

69 Brian McCook, *The Borders of Integration: Polish Migrants in Germany and the United States, 1870–1924* (Athens: Ohio University Press, 2011), pp. 20–21.

70 *Ibid.*, p. 25.

71 *Ibid.*, pp. 70–93.

72 Christoph Rass, 'Temporary Labour Migration and State-Run Recruitment of Foreign Workers in Europe, 1919–1975: A New Migration Regime?', *International Review of Social History 57* (2012), pp. 191–224; *id.*, *Institutionalisierunsprozesse auf einem internationalen Arbeitsmarkt: Bilaterale Wanderungsverträge in Europa zwischen 1919 and 1974* (Paderborn etc.: Schoeningh Ferdinand, 2010).

73 Rudy Damiani, 'Les Italiens: une immigration d'appoint', in Cegarra *et al.*, *Tous gueules noires*, pp. 85–109; *id.* 'Les Italiens du bassin miner du Nord-Pas-de-Calais de 1939 à 1945', in Pierre Milza and Denis Peschanski (eds), *Exils et migration. Italiens et Espagnols en France, 1938–1946* (Paris: Editions L'Harmattan, 1994), pp. 455–464.

74 Caestecker, *Alien Policy in Belgium*, pp. 62–65.

75 *Ibid.*, pp. 221–222.

76 Langeweg, *Mijnbouw en arbeidsmarkt*, pp. 144–150.

77 Damiani, 'Les Italiens', pp. 97–98.

78 Trempé, 'La politique de la main-d'oeuvre', p. 70.

79 Anne Morelli, 'L'appel à la main d'oeuvre italienne pour les charbonnages et sa prise en charge à son arrivée en Belgique dans l'immédiat après-guerre', *Revue Belge d'Histoire Contemporaine 19* (1988), pp. 83–130.

80 René Leboutte, 'Coal Mining, Foreign Workers and Mine Safety: Steps towards European Integration, 1946–85', in: Berger *et al.*, *Towards a Comparative History of Coalfield Societies*, pp. 219–237, 228–230.

81 Tesseltje de Lange, *Staat, markt en migrant. De regulering van arbeidsmigratie naar Nederland 1945–2006* (Hoofdorp: Boom Juridistische uitgevers, 2007), pp. 69–70; Langeweg, *Mijnbouw en arbeidsmarkt*, pp. 186–192.

82 Leboutte, 'Coal Mining, Foreign Workers and Mine Safety'; Christoph Rass and Florian Wöltering, 'Migration und Sozialregion: Wanderungsbeziehungen zwischen europäischen und außereuropäischen Bergrevieren', in Angelika Westermann (ed.), *Montanregion als Sozialregion. Zur gesellschaftlichen Dimension von 'Region' in der Montanwirtschaft* (Husum: Matthiesen, 2012), pp. 59–89, 70.

83 Rass, *Institutionalisierunsprozesse auf einem internationalen Arbeitsmarkt, passim.*

84 Karin Hunn, *'Nächstes Jahr kehren wir zurück . . . ' Die Geschichte der türkischen 'Gastarbeiter' in der Bundesrepublik* (Göttingen: Wallstein Verlag, 2005), pp. 107–109.

85 *Ibid.*, p. 213.

86 *Ibid.*, pp. 219–221.

87 Marie Cegarra, *La mémoire confisquée: les mineurs marocains dans le Nord de la France* (Villeneuve-d'Ascq: Presses Universtaries du Septentrion, 1999), pp. 45–46; *id.*, 'Récession et immigration: les mineurs marocains dans les mines de charbon du Nord/Pas-de-Calais', in Jean-François Eck, Peter Friedemann, and Karl Lauschke (eds) *La reconversion des bassins charbonniers. Une comparison interrégionale*

entre la Ruhr et le Nord-Pas-de-Calais (Villeneuve-d'Ascq: Presses Universtaries du Septentrion, 2006), pp. 157–164.

88   Cegarra, *La mémoire confisquée*, p. 53.

89   Karim Azzouzi, 'Les Marocains dans l'industrie charbonnière belge', *Brood en Rozen. Tijdschrift voor de geschiedenis van sociale bewegingen 9* (2004), pp. 35–53; Tanja Cranssen, 'Marokkaanse mijnwerkers in Limburg, 1963–1975', *Studies over de sociaaleconomische geschiedenis van Limburg/Jaarboek van het Sociaal Historisch Centrum voor Limburg 48* (2003), pp. 121–148.

90   *Ibid.*, pp. 145–146; Cegarra, 'Récession et immigration', p. 127.

91   Piero-D. Galloro, Tamara Pascutto, and Alexia Serré, *Mineurs algériens et marocains. Une autre mémoire du charbon lorrain* (Paris: Editions Autrement, 2011), pp. 45–71.

92   The classic text is Martin Bulmer, 'Sociological Models of the Mining Community', *Sociological Review 23* (1) (1975), pp. 61–92. See also Norman Dennis, Fernando Henriques, and Clifford Slaughter, *Coal is our Life: An Analysis of a Yorkshire Mining Community* (London: Routledge, 1956). Mining communities in Britain are sometimes supposed to be less ethnically diverse than those in Europe and America. For an alternative view, see David Gilbert, 'Imagined Communities and Mining Communities', *Labour History Review 6*(2) (1995), pp. 47–55.

93   Klaus Tenfelde, 'The Miners' Community and the Community of Mining Historians', in *id.* (ed.), *Towards a Social History of Mining in the 19th and 20th Centuries* (Munich: Beck, 1992), pp. 1201–1215, 1207.

94   Leen Beyers, 'Everyone Black?'.

95   *Id.*, 'From Class to Culture: Immigration, Recession, and Daily Ethnic Boundaries in Belgium, 1940s–1990s', *International Review of Social History 53* (2008), pp. 37–61.

96   Aloys Berg, 'Polen und Türken im Ruhrkohlenbergbau. Ein Vergleich zweier Wanderungsvorgänge mit einer Fallstudie über 'Türken im Ruhrgebiet' (PhD Thesis), (University of Bochum, 1990); Diethelm Blecking, 'Polish Community before the First World War and Present-day Turkish Community Formation: Some Thoughts on a Diachronistic Comparison', in: John Belchem and Klaus Tenfelde (eds), *Irish and Polish Migration in Comparative Perspective* (Essen: Klartext Verlag, 2003), pp. 183–200; Leo Lucassen, 'Poles and Turks in the German Ruhr Area: Similarities and Differences', in Leo Lucassen, David Feldman, and Jochen Oltmer (eds), *Paths of Integration: Migrants in Western Europe (1880–2004)* (Amsterdam: Amsterdam University Press, 2006), pp. 27–45; Klaus Tenfelde, 'Schmelztiegel Ruhrgebiet? Polnische und türkische Arbeiter im Bergbau: Integration und Assimilation in der montanindustriellen Erwerbsgesellschaft', *Mitteilungsblatt des Instituts für soziale Bewegungen 36* (2006), pp. 7–28.

97   Leo Lucassen, *The Immigrant Threat: The Integration of Old and New Migrants in Western Europe since 1850* (Urbana: Combined Academic Publishers, 2005).

98   *Ibid.*, pp. 5–8.

99   Roediger, *Wages of Whiteness*.

100   Arnesen, 'Up from Exclusion', p. 164.

101   David R. Roediger and Elizabeth D. Esch, *The Production of Difference: Race and the Management of Labor in U.S. History* (Oxford: Oxford University Press, 2012), p. 89.

102   Upton Sinclair, *King Coal: A Novel* (New York: Hutchinson, 1917), p. 53. 'Dagoes' is contemptuous slang for Italians, Spaniards or Portuguese; 'Hunkies' is ethnic slur used to refer to a labourer from Central Europe; 'Greasers' is a derogatory term for Mexicans.

7 Culture and classed identity in shaping unionisation on mines

Peter Alexander[1]

Introduction

My initial question was 'how did workers' culture contribute to unionisation'? I saw this as a Marxist problem arising from conditions on South African collieries prior to the establishment of a union in 1982, but sensed it had wider implications. Culture and the emergence of trade unions are usually written about separately and I wanted to bring them together. In the process, I found it helpful to conceptualise a mediating term *classed identity*.

Raymond Williams commented: 'Culture is one of the two or three most complicated words in the English language', with this complexity 'not finally in the word but in the problems, which its variations of use significantly indicate'.[2] He distinguished between (a) the 'way of life' of a specific people, perhaps a nation, and (b) practices, such as literature, historically linked to 'high culture', but now including particular kinds of 'mass culture'. Especially in anthropology, the word also refers to material culture, such as food, and, additionally, it can include specific aspects of a way of life, including religion and language, which may define a people, or mark differences among them, or provide connections to others. My interest is in all these aspects of culture, but narrowed to an understanding of mineworkers, with this embodying historical change rather than a description of a particular moment.[3]

Karl Marx wrote – appositely for our discussion – about divisions between 'ordinary' English workers and Irish workers, and the way the former looked down on the latter and aligned themselves with 'the ruling nation', adding: 'The antagonism is artificially kept alive and intensified by the press, the pulpit, the comic papers, in short, by all means at the disposal of the ruling classes'.[4] This example of cultural rupture prefigured both V.I. Lenin's advocacy of proletarian international culture, its opposite and Antonio Gramsci's theorisation of cultural hegemony, in which dominant ideas advanced by schools, churches, the media and so forth, and backed by threat of force, sustained consent, in part through notions of 'common sense'.

'Identity' is often linked with 'consciousness', and Marx, Lenin and Gramsci were more interested in the latter. The first two writers regarded political consciousness as more advanced than trade union consciousness, but did not dwell on pre-union consciousness. Gramsci wrote about:

the active man-in-the mass . . . [who] has two theoretical consciousnesses (or one contradictory consciousness): one which is implicit in his activity and . . . unites him with all fellow workers in the practical transformation of the real world; and one, superficially explicit or verbal, which he has inherited from the past and uncritically absorbed.[5]

Considered in this way, workers had a common identity as workers, but a mix of viewpoints spanning proletarian internationalism and peasant superstition. More than thirty years earlier W.E.B. du Bois theorised a similar contradiction, important in colonies and racially divided societies, that of the 'double consciousness'. He wrote about being an American and a Negro with 'two souls, two thoughts, two unreconciled strivings; two warring ideals in one dark body, whose dogged strength alone keeps it from being torn asunder'.[6] Combining class, consciousness and culture, E.P. Thompson plugged some of the gaps in scholarship on pre-trade union identities and consciousness. He contrasted 'class experience' and 'class-consciousness' proposing that, while the former was 'largely determined' by 'productive relations', the latter was 'the way in which these experiences are handled in cultural terms'.[7] This 'handling' involved adaptation of non-industrial culture to experiences associated with capitalist exploitation, and there are examples of this in the cases that follow.

For international comparison of mining labour, a paper by Clark Kerr and Abraham Siegel, neither of whom were Marxists, proved useful. They hypothesised as follows: 'industries will be highly strike prone when the workers form a relatively homogenous group, which is unusually isolated from the general community, and which is capable of cohesion'.[8] Mining was the prime example of such an industry. However, miners are not necessarily homogenous, and, for instance, the authors acknowledged that 'an isolated mass can be kept from internal solidarity . . . by the turnover of its membership . . . [and] racial, religious, and nationality barriers'.[9] But, what happens when these factors are present, as they often are? Kerr and Siegel also observe that an isolated mass is like a 'race apart', and that its strikes are 'a kind of colonial revolt', but their research was based on Western democracies and predicated on US experience, and one wonders what occurs when workers are, literally, racially excluded or colonial subjects?[10] Further, they imply that strikes are more likely where 'the employer, the landlord, the grocer, and the policemen' are a single entity.[11] Is this valid? Lastly, they assume the existence of unions, so how do we explain cohesion in strikes without unions, as happens in some of our cases. While the authors pose more questions than answers, their questions are stimulating.

In attempting to comprehend 'class' as something more than exploitation in a context where unions were absent, work by a clutch of British sociologists inspired by Pierre Bourdieu also proved valuable. Among them, Wendy Bottero opined: 'The key issue is not self-conscious claims to class identity, but the *classed* nature of social and cultural practices'.[12] She and her colleagues were trying to make sense of a post-modern era marked by growing inequality alongside weakened working-class organisation and, as they saw it, confusion about class labels. Their conceptualisation can be adapted to help us with an earlier era, when – whether or not inequality was rising – working-class organisation was weak and claims of 'working class' identification were unrecorded or marginal. Extrapolating, one can define classed identity as: *practices that some people have in common and distinguish them from other people with different class interests.*[13] This does not negate the importance of exploitation or of consciousness, but it recognises that class is more than an abstract relation to production and it frees consideration of classed practices from assumptions about consciousness. Classed identity, as a concept, can accommodate multiple consciousness and incorporate social and cultural diversity.

In this approach, there is an assumption that cultural practice involves self-expression, rather than merely mutual experience of oppression; so it inserts agency, even if the names of agents are unknown.[14] James Jasper's observation that 'some [social] movements emerge from a pre-existing collective identity' can certainly be applied to labour movements.[15] Classed identities are always present at the birth of a union, but how might culture and the specific form of classed identity influence the personality of the newly born child? This is the question at the heart of our discussion.

This chapter is a tour with a chronological and geographical logic. It takes the reader from the Ruhr, via South Wales, West Virginia, Alabama, Jharia (India), Enugu (Nigeria), the Zambian Copperbelt, to the Transvaal. While not comparative in a systematic sense, it attempts to understand new locations by comparing them to places visited before. We stay in some destinations longer than others simply because I know them better, and while the focus is on coal, which eases comparison, there is a stopover on the Zambian Copperbelt because it is interesting and well researched. Ending with South Africa, the site of my main primary research, was a way of reducing the risk of treating it as a special case. The chapter avoids mere description through its intersection of culture and labour organising, and by placing politics in the frame. Hopefully, its international scope, albeit limited, has helped reduce 'methodological nationalism', thus 'de-centring' analysis and countering exceptionalisms.[16] Inevitably, there are imbalances and absences, with these including production, gender and broader social change (for instance, improved communication).

A number of conclusions emerge. First, classed identity is not, as I initially anticipated, simply part of a transition to trade union consciousness, rather it prepares ground for trade unions, changes with unionisation and survives periods when unions are extinguished. Second, classed identity is sometimes associated with cultural organisations and social institutions that act as proto-unions and are implicated in strikes. Third, it can be an insular barrier to broad class-consciousness, but boundaries are permeable, and there are glimpses of international consciousness that pre-date unionisation. Fourth, variability of ongoing relationships with rural life, as well as relative isolation from urban centres, are factors conditioning cultural practice, classed identity, political participation and unionisation. Fifth, a simple dichotomy between metropolitan and colonised classed identities contributes little to understanding variations of culture and trade unionism revealed in our different cases. Finally, conceptualisation of classed identity focuses attention on commonalities of class experience alongside complexities of class-consciousness.

The Ruhr, South Wales, West Virginia and Alabama

Our tour begins with contrasts between the Ruhr and South Wales, South Wales and West Virginia, and West Virginia and Alabama. As we travel, the clock ticks along. In the Ruhr and South Wales, production took off in the second half the 19th century and was at a roughly similar level. By 1890, the Ruhr employed about 128,000 workers and by 1891 South Wales employed about 115,000.[17] However, at the level of working-class solidarities, differences could hardly be greater. In 1910 two leaders of the South Wales Miners' Federation (SWMF) reported:

> The miners' unions of Germany being split up on four parts have not the power to enforce . . . recognition, and the miners are, therefore, obliged to put up with many indignities which a strong union [like their own] would not suffer for a moment.[18]

The question is: why this difference? For an answer, we must go back to the early years of mining in the two regions, well before unions were formed around the turn of the century (1898 in the case of the SWMF).

In the Ruhr, society was deeply divided along religious lines, with Protestants prominent among owners and Catholics a majority among mine workers. There were many migrants, including a large contingent from Poland, who, whether Catholics (most of them) or Protestants, were derided by other workers and formed their own Polish organisations (including a union). More broadly, Protestants tended to align themselves to the National Liberal Party and Catholics to the Centre Party. Workers became further divided with the advance of the Social Democratic Party (SPD) (and, later still, by support for the Communist Party). Protestant workers' established unions but also clubs that, according to Stefan Berger, aimed to 'enhance workers' allegiance to Kaiser and nation . . . and encourage good relations between employers and employees'.[19] The Catholics and SPD were active in forming, respectively, Christian and Socialist unions, and the SPD, in particular, established a wide array of cultural organisations that broadened its support among workers.

In South Wales, local workers were joined by men from regions of England (including Cornwall) where mining was in decline, and from Ireland, where famine had caused great suffering. There were certainly tensions, but a high proportion of the English miners were, like the Welsh miners, non-conformists, thus easing integration, and although the Irish were mostly Catholics and initially marginalised, they too were tolerated and eventually accepted. To begin with, the Liberal Party attracted a clear majority of the workers. It identified with non-conformism, opposition to the

Tories (regarded as representatives of the bosses), Welsh nationalism and even Irish Home Rule. Workers themselves, rather than parties, developed libraries and institutes and a range of recreational associations, notably choirs and rugby clubs. The SWMF grew out of this cultural activism and religious non-conformity, but it was non-partisan and aimed at inclusivity, recruiting workers whatever their religious or political persuasion. This relatively broad-minded environment allowed the Labour Party to eclipse the Liberals without too much conflict (and later offered a home where communists could prosper).[20]

In explaining the contrast between the Ruhr and South Wales, Leighton James emphasises 'constructed identity', which acknowledges the role of activists, inevitably hidden from a brief account, though certainly important.[21] Berger places more weight on structural differences, a few of which can be mentioned. The Ruhr had a broader range of industry, and mine workers could live in urban areas that were relatively cosmopolitan and socially mixed. There was a greater range of employment opportunities and more contact with politics beyond the coalfield. In South Wales, mining was conducted along narrow valleys, and mine workers were concentrated in small towns and villages. Almost everyone depended on income from collieries, and contact with other classes and social mobility were limited. This context bred a strong sense of 'community' centred on the colliery, the chapel and, later, the union.[22] Ruhr collieries tended to be larger than those in South Wales and capital was more concentrated. In terms of the balance of class forces, owners of Ruhr collieries were relatively more powerful, even without disunity among workers, though the two were related. In Germany, the state was also more present. It owned the mines until the middle of the 19th century and, according to Berger, even at the end of the century workers were addressing complaints to the state.[23] In South Wales there was no such dependency on the state, and so, perhaps, more determination to squeeze concessions from the owners.

In sum, underlying structural differences favoured religious/political/union division on the Ruhr and community and class unity in South Wales. The different make-up of pre-union classed identities contributed greatly to the character of union organisation in the early 20th century. Critically, while unions in the Ruhr were linked to parties with agendas derived from outside the mines, the SWMF was built from the bottom up.

We now turn to the USA, contrasting South Wales and West Virginia and then West Virginia with Alabama. In 1920, South Wales employed about 272,000 workers, which can compared to about 115,000 in West Virginia. Expansion in the latter had been spectacular, with the labour-force growing fourfold in just 20 years, and, moreover, by 1920 it was producing about 60% more coal than South Wales.[24] Alabama was a smaller, less capital-intensive coalfield than West Virginia, employing about 26,000 workers in 1920.[25] Both US coalfields drew in workers from around the world, but in West Virginia, where the local population was sparser and growth more rapid, the foreign-born population was a higher proportion of the total, about a quarter in 1910. In Alabama more than half the workforce was black, with many having been recruited as convicts.[26] The labour history of mining in the two states was similar. The United Mine Workers of

America (UMWA), virtually the only union on the US coalfields, was formed nationally in 1890, but its existence in West Virginia and Alabama was fragile until 1933. That year President Franklin D. Roosevelt passed the National Industrial Recovery Act, after which it took just over a week to organise the whole of West Virginia and only a little longer to organise Alabama.[27] Such rapid unionisation was only possible because of strong classed identities that underpinned and were shaped by earlier bouts of unionisation and brutal defeats of major strikes, notably in 1921 in Alabama and 1922 in West Virginia.

In contrasting West Virginia with South Wales, Roger Fagge emphasises differing power relationships, with these related to cultural disparities. Employers in South Wales could be vicious, but most miners owned their own houses or leased them from private landlords, and shops were numerous and individually owned. Where owners patronised cultural activities, their support was generally benign, and chapels, in particular, provided spaces where meetings could be held. Compare this with West Virginia, where most miners lived in company villages/towns that were run as 'feudal fiefdoms', with the mine operator functioning as employer, landlord and store owner, and as overseer of the police, schools, churches, bars and doctors.[28] The UMWA was excluded from the culture of mining communities, and prevented from organising meetings and other activities. Repressive conditions imposed by West Virginia's mine owners delayed unionisation, but they also created a shared experience for workers. By the time of a 1912 strike workers were already turning against the company churches, and, instead, seeking solace from miner-preachers or ministers in commercial towns, or turning against religious observance altogether.[29] Further, while there was nothing like the associational life that existed in South Wales, there was no shortage of opportunity for drinking, gambling and transactional sex; activities that could be enjoyed regardless of race or nation.[30]

The position in Alabama was not greatly different to that in West Virginia, though it is worth adding some detail. Again, most miners lived in company villages, but racial segregation was starker. The quality of housing for blacks and whites was similar, but there were clearly defined black and white residential areas. All workers ate the same kind of food, but joined racially defined queues in company stores.[31] In interviews I undertook with workers from the 1930s, one of the retirees recalled boyhood stone-throwing: 'sometimes the whites would come home bloody, sometimes we'd come home bloody, [but] we was looking forward to rock battling on Sunday'. Another told me: '[The white workers] decided they'd have . . . a bath-house, so they built a shed . . . [with] a bit for me too. They went in one door . . . and I went in another'. In politics and economics there was white supremacy, with blacks excluded from voting, ownership and better jobs, but among mine workers racial prejudice and status difference were subtler, though still irksome. According to one miner: 'All white folks were 'mister'' . . . [but] they didn't put no handle onto your name'.[32]

Segregation was present in union halls, with whites seated on one side of a central aisle and blacks on the other (a custom still found in 1997).[33] But, interracial unionism made good sense to whites employed in an industry on roughly equal terms and in roughly equal numbers to blacks; and within the union workers called each other 'brother'. One black interviewee remembered a meeting where a naïve white miner complained that 'the colored can make as much money as the white', and a union officer responded, with support from the whites, that if blacks were paid less than whites, 'you would be out of a job'. Yet blacks accepted a tradition where, even where they were a majority, the president of the local branch and the Alabama district would be white, as would the majority of committee members. They calculated that owners had more respect for white representatives and that black leadership would ensure even more public opposition than the union already attracted.[34] In any case, black members could hold sway by insisting on report-backs and by voting for people they trusted.

To the extent that inter-racial and inter-ethnic solidarity prevailed in West Virginia and Alabama, it rested on four pillars. First, while employers' often operated discriminatory hiring, with the exception of convict labour in Alabama (phased out after 1912), all ordinary workers experienced similar conditions of employment. Second, as Ronald Lewis put it in relation to West Virginia: 'a camaraderie among the workers sprouted out of the soil of common grievances against company rule and the common dangers which confront all who go down into the pits'.[35] Third, the two major political parties made little headway among mine workers. Most blacks and many whites were excluded from voting, and, with few exceptions, the Democrats and Republicans identified with bosses who were despised and hated. Fourth, at least in Alabama, activists had built a tradition of inter-racial organising. This was a feature of the Greenback-Labor party in the late 1870s and the Knights of Labor in 1880s, and in 1898 the UMWA secured what may have been the world's first employment contract with an anti-racial discrimination clause (it read, in part, 'all competent colored men are to have an equal chance at all work').[36] Socialist and communist activists reinforced the tradition.

Inter-racialism was a classed identity that recognised common interests among mineworkers, and antagonism with the interests of their bosses, but it was not a developed class-consciousness that rejected all racism and strived for equality across the whole of society. White workers were often racially prejudiced and black workers made pragmatic concessions, but inter-racialism challenged the bigotry of the US South and provided an exemplar for broader working-class unity. The UMWA was the only effective champion of mine workers' aspirations and, in order to succeed, it had no alternative but unite across racial and ethnic divides.

Drawing this account of the four coalfields together, while Kerr and Siegel overgeneralised, juxtaposing the Ruhr and the other cases lends support to their hypothesis about relative isolation. Further, while the employer was not the landlord, the grocer, the police, etc. in South Wales, this was generally true in West Virginia and Alabama, contributing to 'colonial revolt'. Although blacks remained a 'race apart' from whites in the US cases, all miners were a 'race apart' from respectable society and its pro-capitalist political parties. While, there were 'racial, religious and national barriers' in all four cases, only in Germany could these, be 'kept alive' in a sustained way by 'means at the disposal of the ruling class'. Critically, in the Ruhr, the ruling class had a stronger social, cultural and ideological presence among a significant section of workers. Elsewhere, the different character of classed identity created a foundation for counter-hegemonic activity, producing strong trade unions.

Jharia, Enugu and the Copperbelt

Our tour now takes us to British colonies: to India and then to brief stopovers in Nigeria and Zambia. It will be interesting to see whether culture, class identity and unionisation are modified by experiences of colonial rule.

In 1921, coal mining in India employed about 190,000 workers, that is, more than West Virginia; but it produced less than a quarter of the state's coal. At risk of oversimplification, the Indian industry can be divided in two. There were small, labour-intensive mines, sometimes owned by Indians, and larger, capital-intensive collieries, often providing coal for the railways and steel companies. In 1918, about two-thirds of Indian coal was produced by members of the Indian Mining Association, which represented British capitalists concerned with short-term gain based on exploitation of cheap labour. Over the years, there was some shift towards capital-intensive mining, but not much. In 1925, as few as 13.3% of mines were using electricity and 20 years later the figure had risen to just 16.3%. In 1940, machines cut only 11% of the country's coal, and the figure actually declined slightly during the war.[37]

From 1906, India's main coalfield was in Jharia, inland from Calcutta. Here, most mines drew their labour from neighbouring areas, and workers retained links to rural life, returning frequently. These workers were predominantly *adivasis* (India's indigenous population) – hunters and gatherers and small farmers, marginalised by the British and their Indian allies, and derided as 'tribals'. Working and living conditions on the mines were appalling, income was low and debt bondage was common. However, mine work had advantages. It alleviated hunger, families could work together (traditional among *adivasis*) and irregular working was tolerated (especially when mining was impractical due to heavy rain or agricultural commitments were pressing). A high proportion of miners were women but, responding to pressure from social reformers, the British government banned children from underground work in 1923 and it excluded women in 1929. Work teams were mostly paid according to output, often via a contractor who supervised their work. A minority of mine labour, mostly supervisors, clerks and technical staff, were Hindus from higher castes. The more capital-intensive mines often employed oscillating migrants who travelled from further afield.[38]

The first union on the mines, the Indian Colliery Employees Association, formed in 1920, organised employed staff, so mostly Hindus, but it later broadened its membership, and by 1932 had changed its name to the Indian Collieries Labour Union (ICLU). During the 1930s and 1940s, at least a dozen unions organised colliery workers, and an All India Mine Workers' Federation was established in 1943. Like their European counterparts, some unions engaged in upliftment activities, such as literacy classes and lantern lectures. There were various reasons for the multiplicity of unions, some overlapping, and these included: employment level, ethnicity/religion, a focus on colliery/owner/locality/nation, and politics. The largest union was probably the ICLU, which in 1943 recorded 9,779 members. None of them united colliery workers in the manner of the SWMF or UMWA, and the mass of ordinary workers, mainly *adivasi* 'traditional miners', shunned membership. The probable exception was the Hindustan Khan Mazdoor Sangh, a radical nationalist union with substantial *adivasi* membership, but it was only formed in 1946. Traditional miners did not, however, shun strike action. There are records of *adivasis*, women and men, participating in strikes from the 1910s, and we can be sure that dissatisfaction often led to them staying at home. They also joined union-led action, including a 91-day strike in 1938, which mobilised both anti-capitalist and anti-colonial nationalist sentiment.[39]

So why did traditional miners abstain from union organisation? There are doubtless many explanations, but it is reasonable to speculate that, working on small collieries in a labour-intensive industry, they were largely able to defend a way of life by means of localised stoppages (one might even regard it as guerrilla action). In effect, they traded low pay for job control, with this suiting most employers. The capacity of traditional miners to take collective action would have been enhanced by loyalties and practices brought from the countryside and developed on the collieries. Dhiraj K. Nite has pieced together valuable details. Workers lived in *dhowrahs*, long terraces of back-to-back rooms, each about 13 square metres in size, designed for one family, but in practice often accommodating more. Through workers' preferences, contractors' convenience and owners' policy, *dhowrahs* were usually segregated by ethnicity (e.g. Santhal *adivasis* separated from Bauri *adivasis*), caste and village.[40] As with mining elsewhere, there was a strong drinking culture, and grog shops were central to a social life that included women and children.[41] The Khadan-Kali goddess cult was an intriguing feature of colliery life because it merged animist beliefs, prevalent among *adivasis*, and Hindu image worship. Workers made offerings to the colliery-goddess in an attempt to ward off dangers of mine labour.[42] As Nite demonstrates, drinking and divinity were ways of coping, and, while disavowed by modernisers, including some union leaders, they did not prevent workers from fighting to improve their conditions, which they did with increasing force during the 1930s.

In short, *adivasi* miners maintained classed identity rooted in particular cultural traditions and cohesive communities. A divided trade union movement did not provide a viable alternative and, in some measure, the *adivasi* miners' heritage, relations to production and workplace culture militated against strong unions. This is very different to our sketch of classed identity in the European and US cases. Was it a purely Indian picture or part of a colonial landscape? By way of response, we start with two mini case studies from tropical Africa.

Stopover 1: Nigeria. Carolyn Brown writes about 'West Africa's most belligerent working class', the miners employed at the state-owned Enugu Colliery in eastern Nigeria.[43] Their numbers peaked at just over 7,000 in 1945, thus, small compared to India and elsewhere, but locally significant. The colonial government legalised unions in 1939, but the Colliery Workers' Union, formed in 1943, refused to conform to a UK model of proper behaviour, and extended a local tradition of militancy that, arguably, predated the mine's establishment in 1915 and could be traced back to resistance to colonial rule.

Similar to the position in Jharia, a majority of workers lived in villages around the colliery and commuted weekly or daily. They retained a rural foothold that provided influence, alternative sources of income, and spaces for recreation and meetings – all valuable resources. The remainder, mostly clerical and skilled workers, lived in the city. Some joined *nzuko*, improvement associations led by the educated elite, and, at some point, probably before 1920, mine workers developed *nzuko ifunanya*, mutual aid societies organised by job category. These were proto-unions that led strikes, and, following a stoppage in 1937, the colliery management included their leaders in formal structures. Workers developed a range of counter-hegemonic discourse. The first was international. As early as a 1919 strike, they contrasted conditions with those of English miners, and in 1946, the CWU launched a *ca'canny*, the Durham miners' term for a 'go slow'. The second was local Igbo culture. According to this, the village, not the colliery, determined the 'rest day', and there was an expectation of reciprocity, which workers transferred from their chief to their employer, that is, to the colonial state, which owned the mine. Then, third, they raised issues of racism and joined the nationalist revolt against the British. So, Enugu workers' organisation was associated with classed identities that drew on a mix of cultures, not just local ones.

Stopover 2: Zambia. On the Copperbelt of what was then Northern Rhodesia, we meet a workforce that grew rapidly from the start of operations in 1926, and by 1930 included some 30,000 Africans. Most workers were long-distance oscillating migrants, ethnically mixed like those in South Africa, but a growing number had settled in towns, as occurred in Enugu.[44] As in Nigeria, mutual aid associations developed, though here they took the form of ethnically based dance groups. Meanwhile, some of the urbanised workers joined the millenarian Watchtower Society, which, together with dance associations, mobilised a major strike in 1935.[45] Churches, notably the African Methodist Episcopal, advocated racial equality, and sport, especially soccer, broke down ethnic divisions among workers.[46] A high proportion of skilled employees were white, and their 1940 strike encouraged similar action by black workers, now led by the so-called Committee of 17.[47] Companies responded by recognising tribal representatives, supplementing this system with department-based works committees, but by 1946 black workers were joining the white unions.[48] A year later, a representative from the British Trades Union Congress was organising Zambians into unions that restricted demands to pay and conditions, and in 1949 an African Mine Workers' Union was established.[49] As in Nigeria, the union, and Zambian unions in general, disregarded the script, and in the 1950s they engaged in increasingly militant and political action.[50] Once again, classed identities constituted an eclectic mix; this time including Christianity and labourism alongside traditionalism and nationalism; and, once again, there is evidence of proto-unionism.

A number of new issues arise in this section. First, ordinary miners were not entirely dependent on income from mine labour. Whether their ties were local and frequent, as in Jharia and Enugu, or distant and infrequent, as with oscillating migrants, they often retained possibilities for survival in rural areas. This had various implications, including ability to survive work stoppages (at least for short periods); possibly, in consequence, less commitment to trade unions; and cross-fertilisation of cultural practices, ideas and influence between mine and countryside. Classed identity combined relationships with rural life and partial breaks from those relationships, as seen in dance societies. Second, attempts to survive adversity also took the special form of mutual aid societies. Although not discussed here, these likewise played a pivotal role in proto-unionism in early industrialising countries, particularly Britain, where they were termed 'friendly societies'.[51] While not specifically working class, these societies were usually locally organised and membership-based, so had class connotations. Finally, classed identity is not just about a transition from rural pasts towards trade union organisation, it is also about adoption and adaptation of new ideas and cultures in a classed context, for instance, Christianity in the form of Watchtower, knowledge of British trade unionism, and of course, anti-colonial nationalism. The last of these appears more prominently in Enugu and the Copperbelt than in Jharia, perhaps because the mines were larger and more integrated with neighbouring towns, and perhaps because the workers were less marginalised from the majority population. To what extent do these various additions to our account extend to South Africa?

South Africa

From 1902 until 1961, South Africa was part of the British Empire but it was self-governing from 1910. For most of the 20th century, economic and political life in the country was dominated by gold mining, and as late as 1970 South Africa was responsible for more than two-thirds of the mineral's world output. Our focus is on coal mining, but that industry, especially in Transvaal, was closely aligned with the larger gold industry through legislation, ownership, sales and labour supply. South African coal mining was highly mechanised at an early stage, and in 1920 72% of its coal was machine cut (82% in Transvaal); which can be compared to 60% in the US (only 45% in Alabama) and just 13% in the UK.[52] It employed far fewer workers than Indian coal mining (about 34,000 in 1920), but output was only slightly less.[53]

On the collieries, as on the gold mines (referred to simply as 'mines'), racial cleavage shaped all production relationships, culture and unionisation. White workers were exploited and often treated badly, but, unlike the position in Alabama, there were class differences between white and black workers. Most blacks worked under the supervision of a 'miner', always white, and none were employed as mechanics or managers. They were regarded as 'natives' or 'labourers' (and referred to as 'boys'), and, on average, they received about one-tenth the pay of white workers (the 'men').[54] In the first decades of the 20th century, most white mine workers came from the UK, though an increasing proportion was locally born. The majority of black workers were oscillating migrants imported from Mozambique in terms of a 1901 agreement between the British and Portuguese. They were recruited by the Witwatersrand Native Labour Association (Wenela), which was owned by the gold mines through its Chamber of Mines. Oscillating migration was on a different scale to India, partly because, initially, the population in Transvaal's gold- and coal-producing areas was sparse. A large minority of black colliery workers came from Lesotho, then a British colony. Black workers began to settle around the collieries with female partners, a practice opposed by the government, which in 1925 banned collieries from employing more than 15% of these 'married labourers' in their black labour force. Whites lived separately from blacks in much better accommodation, typically in a family house; they wore different and better-quality clothes; ate different, more nutritious food; and benefitted from political and legal rights denied to blacks.

White colliery employees were members of trade unions when they participated in a general strike of white workers in 1914, and, in 1922, they joined the stoppage that culminated in the Rand Revolt (indeed, their strike began a week before that in the gold industry). They made no attempt to involve black workers and were badly defeated (nearly half of them lost their jobs in 1922).[55] 'Class' divided white workers from their bosses but 'race' brought them together. In mining, one element of this was recreation clubs, and in 1949 a local newspaper described the situation at New Douglas as 'more like a popular holiday resort than a colliery' (it included an 18-hole golf course and a yachting lake). In contrast to whites, Transvaal's black colliery workers were not unionised until 1982, when the National Union of Mineworkers (NUM) was formed. There were earlier attempts, but these came to nothing.[56]

However, there were many strikes among black colliery workers. The first was not a workplace stoppage but it did threaten production and had a significant impact. In 1913, a 'nearly absolute mutiny' among workers on the train from Mozambique to Transvaal drew attention to poor conditions on the collieries. The Chamber, concerned that labour supply to gold mines might be jeopardised, insisted the collieries improve pay, reduce hours and abide by basic standards for food and accommodation laid down by the 1911 Native Labour Regulations Act. The Act structured African life on mines and collieries, and underpinned expectations central to their moral economy.[57]

Between 1918 and 1949, Transvaal's colliery workers participated in more than 66 strikes, mostly short, and they were probably the most militant section of South Africa's black working class in this period.[58] In stoppages occurring between 1918 and 1926, collieries with a high proportion of migrant workers were more likely to strike than those with a higher ratio of settled workers. A common tactic was to walk *en masse* to the Wenela compound in Witbank, the main town in Transvaal's principle coal district. With a logic resonant of the Enugu workers' notion of reciprocity, Mozambicans considered this their 'Father's House' and expected a fair hearing, but marches of hundreds of workers could also threaten authority because there were few police in the district. It was usually left to the local native commissioner to resolve matters, and he would generally give the appearance of being impartial and judicious, though could also use force and make arrests. Both sides combined reference to rules with threat of violence.[59] Workers were able to protect the 'moral economy', but there were limits to what they could achieve. This was revealed in a 1919 strike at Tweefontein Colliery, when, for the first time, black workers demanded ten shillings per day. They were then paid about one-fifth this amount and their justification was their ability to maintain production without white miners, who were paid two pounds per day. The strikers were defeated, as were rebellions on the gold mines in 1920 and 1946 around the same issue. 'Ten shillings' threatened South Africa's political economy, and workers only reached this level of pay in 1974, following inflation, labour shortages and a wave of strikes.[60]

Coal had been mined commercially in Transvaal since 1889, yet the region's black colliery workers were the last of the miners considered in this chapter to organise a union. Why was this? The obvious answer, and the most important, is opposition from the owners and repression by the state (both of which were greater in mining than for other black workers in South Africa). But additional factors entered the mix. First, like 'traditional miners' in Jharia – and, indeed, many colliery workers elsewhere – Witbank's labourers had the capacity to hold short, effective strikes. In part, this derived from the possibility of rural survival, albeit at a very meagre level. Second, mine culture provided a foundation for common classed identity, and we now turn to this.

Migrant workers lived in compounds divided into rooms that usually housed 16–20 single men. As in Jharia, through a mixture of self-selection and management policy, they tended to share accommodation with workers speaking the same language and sometimes from the same locality. Each room had an *isabonda*, a man chosen by roommates to resolve petty disputes and represent them to colliery authorities. It is likely that on South African mines and collieries it was the room rather than mutual aid societies or dance groups that provided a basis for proto-unionism. Locality and ethnicity enhanced trust and solidarity and there is evidence of strikes being mobilised by men from one room spreading a call to other rooms, initially through an ethnic network. *Izabonda* also played a critical role in mobilising the 1946 strike.[61]

Access to compounds was restricted, and together with married quarters and neighbouring shacks, they provided space for relatively closed communities. Colliery workers did visit Witbank or other towns but these were a long distance from some collieries and township residents generally looked down on their migrant cousins with distain.[62] Compounds were the place where the workers drank, smoked and gossiped together, where they had sex (sometimes with other men, sometimes with female visitors) and where, for the most part, they participated in religious and recreational activities.[63] By law, Sundays were a day of rest, and this contributed to shared experiences. Research on colliery churches is under-developed, but I have not come across Watchtower or similar movements.[64] More is known about the two main recreational activities: dancing and soccer.

According to Hugh Tracey, writing in 1952, 'Dancing is the main recreation of over a third of all the men who flock to Johannesburg ... over one hundred thousand men'.[65] With regard to the collieries, where there were competitions by 1921, an old miner recalled that dancing was 'the main thing'.[66] Mine dancing, which might be described as 'break-dancing meets line-dancing', was endogenous to the mines, but drew on rural influences (see Figures 7.1 and 7.2). Watching old films, I was struck by its complexity, artistry, humour and athleticism and by black leadership (albeit with white judges). Despite some opposition from moralists and occasional concerns about disorder at large events, the collieries and mines were enthusiastic benefactors. Managers were probably motivated by a desire for fitness, harmless outlets of energy, colliery/mine pride and by their own egos. Especially after the 1946 strike, dancing was also seen as a means of encouraging tribal allegiance. In addition, it was developed as a tourist attraction (see Figure 7.1). But in 1984 public performances were summarily halted, perhaps because the mines regarded them as anachronistic, but they were also opposed by NUM, which saw them as divisive and competition for miners' loyalty and time.[67]

Soccer was the other popular sport on the collieries. By 1935, it was being played on colliery fields, and, in 1940, colliery sides broke away from a league that covered all teams in the Witbank area.[68] While dance teams were divided according to 'tribe', soccer cut across ethnic divisions and was played according to international standards. While dancing was seen as primarily Mozambican, soccer was especially prevalent among men from Lesotho, some of whom were recruited as workers because of their prowess on the field. As with dancing, and indeed churches, soccer provided scope for nurturing leadership skills, important in union organisation, and Gwede Mantashe, who became general secretary of NUM, was a well-known Witbank district footballer. Nevertheless, according to one NUM colliery worker, as an NUM activist he was expected to discourage '*obsession* with soccer'.[69]

Drawing threads together, classed identity among Transvaal's black colliery workers bore some similarity with that in South Wales and elsewhere, but isolation was more extreme. Blacks were segregated from whites, barred from their occupations and incomes, and excluded from their freedoms, social world and labour unions. Their lives and geographical location also separated them from black workers and professionals living in towns. While there were differences between single migrants and settled, married workers, they worked together, lived in close proximity, interacted socially and shared racial oppression. Compounds, with their rules and rooms, provided a core to colliery life that stretched out to include recreational and religious activity. Nationalism appears less significant in pre-trade union Transvaal than in Enugu and the Copperbelt, perhaps because of greater distances from urban areas. Even here, some nuance is required: most workers came from outside South Africa, and Mozambican migrants played an important part in the long struggle against Portuguese rule.[70]

Colliery culture had two major frames. The first was a high degree of regulation shaped, in particular, by centralisation of capital in gold mining, requirement for lots of low-paid workers, and alignment of interests between capitalists and the colonial state. The second was related: a large majority of workers were oscillating migrants. Combining these two structures produced constraints on workers, but also obligations on owners and, for workers, expectations and some capacity to defend the *status quo*. In practice, classed identity integrated resilience with reluctance to organise trade unions.

Migrants had different home cultures that were refreshed by continuing contact with rural areas and salient distinctions on the collieries, notably allocation of rooms. Ethnicity was double edged. On the one hand, divisions were associated with bloodshed and acted as a barrier to unionisation; on the other, it could bolster strike solidarity. On the collieries, as elsewhere, ethnic identification was weaved into adjustment to modernity, thus contributing texture to classed identity. Its significance was celebrated in dancing and challenged by soccer, but both were part of a common colliery culture. Once economic and political certainties were thrown into turmoil in the 1970s, and NUM showed it was a force to be reckoned with, classed identity provided a foundation for union growth and construction of a new order. The irony was that, as NUM gained ground, it undermined colliery culture and gradually fractured the identity. It opposed sex between male workers, heavy drinking and mine-dancing competitions; took control of compounds in the pivotal 1987 strike; was centrally involved in breaking racial barriers and the overthrow of apartheid; and championed living-out allowances, employment of women underground, dismantling of single-sex hostels and reduced dependence on oscillating migration. While contradictory, mining culture helped workers survive the misery of daily life and classed identity eventually contributed to an ecology conducive to unionisation and reframing of miners' existence.

Figure 7.1 Dancing at Robinson Deep Mine, c. 1951. Probably Ndau-Tswa miners from Mozambique, whose style was noted for its acrobatic prowess and humour. See Hugh Tracey, *African Dances of the Witwatersrand Gold Mines* (Johannesburg: African Music Society, 1952), p. 17. Photographer: Merlyn Severn. Reproduced with permission from the International Library of African Music (located at Rhodes University, Grahamstown).

Figure 7.2 Dancing at Robinson Deep Mine, c. 1951. Probably Mpondo miners. Photographer: Merlyn Severn. Reproduced with permission from the International Library of African Music (located at Rhodes University, Grahamstown).

Conclusions

Modern industry created a context where one 'way of life' was superseded by another. A new classed identity emerged, which, in the main – though not without contradictions and complications – provided a foundation for trade union organisation.

Mine workers drank beer to get drunk and socialise; they went to church because they were Christians; and they were enthralled by the artistry, teamwork and excitement of dancing and soccer, but, for the most part, they drank beer, prayed and participated in sports with people like themselves, other miners. They developed a common culture where grievances were formulated, foes were identified and leaders emerged. Aspects of the old ways persisted and some enhanced worker solidarity through expectations of reciprocity from employers, material resources to survive lost income and, in the case of South Africa, ethnic networks that mobilised strikes. In the new world, religion might placate workers, as with Protestantism on the Ruhr, or contribute to a distinctive colliery culture, as with the Khadan-Kali cult, but it could also stimulate anti-establishment dispositions, such as Welsh non-conformism and, more dramatically, Watchtower. Critically, miners developed proto-unionism in the form of mutual aid societies, dance groups and *izabonda*.

However, workers often retained an expectation of returning 'home', so the past was also conceived as a future, and thus important for present identity. This was particularly the case with oscillating migrant workers. Over time, the significance of older identifications tended to wane – miners worked alongside people with different features and cultures, they found common languages and had children born locally – but there were also countervailing pressures. Importantly, ethnicity and race often overlapped material differences and provided markers of inequality. In varying degrees, this was true of all cases represented here. Antagonisms were often 'kept alive' by dominant interests, but this was only possible because, in some measure, they had a life of their own; that is, they were not just a product of 'consent'. In Jharia, Hindu supervisors joined unions prior to ordinary *adivasi* miners, in Transvaal white miners excluded black labourers, and so forth.

While colonialism added a layer of antagonism between master and servant, it would be wrong to create stereotypes of metropolitan and colonised classed identities. In the Ruhr and South Wales, miners were active participants in political parties, but in the US cases – where there was greater repression and the employer was also 'landlord and grocer' – recognition of miners as a 'race apart' who engaged in 'colonial revolts' was perceptive. In the colonies, there were also variations, including some important contrasts within cases, for instance, between traditional and employed, rural and urbanised, and foreign and local. Racial segregation and colonial exclusion produced extra contradictions of consciousness, but also awakened a politics of humanity and democracy that coloured classed identity and existed alongside growing unions.

The relationship between classed identity and trade unionism was not straightforward. Unions mostly came from outside the mines and could be associated with party rivalries that divided workers; and strikes might succeed if localised, but attract repression if allied with union mobilisation. Moreover, while unions brought knowledge of wider society, workers sometimes gained international insights even without the presence of unions (as in the Enugu case). However, there can be no doubt that trade unions generally united workers, improved their living standards and helped win democratic rights. In the process of achieving broader working-class unity, unions sometimes came into conflict with localised classed identities, and recreational pursuits could be a negative pull on activists' limited time. However, classed identities appeared before unionisation, continued through union eras and survived crushing defeats. One cannot comprehend the wildfire spread of the UMWA in 1933 without awareness of the inter-racial tradition that was already well established.

Too often, the history of mining culture is written separately from that of industrial relations, and this chapter brings them together. In the process, it emphasises the importance of workers as actors in pre-union labour history. They created intersecting networks of solidarity and classed identity through cultural activity. By drawing examples from the global South into conversation with some from the North, I have expanded an understanding of mine culture to include important rural relationships, proto-unionism and anti-colonial politics. More broadly, the approach adopted here pushes us to reconsider oversimplified associations between organisation, consciousness and identity. Unions cohere a dominant consciousness among workers, but old differences can be reimagined, as nationalism for instance. Hopefully this chapter will encourage research that, by integrating culture and class struggle, offers a more complex understanding of consciousness. The notion of *classed identity* helps take us in that direction.

Notes

1   Stefan Berger encouraged me to write this chapter, and I am profoundly grateful for his patience. I am also obliged to the Department of Science and Technology and the National Research Foundation, which support my South African Research Chair in Social Change; the Institute for Human Sciences, Vienna, which provided a visiting fellowship; and the many individuals who commented on earlier drafts.
2   Raymond Williams, *Keywords. A Vocabulary of Culture and Society* (London: Fontana Paperbacks, 1976), pp. 87–92.
3   Even here there are problems, Stefan Berger commented: 'working-class culture can take on . . . a chameleon-like quality'. See 'Working-Class Culture and the Labour Movement in the South Wales and the Ruhr Coalfields, 1850–2000: A Comparison', *Journal of Welsh Labour History/Cylchgrawn Hanes Llafur Cymru 8*(2) (2001), p. 6.
4   Karl Marx to Sigfrid Meyer and August Vogt, Letter, 9 April 1870, *Marx and Engels Selected Correspondence*, downloaded from Marxists.org, www.marxists.org/archive/marx/works/1870/letters/70_04_09.htm, on 21 December 2018.
5   Antonio Gramsci, *Selections from Prison Notebooks* (London: Lawrence and Wishart, 1971), p. 333.
6   W.E.B. Du Bois, 'Strivings of the Negro People', *The Atlantic*, August 1897. Downloaded from www.theatlantic.com/magazine/archive/1897/08/strivings-of-the-negro-people/305446/ on 25 December 2018. Du Bois was not yet a Marxist, and the implication of his later writing was that black workers could also develop a socialist class-consciousness. Mosa Phadi, in 'What it Means to Be Black in Post-Apartheid South Africa' (D Litt et Phil) (University of Johannesburg, 2018), shows that black South Africans now have a multiple consciousness of 'blackness'.
7   E.P. Thompson, *The Making of the English Working Class* (Harmdonsworth: Penguin, 1968), p. 9.
8   Clark Kerr and Abraham Siegel, 'The Interindustry Propensity to Strike: An International Comparison', in: Arthur Kornhauser, Robert Dubin and Arthur M. Ross (eds), *Industrial Conflict* (London: McGraw Hill, 1954), p. 195.
9   *Ibid.*, p. 193.
10   *Ibid.*
11   *Ibid.*, p. 192.
12   Wendy Bottero, 'Class Identities and the Identity of Class', *Sociology 38*(5) (2004), pp. 985–1003. My emphasis. See also, Fiona Devine and Mike Savage, 'Conclusion', in: Rosemary Crompton, Fiona Devine, Mike Savage and John Scott (eds), *Renewing Class Analysis* (Oxford: Blackwell, 2000).
13   I first used the term 'classed identity' in an unpublished 2010 paper 'Regulations, Recreation and Resistance: Everyday Life on Witbank Collieries'.
14   In this sense it was 'active experience'. See Lawrence Cox and Alf Gunvald Nilsen, *We Make our Own History: Marxism and Social Movements in the Twilight of Neoliberalism* (London: Pluto Press, 2014), p. 7. Having defined culture in a broad way, I have simplified matters by writing about 'cultural practices' rather than 'social and cultural practices'.
15   James M. Jasper, *Protest: A Cultural Introduction to Social Movements* (Cambridge: Polity Press, 2014), p. 113.
16   On countering 'methodological nationalism' see Marcel van der Linden, *Workers of the World Unite: Essays Toward a Global Labour History* (Leiden: Brill, 2008).
17   Leighton S. James, *The Politics of Identity and Civil Society in Britain and Germany: Miners in the Ruhr and South Wales, 1890–1926* (Manchester: Manchester University Press, 2008), p. 24.
18   Robert Smillie and Robert Onions, in *ibid.*, p. 2
19   Berger, 'Working-Class Culture', p. 16.

20 Berger, 'Working-Class Culture'; James, *Politics of Identity*.

21 James, *Politics of Identity*.

22 In South Wales, but not the Ruhr, the unemployed could join the union, which further strengthened community bonds and union dominance. Berger, 'Working-Class Culture', p. 23.

23 *Ibid.*, p. 11.

24 Roger Fagge, *Power, Culture and Conflict in the Coalfields: West Virginia and South Wales, 1900–1922* (Manchester: Manchester University Press, 1996), p. 12. West Virginia produced 81.2 million tons in 1921, compared with 50.3 million tons in South Wales in 1922.

25 Daniel Letwin, *The Challenge of Interracial Unionism: Alabama Coal Miners, 1878–1921* (Chapel Hill: University of North Carolina Press, 1998), p. 23. Assessment of Alabama draws largely on my own work. Peter Alexander, 'Rising from the Ashes: Alabama Coal Miners, 1921–1941', in Edwin L. Brown and Colin J. Davis (eds), *It is Union and Liberty: Alabama Coal Miners and the UMW* (Tuscaloosa: University of Alabama Press, 1999); Peter Alexander, 'Race, Class Loyalty and the Structure of Capitalism: Coal Miners in Alabama and the Transvaal, 1918–1922', *Journal of Southern African Studies 30*(1) (2004), pp. 114–132. See Brian Kelly, *Race, Class, and Power in the Alabama Coalfields, 1908–21* (Urbana and Chicago: University of Illinois Press, 2001).

26 For 1910, the West Virginia figures were: 42.6% American-born whites, 24.3% foreign-born white, and 9.3% unknown or other. By 1920, only 3.6% of Alabama's miners were foreign-born. Letwin, *The Challenge of Interracial Unionism*; Alexander, 'Race, Class Loyalty', pp. 117–118.

27 Fagge, *Power, Culture and Conflict*, p. 110; Peter Alexander, 'Rising from the Ashes', p. 70.

28 Fagge, *Power, Culture and Conflict*, pp. 29–48 and 64–77.

29 David Alan Corbin, *Life, Work, and Rebellion, in the Coalfields: Southern West Virginia, 1880–1922* (Urbana: University of Illinois Press, 1989), pp. 149–158.

30 Fagge, *Power, Culture and Conflict*, p. 75.

31 By the 1920s many miners were virtually self-sufficient in food (from hunting, fishing, gathering fruit, cultivating plots and rearing animals), and this helped some of them survive strikes.

32 Interviews were conducted by me in Alabama in 1997, and are quoted in Alexander, 'Rising from the Ashes', pp. 64–68.

33 Alexander, 'Race, Class Loyalty', p. 125, n. 74.

34 See also, Ronald L. Lewis, *Black Coal Miners in America: Race, Class and Community Conflict 1780–1980* (Lexington: University of Kentucky Press, 1987), pp. 44–45.

35 *Ibid.*

36 Catholic University of America, John Mitchell Papers, *Contract between the Tennessee Coal, Iron and Railroad Co. and its Pratt Miners*, 30 June 1898. See Alexander, 'Race, Class Loyalty', p. 125.

37 Fagge, *Power, Culture and Conflict*, p. 12; Peter Alexander, 'Women and Coal Mining in India and South Africa, c1900–1940', *African Studies 66*(2–3) (2007), p. 204. In 1930, the average Indian miner produced 130 tons, which can be compared to 270 in the UK, 311 in Germany and 831 in the USA.

38 Alexander, 'Women and Coal Mining'; Dhiraj K. Nite 'Family and Work: An Exploration into the Displacement of Family Gangs and the Politics of Wages in the Jharia Coalfields 1890s–1940s', manuscript (unpublished, 2005); A.B. Ghosh, *Coal Industry in India: An Historical and Analytical Account. Part 1 (pre-Independence Period)* (New Delhi: Sultan Chand, 1977); Dilip Simeon, *The Politics of Labour Under Late Colonialism: Workers, Unions and the State in Chota Nagpur 1928–1939* (New Delhi: Manohar, 1995); Lindsay Barnes, 'Women, Work and Struggle: Bhowra Colliery 1900–1985', (PhD thesis) (Jawaharal Nehru University, 1989); Dhiraj K. Nite, 'Work and Culture in the Mines: Jharia Coalfield, 1890s–1940s', (MPhil thesis) (Jawaharal

Nehru University, 2004); Kuntala Lahiri-Dutt, *From Gin Girls to Scavengers: Women in Indian Collieries* (Canberra: Australian National University, 2001), p. 7. By 1920, for every ten men working underground in Indian coal mines there were seven women; by 1935, the ratio had fallen to 10:1, but it rose again slightly during World War Two.

39 I am indebted to Dhiraj Nite for sharing his knowledge and insights about Indian mining, and in particular for responses to questions in emails headed 'Hardcore Labour' dated between 24 October 2018 and 12 November 2018. This paragraph benefits from that correspondence and also from Dhiraj Kumar Nite, 'Employee Benefits, Migration and Social Struggles: An Indian coalfield, 1895–1970', *Labor History* (downloaded from <www.tandfonline.com/doi/full/10.1080/0023656X.2019.1537038> on 14 June 2019), and Dhiraj Kumar Nite, 'Familist Movement and Social Mobility: The Indian Colliers (Jharia) 1895–1970', *Indian Historical Review 41*(297) (2014), p. 309. Some details are drawn from Anonymous, 'Emergence of Trade Unions in the Coalmines', manuscript (N.p., N.d.) downloaded from <http://shodhganga.inflibnet.ac.in/bitstream/10603/63959/9/09_chapter%203.pdf> on 14 December 2018. The phenomenon of strikes without unions is examined in Susan Wolcott, 'Strikes in Colonial India, 1921–1938', *ILR 61*(4) (2008), pp. 460–484.

40 Nite, 'Employee 'Benefits', and his manuscript for this article, which includes additional detail.

41 Dhiraj K. Nite, 'Drinking Culture and Class Conflict: Jharia Coalfields, 1920–1940', manuscript (2008). I am grateful to Dhiraj for providing a copy of this paper.

42 Dhiraj K. Nite, 'Worshiping the Colliery-goddess: Religion, Risk and Safety in the Indian Coalfield (Jharia), 1895–2009,' *Contributions to Indian Sociology 50*(2) (2016), pp. 163–186.

43 Carolyn A. Brown. *'We Were all Slaves': African Miners, Culture, and Resistance at the Enugu Government Colliery* (Portsmouth NH, Heinemann, 2003), p. 3. For this paragraph and the next, see particularly pp. 126–28, 142–49, 245, 265, 295. I am grateful to Carolyn for discussions about African mine workers conducted over more than 20 years.

44 A.L. Epstein, *Politics in an Urban African Community* (Manchester: Manchester University Press, 1958), pp. 1–4.

45 Charles Perrings, *Black Mineworkers in Central Africa: Industrial Strategies and the Evolution of an African Proletariat in the Copperbelt 1911–41* (New York: Heineman Educational Publishers, 1979), pp. 213–217.

46 Jane Parpert, *Labour and Capital on the African Copperbelt* (Philadelphia: Temple University Press, 1983), p 68; Epstein, *Politics in an Urban African Community*, p. 10. On soccer see Ridgeway Liwena, *The Zambian Soccer Scene* (Lusaka: Liwena Publishing, 1984).

47 Epstein, *Politics in an Urban African Community*, p. 65; see also Miles Larmer, *Mineworkers in Zambia: Labour and Political Change in Post-colonial Africa* (London: I.B. Tauris, 2007), p. 32.

48 Epstein, *Politics in an Urban African Community*, p, 89; Parpert, *Labour and Capital*, p. 2.

49 Larmer, *Mineworkers in Zambia*, p. 33. Hortense Powdermaker, *Copper Town. Changing Africa: The Human Situation on the Rhodesian Copperbelt* (New York: Harper & Row, 1973), p. 120, mentions a union branch chair who was also chair of the football club.

50 Larmer, *Mineworkers in Zambia*, pp. 34–38.

51 See Simon Cordery and Monmouth College, *British Friendly Societies 1750–1914* (Basingstoke: Palgrave Macmillan, 2003).

52 Peter Alexander, 'Coal, Control and Class Experience in South Africa's Rand Revolt of 1922', *Comparative Studies of South Asia, Africa and the Middle-East 19*(1) (1991), p. 34. According to Fagge, *Power, Culture and Conflict*, p. 144, 76% of West Virginia's coal was machine-cut in 1922.

53 Alexander, 'Women and Coal Mining', p. 211. Gold mines affiliated to the Chamber of Mines, which were most of them, employed about 200,000 workers at the same date; Francis Wilson, *Labour in the South African Gold Mines 1911–1969* (Cambridge: Cambridge University Press, 1972), p. 157.

54 Miners in India were paid even less than black South Africans; Alexander, 'Women and Coal Mining', pp. 206, 213–214. On the collieries, black workers out-numbered white workers by about 20 to 1.

55 Peter Alexander, 'Challenging Cheap-labour Theory: Natal and Transvaal coal miners, ca 1890–1950', *Labor History 49*(1) (2008), pp. 47–70; Jeremy Krikler, *The Rand Revolt: The 1922 Insurrection and Racial Killing in South Africa* (Johannesburg: Jonathan Ball, 2005). The employers provoked the 1922 strikes, and it is likely they were encouraged by knowledge of the miners strike in the USA; Alexander, 'Race, Class Loyalty', pp. 128–129.

56 Peter Alexander, 'Paternalised Migrants, Policing and Political Economy, Highveld Colliery Strikes, 1925–49,' *Social Dynamics 20*(1) (2003), pp. 56–57.

57 Peter Alexander, 'A Moral Economy, an Isolated Mass and Paternalized Migrants: Transvaal Colliery Strikes, 1925–49,' in: Stefan Berger, Andy Croll and Norry LaPorte (eds), *Towards a Comparative History of Coalfield Societies* (Aldershot: Ashgate, 2005).

58 *Ibid.*

59 Alexander, *'Paternalised Migrants'*, p. 60.

60 Peter Alexander, 'Oscillating Migrants, "Detribalised Families" and Militancy: Mozambicans on Witbank Collieries, 1918–1927', *Journal of Southern African Studies 27*(3) (2001), pp. 520–521.

61 Alexander, 'Paternalised Migrants', p. 63; Dunbar Moodie, *Going for Gold: Men, Mines, and Migration* (Berkeley: University of California Press, 1994), pp. 96–102, 237–238; P.L. Bonner, 'The 1920 Black Mineworkers' Strike: A Preliminary Account', in: Belinda Bozzoli (ed.), *Labour, Townships and Protest: Studies in the Social History of the Witwatersrand* (Johannesburg: Ravan Press, 1979), pp. 273, 284.

62 See, for instance, Hugh Masekela and D. Michael Cheers, *Still Grazing* (New York: Three Rivers Press, 2004), pp. 3, 12, 15.

63 Alexander, 'Regulations'; see also, T. Dunbar Moodie, 'Migrancy and Male Sexuality on the South African Gold Mines', *Journal of Southern African Studies 14*(2) (1988), pp. 228–256. 'Unnatural vice', as it was called in official discourse, typically involved a senior man taking a young recruit as his wife and practicing 'thigh sex' rather than anal intercourse.

64 But, see Alexander, 'Regulations', pp. 14–16. Ancestor beliefs and the power of *sangomas* also require research.

65 Hugh Tracey, *African Dances of the Witwatersrand Gold Mines* (Johannesburg: African Music Society, 1952), p. 1. This book is richly illustrated with photographs by Merlyn Severn and some editions include CDs with recordings of music that accompanied the dancing. It is available from the International Library of African Music at Rhodes University.

66 Daniel Langa interviewed by Rudzani Mudau, 2002. Langa was born in Witbank town in 1934 and began work at Navigation Colliery in 1951.

67 Alexander, 'Regulations', pp. 16–18. See also, Patrick Harries, *Work, Culture, and Identity: Migrant Laborers in Mozambique and South Africa, c.1860–1910* (Portsmouth: Heinemann, 1994), including photographs; Cecile Badenhorst and Charles Mather, 'Tribal Recreation and Recreating Tribalism: Culture, Leisure and Social Culture on South Africa's Gold Mines 1940–1950', *Journal of Southern African Studies 23*(3) (1997), pp. 473–489. Mine dancing survived, and in 2012, NUM kindly organised a show for the International Mining History Conference. Although the dancers were skilled and entertaining, they were a shadow of virtuoso performers of yesteryear.

68 Alexander, 'Regulations', pp. 19–22. See references to literature on soccer in this paper, particularly, Rudzani Mudau, 'Sport and the Development of New Mining Communities' (MA thesis) (University of Johannesburg, 2006).

69 My emphasis. Dhiraj Nite and Paul Stewart, *Mining Faces: An Oral History of Work on Gold and Coal Mines in South Africa, 1951–2011* (Auckland Park: Jacana, 2012), p. 36. Another famous soccer-playing activist was Mgcineni 'Mambush' Noki, leader of the massacred Marikana strikers. His nickname came from Mambush Mudau, then all-time top scorer in South Africa's Premier Soccer League. See Luke Sinwell with Simphiwe Mbatha, *The Spirit of Marikana: The Rise of Insurgent Trade Unionism in South Africa* (London: Pluto Press, 2016).

70 See 1919 leaflet produced by 'African Union of Natives of the Province of Mozambique' calling for participation and a meeting at Simmer and Jack Compound; in Alexander, 'Oscillating Migrants', p. 521.

8 Feminising an ancient human endeavour

Gendered spaces in mining

Kuntala Lahiri-Dutt

Introduction

Social science texts have conventionally seen mineral resources as 'gifts of nature', analysing why certain industries develop in certain locations to theorise about them. Gender did not enter into these locational conversations on mineral resource extraction. Consequently, in much of the literature on mining, gender was, to use a technical, engineering term, turned into an 'overburden'; that is disposable material treated as waste, and material that is only tangential to the major purpose of extracting the minerals that are valuable and worthy.

In spite of this neglect by mainstream social scientists, feminists from almost all disciplines have explored through this avowed gender-blind field in order to illuminate mining as a heavily gendered human endeavour. They have represented mining as gendered work where gendered labour is performed, highlighted the gendered nature of places where these various kinds of mining work are performed, highlighted the gendered contributions to working-class struggles against the capitalist exploitation of large-scale, industrial and capitalised extractive operations, the gendered nature of the impacts of mining on the host communities, and the heavily gendered nature of the organisation of mining as an industry and its role in building the gendered communities that sustained the masculine industry.

A historiography of this literature illuminates the fact that industrial production and processing of the minerals was, and remains, a gendered labour process that has distinctly gendered effects on communities, and that creates gendered places of production and reproduction. Yet, this body of literature has been lying largely hidden, unexplored and untapped, under the heavy weight of hegemonic masculinist knowledge propagated by conventional social science research.

Drawing on this feminist literature, this chapter presents the main strands of feminist arguments and brings the political economy approaches closer to the emergent threads of political ecology studies. In other words, the chapter offers an overview of the 'industrial' studies of labour by feminists coupled with feminist writings on the 'gender selective impacts of mining' emerging primarily from the studies in less-affluent nations where mining has been expanding rapidly since the 1980s. It argues that the masculinity of mining as an industry cannot be divorced from the gendered impacts of the industry; a masculine workplace with masculine labour processes is intimately linked to situations where women bear disproportionately heavy burdens of environmental degradation and social disruptions caused by mining activities.

Feminising ancient human endeavour: gendered spaces in mining

Gender concerns: in both the industry and in communities it builds

Social scientists from various disciplines have written extensively about mineral resources, their extraction and the mining industry. The conventional approach was to see the mineral resources as 'gifts of nature'; this approach was predominant among early economic geographers who engaged in deep locational conversations on how the natural endowments led to the development of mining industries in certain locations. These locational discussions were in exclusion to those amongst the labour studies experts who examined the class-based labour processes unfolding in capitalist developments of mineral extraction. In the volumes of literature generated by both these groups of experts of mining and society, gender figured only marginally. Undoubtedly, the spatial debates excluded human society as an active agent, but what is even more interesting is that even the debates on labour were ungendered, generally presenting the labour force as a class with a unity of purpose and similarity of interests. Experts who explored the heterogeneities within it only rarely went beyond seeking race and ethnic diversities. If mining labour were bound to their camps and compounds, so were the views of the social researchers who wrote the historiography of mining limited to a masculine manner of envisioning the world. The scholarly world was dominated by men at that time, men who were trained in the use of such words as 'mankind', and used to see men as the *natural* workers. One might say, using a mining analogy, that women and gender were thus treated by scholars as a social 'overburden', waste materials that are disposable and marginal to the main purpose of extracting the multitude of social, political and economic meanings and procedures of mineral extraction, processing and the distribution of finished products. Women's stories and knowledge lay hidden under layers of rock-solid masculinist normativity in thinking about mining, waiting to be uncovered by feminists from all disciplines with their eclectic methods and diverse tools.

The hegemonic masculine view of mining led to the reaffirmation, by popular literature, of the simultaneous cruelty and hellish romance of the miner's occupation, the manual character of the masculine work, the dirt and the risk. These portrayals reinforced the approaches portrayed by social scientists at the time. Such mutually supporting depictions further turned women invisible in the literature on mining. A commentary by Allen about such masculinist analysis is:

Mining evokes popular images of hard unrefined men, distinct and separate from other workers, hewing in mysterious dungeons of coal: dirty, strange men, in some ways frightening and for this reason repellent, yet attractive because they are masculine and sensuous.[1]

Not just a masculine normativity in social sciences, the other reason of such gender neutrality, no, gender blindness, is rooted in a kind of 'othering' of the act of mining as a whole. Orwell's (1937) words represent this best: 'all of us owe the comparative decency of our lives to poor drudges underground, blackened to the eyes with their throats full of coal dust, driving their shovels forward with arms and belly muscles of steel'.[2] This othering led to, as Campbell observed, turning the male miner into an iconic figure:

> Miners are men's love objects. They bring together all the necessary elements of romance . . . It is the nature of this work that produces a tendency among men to see it as essential and elemental, all those images of men down in the abdomen of the earth, raiding its womb for the fuel that makes the world go round. The intestinal metaphors foster the cult of this work as dark and dangerous, an exotic oppression . . . it constructs the miner as earth-man and earth-man as true man. And it completed the equation between some idea of elemental work and essential masculinity. This romance is duly mirrored in working class politics – miners are the Clark Gables, the Reds of class struggle.[3]

This chapter steps away from such avowed gender neutrality and illuminates the extractive industries as gendered places of work at various scales. In addition, it also illuminates the industrial production and processing of the minerals as a gendered labour process having gendered effects on communities. It draws on the growing body of literature that has been built by feminist researchers over the years, literature that has been lying hidden, unexplored and untapped, under the heavy weight of hegemonic masculinist knowledge propagated by conventional social sciences. In this context, it is imperative for feminists to ask two central questions to expand the discourse. The first question is why there exists a need for feminists to consider mineral extractive industries such as mining, and the second is through what approaches we should study mining.

My objective is to present the major strands of feminist arguments arising from approaches that are based on both political ecology and political economy. Thus the chapter presents an overview of the 'industrial' studies of labour by feminists, coupled with feminist writings on the 'gender selective impacts of mining' emerging primarily from the studies in less-affluent nations where mining has been expanding rapidly since the 1980s. Indeed, the masculinity of mining as an industry cannot be divorced from the gendered impacts of the industry; a masculine workplace with masculine labour processes are intimately linked to situations where women bear disproportionately heavy burdens of environmental degradation and social disruptions caused by mining activities.[4]

Images of mining that have conventionally been invoked by social researchers dealing with extractive industries continue to be masculine, illustrating not just the rough, dangerous and manual nature of the work as it used to be but also a new kind of masculinity that is associated with power that emanates from the capital that such industries represent. An in-depth exploration of these representations is impossible within the limited space of this chapter. Suffice it to say that it is possible that the equation of modern mining with the 'death of nature' by ecofeminists such as Merchant[5] came from an association of women and nature, and the obvious analogy of mining violating a sacred mother's womb. This association remains of feminist interest and may require further exploration because for most of human history, as Mircea Eliade[6] shows, people actually believed that minerals grow organically inside the earth, drawing an analogy between obstetrics and mining, equating the mineral ore to the embryo, a mine to the uterus, a shaft to the vagina, and the miners to obstetricians. Such gendered symbolism about mining can also be detected in the work of anthropologist Michael Taussig[7] (in interpreting the changing belief systems surrounding mining in Latin America). Taussig documented that as Bolivian peasants were transformed into industrial proletariats, their incorporeal mother earth and the benevolent spirit of *pachamama* changed into the idol of an unpredictable male devil that inhabits the underground and who, instead of being worshipped, is feared by all.

Traditional social research on mining has conventionally explored less of this symbolic domain and more of the *real* or the absolute space that minerals and mining occupy. The thematic areas that have been explored by conventional social science research until now have included the 'special' nature of mining settlements that defy any spatial modeling, the widespread in-migration into mining areas leading to rapid and widespread urbanisation, and the 'place identity' that extensive mining of minerals such as coal creates – whether in mining camps or over extensive regions. As a feminist researcher, one cannot help but observe that each of these areas of research can be gendered. Indeed in recent years, feminist contributions to labour studies and to the examinations of economic work have brought to light women's key roles in building the mining industry and constructing coherent mining communities.

This chapter summarises these feminist contributions under *four* broad (and somewhat overlapping) heads: mining as a gendered industry or a workplace (or women and men *in* mining); the working class struggles by mining workers as gendered which studies the gender interstices of home and work in mining communities; and mining as a global/national agent of capital accumulation and dispossession of the poor in the contemporary world. At the end of the chapter, I offer some possible directions for future feminist research on mining.

Women as economic agents: mining as a gendered industry

Influential authors such as Lewis Mumford[8] and Carolyn Merchant,[9] who were familiar with the extractive industries that came into existence in Europe and America after the Industrial Revolution, saw these industries as one of the ultimate expressions of modernity – a metaphor of the modern world – representing capitalist exploitation of labour and the gross commodification of nature by the hegemonic corporate enterprise. The extraction of mineral commodities was instrumental to capitalist exploitation as coal and iron ore formed the backbone of modern industries that built themselves on the manual labour performed in the mines (and factories). Humphries offers a Marxist explanation, pointing out that the rural poor in the late eighteenth and early nineteenth century England enjoyed non-legal but insuperable customary and common rights to local agricultural resources.[10] The loss of these resources through enclosures changed women's economic positions adversely within the families as their dependence on wage (and wage earners) increased.

No matter how much the ecofeminists such as Carolyn Merchant equate nature with women, there is no doubt that women have been in the mines with men, as part of a family labor unit, from early times. In between the early days and contemporary times that have seen women return to mining work, there was a brief period – following the 1842 Mining Act of the UK and measures by the International Labour Organization (ILO) to stop women's work at night shifts and in underground – when women were prevented from working in the mines. In early mining, there was hardly any difference between it and farming in terms of peasant production strategies. Both were family-based production, carried out primarily to ensure the subsistence and survival of the household unit. This is evident from the depictions in early treatise such as *De Re Metallica*[11] that portray women as taking on the tasks of breaking and sorting of ores, hauling and transporting them, participating in the smelting and processing activities, and sometimes even undertaking the physically demanding job of working the windlasses. Vanja outlines women's contributions in preindustrial mining of an artisanal nature in Europe, where women worked in small pits and smelted ore.[12] This tradition of women accompanying men into mines continued into the colonial period when the major gold rushes took place in the 'New Worlds' of America and Australia. Some women, however, went into the mines on their own volition and not just to 'service' the men or as supplementary labour; the fact that some women prospected for gold during major rushes such as in the American West has been well-documented by Zanjani.[13]

The transformation of Europe and Anglo-America into industrial societies was possible because of a mineral extractive boom. This boom led to a mushrooming of collieries and iron ore mines, in addition to silver and gold mining. Early feminist scholars[14] noted that women played a key role in Industrial Revolution as the home and the workshop became separated and a great number of them turned into wage earners in the outside world. The contributions of women, although apparently wiped out from the mainstream consciousness, become visible in odd places and memorialised as a forgotten past; for example, the museum (Musée D'histoire Naturelle et de Géologie) in Lille, in northern France, has statues of women colliery workers in their work clothes. Similarly, in the Kohle Global exhibition in the Ruhr Museum in Essen, the contributions of women colliery workers of the time are celebrated. These early modern or capitalist industrial mines also saw the participation of women in large numbers. In the mines in Belgium until the early twentieth century, 'numbers of women working underground in Belgian coal mines actually grew . . . [and] . . . Belgium's women coal-miners earned some significant portion of the public respect and reverence elsewhere given so readily to male coal-workers'.[15]

Women coal mine workers came to be known as *hiercheuse*, a proud title connoting the feminine version of *mineurs*, the male miners. Hilden further goes on to narrate that while the male miners wore hard leather helmets, trousered suits and wooden clogs, the *hiercheuses* were dressed in white linen suits of knee-length trousers and a buttoned, long-sleeved jacket. Groups of women worked in shifts, going down the mine at 5am, and returning at about 9pm, loading between 60 and 70 chariots during these hours.

One might expect that European colonial powers would also hire women when they were establishing mines in the colonies, but that was not the case. In contrast, early Indian collieries initially employed women as part of a 'family labour' system that kept the workers, often drawn from local tribal communities, tied to the modern industrial mines.[16] With respect to South African collieries, Alexander has shown that the colonisers hired male black men and kept them in barracks, strictly separate from those of the white workers.[17] The job division in the pits meant that *kamins* (female workers) performed quite different tasks from those performed by the *coolies* (male workers) in India.[18] The *kamins* worked not only as the 'gin girls' (from the term 'engine') to lift coal from the shafts, but also as loaders of coal (cut usually by their male partners – father, brother or husband) in both surface and shallow underground or open cast mines that were locally known as *pukuriya khads*. A similar pattern of sex-based labour division existed elsewhere in modernising Asia as well; Nakamura describes the *naya* ('stable') system of work in Japanese coal mines:

> A working pattern in which a married couple worked as a unit, with the husband (*sakiyama*) digging out the ore and the wife (*atoyama*) assisting him by carrying away the coal, became widespread [during the early twentieth century]. Married women comprised most of the female workforce in the coal mining industry.[19]

The *naya* system was an indirect labor management system with the *naya* chief as an agent, who recruited the miners, and then made them live in the bunkhouses he provided, and supervised their daily lives. Monies borrowed from him kept the miners in bondage to him.

However, in spite of this long and impressive history, the issue of women's labour in the mines remains an area fraught with controversies owing to four main reasons: sex-segregation of jobs that pushed women into relatively lower-status jobs compared to those performed by men; the marking of tasks as male and female tasks which created spatial segregation *within* the mine or even the pit; the prevalence of a family wage that meant women hardly ever received the full recognition for their component of labour; and the various protective legislations, largely initiated by the International Labour Office during the 1920s, around women's labour in mining. Women's labour in mines within a strict sex-based division of tasks was (and still remains) subject to gender ideologies – situated in the home as well as those propagated by the state. Protective legislations define women as 'special workers'. This 'special' nature of women workers is conceived by labour experts in three ways: women's 'natural' task is reproductive labour; their physical frailty circumscribes the kinds of work they can do; and as dependents unable to uphold their own interests, they require the protection of the state. Discourses on the nature of women workers as 'special' build, reinforce and perpetuate a 'maternal wall', bringing women's domestic and reproductive roles to the fore, rather than their productive roles. Populist and universalist conceptions of femininity and womanhood tend to naturalise contested gender norms through these protective legislations that are problematic and that eventually operate against women's interests. Their long-term policing effects are evident, as in the mines, where masculinities became inscribed onto the bodies of miners and into the spaces of mining.

Not only in the workforce, the relationship between gender ideologies and the sex-based division of labour has meant that women's unpaid labour as part of the family has also remained largely invisible. This is more than evident in mining communities where men were seen as the primary workers and women were seen simply as supplementing their labour. Describing women labouring in early modern collieries in England, John argued that since the male members utilised the labour of their female relatives, those women who worked as 'pit-brow lasses' (those who worked at the brow of coal pits) were not usually recorded in colliery accounts by the government officials who were in charge of keeping records. She observes:

> In the eighteenth century the employment of women was still part of a family concern, male members utilising the help of their female relatives wherever possible. Since the hiring and payment would be the responsibility of the male collier, women were not usually recorded in colliery accounts.[20]

This invisibility that ensured the maintenance of masculine dominance within the household also meant that the division of able-bodied and the disabled was not sharply drawn and the male was presented as the head even when physically ill because the central concern to mining families was to ensure the physical well-being and continued wage-earning capacity of male bread winners.[21]

The relationship between state gender ideologies and economic imperatives meant that women were encouraged to join the mining industry at the convenience of the state. For example, Ilič has documented that despite the fact that the Soviet Labour Code included a ban on the employment of female labour in underground work, women continued to work extensively and in increasing numbers throughout the interwar period in response to the need for additional labour.[22]

These factors made women's labour invisible. Clearly, official employment records do not show the full extent of women's participation in mining.

When government officials visited an asbestos mine in South Africa in 1950, they found the records showed that the mine employed 100 males but made no mention of women. A month later, a health inspector, also from the government, visited the same mine and noted that there were 102 male employees and 40 women.[23] Such discrepancies occurred because of women's invisibility as workers, but not necessarily as individuals with health-related problems.

In retaliation for the invisibility of women mine workers in government records, feminist labour historians have resorted to alternative sources of documentation such as oral histories. For example, Gier and Mercier[24] and Tallichet[25] have used oral histories of women mine workers in the mid-Western regions of the USA and in central Appalachia respectively in order to unearth their hidden histories.

The most remarkable phenomenon in the contemporary world is a 're-feminisation' that the extractive industries are experiencing. After decades of being known for its hyper-masculine labour regimes, the awareness of the differences of women's issues compared to men's – low participation in formal, industrialised mining and its excessive impacts on women in mine-affected communities – have enhanced the visibility of women/gender in extractive industries. Equal employment opportunity principles have opened up hitherto out-of-bounds jobs for women in countries such as the US and Canada, and as large-scale industrialised capital charts out its new territories in the countries of Global South, it is absorbing women into some of the mining operations.[26] Some countries with large mining sectors, such as South Africa, have created new rules for hiring women, primarily in order to ensure that some benefits, such as cash incomes, flow to women in mining communities. Increased awareness of the need for greater equity between genders has forced large mining corporations, irrespective of actual position of women in these companies, to highlight women in their public profiles. Together, one can say that these are driving the re-feminisation of mining, posing a challenge to scrutinise the process to understand multiple aspects of women's engagements with diverse forms of extractive processes, and link them to gender equality. One can detect increased inclusion of 'gender' in policies and processes of formal, capitalised mining such as in social impact assessments,[27] community engagement,[28] mining companies' negotiations and agreement-making,[29] sustainability reporting,[30] women's access to local economic and community development, services and infrastructure,[31] inclusion of Indigenous women[32] and provision of job opportunities.[33] Responses to women's concerns in industry initiatives have been much slower than other natural resource management sectors.[34] Progress on the ground has been limited because mining company personnel rarely question patriarchal systems in mine-affected communities for fear of offending 'custom'.[35] However, in terms of numbers alone, the proportion of women remains low in large, industrial mines.

Moreover, a number of issues remain poorly debated; for example, even in countries with affirmative actions in employment opportunities, women continue to experience discrimination in various ways and forms that are not clearly understood from numbers alone.[36] The vulnerabilities that women experience do not become apparent from the win-win arguments proposed by mining corporations on their glossy pamphlets and professional websites.[37] Yet, cumulatively, all these indicate the fact that both international pressure and gender awareness are building, and the awareness of women's contributions to mining as an industry and as a workplace is increasing.

Women as reproductive agents: mining wives building up the community and family

Linda Rhodes lived in mining camps as the wife of a mining engineer. Based on her personal experiences, she shows how unpaid labour by wives – at home and in the community – helps sustain a flourishing social life around mines.[38] Her observations are supported by the earlier observations of Robinson who documented that the managers' wives in a mining town in Indonesia were expected to take on leadership and welfare roles in the community through involvement in the Association of Inco Families, an organisation in which their position paralleled that of their husbands.[39] In other words, they were constructed as an auxiliary branch of the capitalist enterprise of the mining company. Broadly speaking, this illustrates the meaning of *mining culture*, a culture that attributes domesticity to miners' wives by socially constructing and locating women within the home in mining communities. These women can then be described as 'the hewers of cakes and drawers of tea', and relegated to their place at home for men to gather in union halls (or local pubs) in order to form their class-solidarity.[40] Marxist geographers McDowell and Massey in their research on the colliery settlements of Durham, England, analysed this phenomenon as one of gender divisions of labor creating a spatial division between the home and workplace (the mine).[41]

Mining settlements are not organic and can be seen as 'remote' and a kind of exception to ordinary settlements.[42] More importantly, the communities who inhabit these settlements comprise of new migrants, building up a 'new' community, whose very existence is ruled by the state, the company and labour organisations – all male-dominated institutions.[43] Klubock's study of a copper mine community in Chile shows how transnational mining capital and the state ignore the complexities of gender within the community in monitoring men's and women's roles and behaviours, thus engineering a shift to the male head-of-household model.[44] Anthropologist June Nash put women in Bolivian mining communities within the context of the home as wife of a male miner, subjected to the limitations of the house, to the dominance by the man whose needs she must dedicate herself to and to almost continuous childbearing: 'Male and female roles are dichotomized in the mining community, and there is still a mystique about women not entering the mine'.[45] Women in mining communities are therefore seen as belonging to the working class *because* of their men, the 'male contoured social landscape' burgeoning with the tacit as well as overt support from the company (for example the Anaconda Copper Mining Company in Butte, Montana[46]).

This statement continues to be broadly true of all contemporary mining settlements. Not just the larger *company town* but also the camps meant for mine workers living close to the pits are highly structured along class hierarchies.[47]

In these settlements, company hierarchies are reproduced within and between social spheres to provide an informal instrument of subjugation for those occupying a lower status at work. Managers' wives have relatively higher social and economic status than the wives of staff members and much more so than the local or indigenous women, who in turn, then develop a heightened awareness of class and wealth. Also, in multinational operations, race too can play an important role in segregating places according to the complex intersection of gender, race and class.[48]

Undoubtedly, the mining communities are where everyday lives are performed and gender is constructed and enacted through the processes of daily production and consumption. Feminists' interests emerge from the fact that gendered people from these communities occupy the interstices of home and work. The contrasting female identities – of the domestic woman versus the political woman (or the economic woman) – become a key force in reinforcing normative gender roles by ascribing certain tasks to the individuals from each sex. The majority of domestic women shoulder onerous maternal and domestic burdens of unpaid domestic and emotional work, and accept male subordination to fit the picture. This is fast becoming the norm. To what extent women have a degree of autonomy in producing their identities is still unclear.[49]

Women as political agents: active in mining struggles

It is important to note that women are not only workers in the mining industry. They are also political agents who are active in mining struggles. Throughout the history of modern industrial mining development, they have fought (and continue to do so) against exploitation, either standing side-by-side with men or in the wings by providing critical support and logistical assistance. Their efforts have helped in the maintenance of solidarity as part of the working class. Mining and gender literature also uncovers the histories of, and presents, women who helped the male (and female) workers to protect both their families and their collective interests. The archetypal images of women protesting against capitalist exploitation in mines are contained in Emil Zola's masterpiece, *Germinal*,[50] set in the context of the coal miners' strike in northern France in 1860 and in particular, that of the inspirational Mother in Maxim Gorky's 1906[51] novel about Russian factory workers. It is important to remember, however, that the struggles of industrial mine workers, which such images represent, are somewhat different from the contemporary struggles portrayed in the social movements and resource conflicts literature.

Throughout the industrial world, coal mine workers often led the demands for improvements in working conditions, and women were an auxiliary force in these class movements. While women supported workers' struggles against capitalist exploitation and mine closure, and even confronted management directly over issues like food and housing, they also adopted an impressive number of strategies to ensure their own position. Their strategies often pitted gender against gender and even occasionally transcended class lines. Women activists, even though working for auxiliaries, sometimes acquired considerable power by refashioning the rhetoric surrounding motherhood and through the deployment of strategies to organise themselves.[52] Another strategy involved closer self-observation of women and their relationship with the trade union activists.[53]

The crucial roles women played in miners' strikes were invaluable, as they not only supported men's struggles, but also took up activist roles themselves to be the 'voice' for their own concerns. Describing the more recent strikes in the Bowen Basin in Australia in the 1980s or the dispute with the Australian mining company, BHP in 2001, Murray and Peetz comment on the support that women gave to striking miners and to each other: 'without this support it is questionable whether any of the actions would have succeeded.'[54] Similar stories have been documented of miners' strikes in other countries. Stead for example, described women's roles in the miners' strike in Britain during 1984–85.[55] The key areas that emerged include not just the unpaid work by wives and the economic support they provided to families through the crisis by adopting diverse economic activities, but also their active political roles as protesters representing class interests.

The supportive roles played by women in mining struggles are legendary. However, some feminists have critiqued this view based on the conflicts in class and gender identities that never make women an integral part of mining unions anywhere in the world. Indeed, the politics of socialism, in which miners have historically played a central part, tends to generalise and present a fictitious character of 'the miner's wife' who supports men's struggles in solidarity.

Trade unions were most often insensitive to the needs of women as workers and held working-class interests as representing the interests of all workers, irrespective of their gender. For example, in Indian collieries, the need to improve working conditions was advocated by male-dominated trade unions, which could only bargain for higher wages on the grounds of a shortage of sufficiently skilled labourers to operate the machines. The camaraderie of male-dominated trade unions, a vital asset in the struggle against labour exploitation, has compounded the marginalisation of women workers in Indian coal mines, where most women mine workers are illiterate and from lower castes or indigenous groups, turning the male miner as the archetypal 'working class', subsuming women's interests in the mining industry.[56] In some instances, the ideal of dual unionism (separate unions for female and male workers) meant that women's unions were always lower in status, poorly staffed and had a largely ineffective voice. The All India Trade Union Association, founded in 1920, dealt with 'women workers' issues' by forming separate trade unions for women. The separation of women's interests as 'special' and different from 'general workers' issues' meant that men in trade unions dealt with 'bigger' and more pressing agendas. This separation, though changed later, left a legacy by segregating Indian working-class interests on a gender basis, and by encouraging a view of the industrial labour force as the domain of male workers. In 1960, Sengupta observed that 'the question of women in Indian industries is still considered in many circles as a minor problem compared to other labour problems'.[57] Little had changed by 1997, when Fernandes observed that 'the question of gender has been viewed by unions as a marginal concern'.[58] This is because the conventional wisdom of trade union leaders is to deal with what they see as the most important problems such as lockouts and mass retrenchments. These so-called 'bigger' problems continue to subsume issues relating to the position of women workers within the trade unions in Indian collieries.[59]

Mining corporations are aware of this social need of mining communities to conform to gender ideologies and being gendered organisations, they exploit this social need and themselves push women into domestic spaces. In the past, in mining towns of the Northern Rhodesian copper belt, mining corporations encouraged more mine workers to live with their wives at home to enhance the 'stability' of the labour force, while women wrestled with their 'male protectors' for access to family incomes.[60] Although a woman's potential prosperity largely depended on her husband's wages, higher wages alone did not guarantee a woman's financial position. In contemporary times, large mining corporations take up other strategies such as building women's clubs to create and reinforce a strong occupational identity that overall emphasises women purely as reproductive agents by placing them as part of the home and the community.

Women as victims: dispossessed by mining

In contrast to the historical literature, contemporary feminist studies have primarily used the lens of political ecology to analyse the literature on mining. The reason for this may possibly be attributed to the fact that since the 1980s, most new and large mining projects have been established in less-affluent countries; very few nations of the industrial world can still claim to have a thriving mining sector, except the USA, Australia and Canada. It is also possible that this selective focus is reminiscent of the bias expressed by Mumford and Merchant, which I referred to earlier in this chapter.

Collectively, this genre of research has emphasised that the introduction of large-scale mining affects women disproportionately. The impacts on women are more severe than that on men. These impacts range from the erosion of the physical subsistence base to social and cultural changes such as changed notions of authority and interpersonal power equations both at home and within the communities. Women are affected both by lack of access to assets and resources, as well as increased cash flows into local economies which predominantly fall into the hands of men. The gendered impacts often cut across class and race, but women (and men) who are already at a disadvantaged position within the socio-economic hierarchy are more adversely affected by large-scale mining than their more privileged counterparts. Again, the gender-selective impacts have been noted both in economically well-developed countries such as Canada as well as in less-developed countries with smaller economies. For example, Hipwell et al. noted three broad categories of gendered impacts of mining on the Indigenous populations in Canada: those affecting women's health and well-being, those affecting women's work and traditional roles and those affecting the (unequal) distribution of the economic benefits from mining.[61]

Equally important to their discussion were the loss of traditional autonomy and the changes in the productive roles of women.

In less-affluent countries, the family unit's food security primarily burdened already fast-depleting subsistence bases combined with environmental degradation for women.[62] Degradation in the physical environment in mining areas occurs due to the contamination of air, dust and water from activities related to mineral extraction and processing. Positioning the problem in the environmental justice paradigm[63] leads to the observation that the alienation or loss of productive agricultural lands, reduced access to water sources, the loss of livelihood opportunities and the loss of forest cover negatively affect women more than men. Bhanumathi also observed the decreased ability of women to work on remaining land as men tend to migrate out of villages to cities in search of cash incomes.[64]

In her early ethnographic work on the political economy of development in a mining town in Indonesia, Robinson observes that 'The fundamental change in Soroako has been the loss of the village's most productive agricultural land to make way for the mining project'.[65] Therefore, the loss of productive land rapidly transforms a large proportion of male villagers dependent on wage labour for the company, reducing them to a semi-proletariat. Generally, women tend to become less active economically due to a sudden change in the production systems, relations and spatial orientations according to Rothermund.[66] Consequently, women in mining tracts in India sometimes resort to a variety of activities in the informal sector, or become more homebound than before. As cultural expressions of gender change, both women's autonomy and personal freedom which they may have enjoyed earlier either become restricted or decline altogether. As the community changes from peasant agriculture to wage labour for subsistence, women generally experience a decline in their economic independence.

Mining projects are highly capitalised; their heavy equipment, offices, roads and other infrastructure, salaried workers and their residences introduce a new cash-based economy into a rural area. Often these cash incomes actually play significant roles in reducing women's autonomy. A variety of social and cultural impacts occur when large numbers of migrant population disrupt the social fabric of a mining area. The loss or changes to traditional culture remove some of those older cultural norms that might have attributed some authority and power to women. Thus, the loss of older values often more adversely affects women than men; the effect is particularly evident in the devaluation of women's productive work at home and by undermining their status as decision makers and landowners. In the Pacific islands, a sudden influx of mining revenues within local communities generally marginalises women. In general, the monetised economy introduces a different 'mining culture'. This external culture is reinforced by the mining company which attributes new notions of authority to men, putting women either in lower-status jobs or rendering them invisible by involving men in decision-making processes relating to issues of compensation and agreement making. When a mining corporation negotiates with the community, the latter is most often represented by traditional leaders who are usually elders and of the male gender. This happens because being gendered organisations themselves, the mining corporation personnel are usually also male, who carry with them false assumptions about who might be the 'head of the family', or how household resources are allocated among the members. Usually, the land belongs to men, and more often than not, men receive monetary compensation for the loss of land, making it difficult for women who have little formal political authority to be able to influence how the mine shapes their autonomy in decision making and, in a broader perspective, their lives.

One interesting dimension in determining the winners and losers of the mining projects among women is their geographical distance from the mining operation. In a study carried out in eastern Kalimantan in Indonesia, it was found that generally, physical proximity to the mines leads to the direct experiencing of noise and vibrations, and the visibility of gigantic machines arouses fear and a sense of insecurity creating a heightened sense of negative impacts for women according to Lahiri-Dutt and Mahy.[67]

The entry of a cash-based economy with mining also affects women indirectly. The extra cash earned by men being spent on sexual promiscuity, on pubs, karaoke bars and brothels that come up overnight in the most remote places. Lahiri-Dutt and Mahy observe that generally speaking, the negative gender impacts of mining are related to the shifting power equations within communities and families. These shifts can be specific to a particular person or groups, and they also include the increased cost of living, the lack of direct employment opportunities in the mine for women, environmental impacts and women's lack of decision-making power at the community level. All these shifts translate into a dependency on male relatives.[68]

Reflecting on the literature that relates to the impacts of mining on women, one observes that it has reaffirmed the role of women as victims of multinational capital that destroys the environment and adversely affects the lives of the poor. Of the several victim figures, the most convincing and lasting one has been that of 'the prostitute' who has been turned into one by mining projects and who makes a living selling sex around the mine site. Again, as a metaphor, this has been interpreted as equal to the vandalisation of nature by mining, thereby causing the degradation of women by degrading the land. Indeed, the attitudes towards both sexuality and towards women change with the rapid influx of money and new social-cultural values.[69] For example, while many societies in Papua New Guinea incorporated long periods of male sexual abstinence, there is evidence that in mining towns this is being eroded. Macintyre notes that mining communities report a growing incidence of alcoholism, rape and other forms of violence against women and an increasing incidence of teenage pregnancy.[70] However, the theoretical positions of much of this evidence have been questioned in recent years.[71]

O'Faircheallaigh[72] and Lahiri-Dutt[73] have critiqued the theoretical basis for viewing women only as victims of social and economic change. Overall, attention to gender can lead to more sustainable livelihoods for mining communities.[74] This discussion is the derivative of the highly contested literature on women, environment and development (WED) which emphasises the affinity of women with their environments, is reminiscent of biological determinism and essentialism and characterised by the absence of social, material or historical context. This literature homogenises all women as a single category, and romanticises their special closeness to the environment at the cost of exploring the intersection of race, ethnicity and class relations. This scenario leads to the danger of depoliticising the environmental and community politics in mining and creates a dualism between women and men with separate spheres and spaces of production.

Conclusion

As is apparent from the discussion above, I clearly needed to rely more heavily on certain specific disciplines within the social sciences in writing this chapter that is intended to address a broad audience generally interested in mining. While feminist historians have contributed generously in 'uncovering' the hitherto hidden histories of women in mining, more contemporary studies have focused on the political ecology of mining as the industry has moved into the less-affluent countries, impacting on the poorer and the disadvantaged in general but more specifically women from these communities. It turned into quite a challenge to find contemporary studies of labour processes within the mining industries of these countries. In a way, I see this as an important turn within feminist studies of mining: from the earlier pure emphasis on labour processes in the mining industry and its associated spaces, feminists are using a combination of political economy and political ecology to deepen their understanding of gender both in the industry and how it impacts on the mining communities.

Tentatively, I can only speculate and offer some broad and general views for this paucity in the creation of new knowledge on gendered perspectives on mining and vice versa. In my view, one of the reasons lies in how most social researchers see mining: quintessentially as the 'other', as a human endeavour that is 'not natural' and one that competes with farming for land; that is physically remote and scattered; that is associated with chance or luck; that does not yield itself easily to modeling; and finally, one that represents a 'special' kind of human project in its disregard for preserving nature/the environment. More significantly, the paucity of information may also have to do with the subjectivities of many feminists whose works dominate mining literature: their general location in industrial nations in many of which mining has become a part of history, and their position as second-wave feminists with left-leaning politics which makes them uncomfortable with the fact that capitalised mining has expanded rapidly into the less-affluent nations since the 1980s where the gender and class equations are somewhat different.

Therefore, in thinking about contemporary mining in these countries, some feminists put aside their historical understandings of race/class/gender, and find it easier to adopt a lens of political ecology to show that poor women are leading the protests against mining operations that undermine the very basis of their survival. These contributions are significant to the social studies of extractive industries, but they in a perverse manner naturalise men as mine workers usurping benefits and therefore reaffirm women as either the victims or as agents resisting against mining, further reaffirming women as 'the other' of mining and the miners, equivalent to *mother nature*.

From my position at the margin of margins, I can only indicate the possible directions of future research on gender *in* and gender *and* mining through a feminist lens. The body of knowledge that developed in the context of modern industrial mining remains invaluable and will remain so, but tomorrow's feminists would do well to engage more with four main sets of theories: postcolonial feminist perspectives that critically reflect on power relations, intersectionality, feminist political ecology and related but not quite the same, gender and development (GAD) theories. They would explore how mining impacts not just on all women as a homogenous group but how gender creates advantages as well as disadvantages selectively. They would explore deeply how gender can be integrated within the mining 'project cycle', that is, from the stage of exploration to that of closure, and by engaging with the mining communities, companies and policy makers, would offer new solutions to the continued invisibility of women and gender. They would also show how the new working classes brought into existence in less-affluent nations by contemporary capitalised mining projects are essentially differentiated not just in their rural roots but also in the new, flexible, work arrangements (such as Fly In Fly Out or FIFO), and in how gender is (re) negotiated in the fluid spaces and selves created by these arrangements. They would explore what race and age mean for women and men in mining camps and towns, thereby focusing on how the gendered bodies of individual men and women perform certain kinds of work, how masculinity (and femininity) of the enterprise is transmitted into the communities, and the gendered social lives within these communities, as well as the implications of the intricate sexually based division of labour within mining organisations.

Acknowledgement

I prepared this chapter during my stay as a Senior Visiting Fellow at the Asia Research Institute (ARI) of the National University in Singapore (NUS). I thank both ARI and NUS for providing me with an engaging academic environment that immensely facilitated this writing project.

Notes

1 Vic L. Allen, *The Militancy of British miners* (Shipley: Moor Press, 1981), p. 4.
2 Orwell, George, 'Down the Mine', in *Collected Essays* (London: Secker and Warburg, 1937).
3 Beatrix Campbell, *Wigan Pier Revisited: Poverty and Politics in the Eighties* (London: Virago, 1984), p. 97.
4 A caveat is better mentioned up front: this chapter does not explore informal mineral resource extraction that has been a central plank of my recent feminist critique of mining: Kuntala Lahiri-Dutt, 'The shifting gender of coal: Feminist musings on women's work in Indian collieries', *South Asia: Journal of the South Asian Studies Association* 35(2) (2012), pp. 456–476.

5 Carolyn Merchant, *The Death of Nature: Women, Ecology and Scientific Revolution* (San Francisco, CA: Harper, 1990).
6 Mircea Eliade, *The Forge and the Crucible: Origins and Structure of Alchemy* (Chicago, IL: University of Chicago Press, 1962).
7 Michael T. Taussig, *The Devil and Commodity Fetishism in South America* (Chapel Hill, NC: University of North Carolina Press, 1980).
8 Lewis Mumford, *Techniques and Civilisation* (New York: Harcourt Brace and Company, 1934); and Lewis Mumford, *The Myth of the Machine, Vol 2: The Pentagon of Power* (New York: Harcourt Brace Jovanovich, 1967).
9 Merchant, *The Death of Nature.*
10 Jane Humphries, 'Enclosures, common rights, and women: The proletarianization of families in the late Eighteenth and early Nineteenth centuries', *The Journal of Economic History 1*(3) (1990), pp. 17–42.
11 Agricola Georgius, *De Re Metallica*, translated from the Latin by H. and L. Hoover, (New York: Dover Publications Inc., 1556). This was the first European mining text-book to contain numerous woodblocks depicting women carrying out these tasks. Also see Pamela O. Long, 'Of mining, smelting, and printing: Agricola's "De re metallica"', *Technology and Culture 44*(1) (2003), pp. 97–101.
12 Christina Vanja, 'Mining women in early modern European society', in: Thomas Max Safley and Leonard N. Rosenband (eds), *The Workplace Before the Factory: Artisans and Proletarians, 1500–1800* (Ithaca and London: Cornell University Press, 1993), pp. 100–117.
13 Sally Zanjani, *A Mine of her Own: Women Prospectors in the American West, 1850–1950* (Lincoln, NE and London: University of Nebraska Press, 2006).
14 Such as Ivy Pinchbeck, *Women Workers and the Industrial Revolution, 1750–1850* (London: Frank Cass, 1930).
15 Patricia Penn Hilden, *Women, Work and Politics, Belgium, 1830–1914* (Oxford: Clarendon Press, 1993), pp. 110–111.
16 Kuntala Lahiri-Dutt, 'Digging women: Towards a new agenda for feminist critiques of mining', *Gender, Place and Culture: A Journal of Feminist Geography 19*(2) (2012), pp. 193–212.
17 Peter Alexander, 'Women and coal mining in India and South Africa', *African Studies 66* (2007), pp. 201–22.
18 Kuntala Lahiri-Dutt, and Martha Macintyre, 'Introduction', in: Kuntala Lahiri-Dutt and Martha Mcintyre (eds), *Women Miners in Developing Countries: Pit Women and Others* (Aldershot: Ashgate, 2006), pp. 1–18.
19 Masanori Nakamura, *Technology Change and Female Labour: Manufacturing Industries of Japan* (Tokyo: United Nations University Press, 1994), pp. 15–16.
20 Angela John, *By the Sweat of their Brow: Women Workers at Victorian Coal Mines* (London: Croom Helm, 1980), p. 30.
21 Nancy Forestell, '"And I feel like I'm dying from mining for gold": Disability, gender, and the mining community, 1920–1950', *Labor Studies in Working-Class History of the Americas 3*(3) (2006), pp. 77–93.
22 Melanie Ilič, 'Women workers in the Soviet mining industry: A case study of labour protection', *Europe-Asia Studies 48*(8) (1996), pp. 1387–1401.
23 Jock McCulloch, 'Women mining asbestos in South Africa', *Journal of Southern African Studies 29*(2) (2003), pp. 413–432.
24 Jaci Gier and Laurie Mercier (eds), *Mining Women: Gender in the Development of a Global Industry, 1670–2005* (New York: Palgrave Macmillan, 2006).
25 Suzanne E. Tallichet, *Daughters of the Mountain: Women Coal Miners in Central Appalachia* (University Park, PA: The Pennsylvania State University Press; Urbana and Chicago, IL: University of Illinois Press, 2006).
26 Lahiri-Dutt and Macintyre, 'Introduction'.

27 Ginger Gibson, and Deanna Kemp, 'Corporate engagement with Indigenous women in the minerals industry: Making space for theory', in: Ciaran O'Faircheallaigh and Saleem Ali (eds), *Earth Matters: Indigenous Peoples, the Extractive Industries and Corporate Social Responsibility* (Sheffield: Greenleaf Publishing, 2008), pp. 104–122.

28 Deanna Kemp, Carol Bond, Daniel Franks, and Claire Cote, 'Mining, water and human rights: Making the connection', *Journal of Cleaner Production 15*(18) (2010), pp. 1553–1562.

29 Ciaran O'Faircheallaigh, 'Women's absence, women's power: Indigenous women and negotiations with mining companies in Australia and Canada', *Ethnic and Racial Studies 36*(11) (2013), pp. 1789–1807.

30 Kate Grosser, and Jeremy Moon, 'Developments in company reporting on workplace gender equality? A corporate social responsibility perspective', *Accounting Forum 32*(3) (2008), pp. 179–198.

31 David Trigger, Julia Keenan, Kim de Rijke, and Will Rifkin, 'Aboriginal engagement and agreement-making with a rapidly developing resource industry: Coal seam gas development in Australia', *The Extractive Industries and Society 1*(2) (2014), pp. 176–188.

32 Joni Parmenter, 'Experiences of Indigenous women in the Australian mining industry', in: Kuntala Lahiri-Dutt (ed.), *Gendering the Field: Towards Sustainable Livelihoods for Mining Communities* (Canberra: ANU E Press, 2011), pp. 67–86.

33 Lahiri-Dutt and Macintyre, 'Introduction'.

34 Martha Macintyre, 'Money changes everything: Papua New Guinean women in the modern economy', in: Mary Patterson and Martha Macintyre (eds), *Managing Modernity in the Western Pacific* (St Lucia: University of Queensland Press, 2011), pp. 90–120.

35 Martha Macintyre, 'Women and mining projects in Papua New Guinea: Problems of consultation, representation, and women's rights as citizens', in: Ingrid MacDonald and Claire Rowland (eds), *Tunnel Vision: Women, Mining and Communities* (Fitzroy: Oxfam Community Aid Abroad, 2002), pp. 26–30.

36 Minerals Council of Australia (MCA), *MCA Workforce Gender Diversity Review* (Canberra: White Paper, 2013).

37 Robyn Mayes, and Barbara Pini, 'The Australian mining industry and the ideal mining woman: Mobilizing a public business case for gender equality', *The Journal of Industrial Relations 56*(4) (2014), pp. 527–546.

38 Linda Rhodes, *Two for the Price of One: The Lives of Mining Wives* (Perth: Curtin University Press, 2006).

39 Kathryn Robinson, *Stepchildren of Progress: The Political Economy of Development in an Indonesian Mining Town* (Albany, NY: State University of New York Press, 1986).

40 Julie-Kathy Gibson-Graham, '"Stuffed if I know!" Reflections on post-modern feminist social research', *Gender, Place and Culture 1*(2) (1994), pp. 205–224.

41 Linda McDowell, and Doreen Massey, 'Coal mining and place of women: A case of nineteenth century Britain', in: Doreen Massey and John Allen (eds), *Geography Matters! A Reader* (Cambridge: The Open University, 1984), pp. 128–147.

42 Catherine Pattenden, 'Shifting Sands: Transience, Mobility and the Politics of Community in a Remote Mining Town' (Unpublished PhD thesis) (Discipline of Anthropology and Sociology, School of Social and Cultural Studies, University of Western Australia, 2005).

43 Katy Jenkins, 'Women, mining and development: An emerging research agenda', *The Extractive Industries and Society 1*(2) (2014), pp. 329–339.

44 Thomas Miller Klubock, *Contested Communities: Class, Gender and Politics in Chile's El Teniente Copper Mine, 1904–1951* (Durham, NC: Duke University Press, 1998).

45 June Nash, *We Eat the Mines and the Mines Eat Us: Dependency and Exploitation in Bolivian Tin Mines* (New York: Columbia University Press, 1979), pp. 12–13.

46 Mary Murphy, *Mining Cultures: Men, Women and Leisure in Butte, 1914–41 (Women in American History)*, (University of Illinois Press, IL: Illinois, 1997).

47 Jessica Smith Rolston, *Mining Coal and Undermining Gender: Rhythms of Work and Family in the American West* (New Brunswick: Rutgers University Press, 2014).

48 Kuntala Lahiri-Dutt, 'Gender (plays) in Tanjung Bara mining camp in Eastern Kalimantan, Indonesia', *Gender, Place and Culture 20*(8) (2013), pp. 979–998.

49 Valerie Gordon Hall, 'Contrasting female identities: Women in coal mining communities in Northumberland, England, 1900–1939', *Journal of Women's History 13*(2) (2001), pp. 107–131.

50 Zola, Emil, *Germinal* (G. Charpentier Publishers. Translated into English from the Original French (1885) by Eldritch Press, 1985).

51 Gorky, Maxim, *Mother* (Translated into English from the Original (1906) Russian by Mahaveer Press, New Delhi, 2012).

52 Caroline Waldron Merithew, '"We were not ladies": Gender, class and a women's auxiliary battle for mining unionism', *Journal of Women's History 18*(2) (2006), pp. 63–94.

53 Karen Beckwith, 'Collective identities of class and gender: Working class women in the Pittston coal strike', *Political Psychology 19*(1) (1998), pp. 147–167.

54 Georgina Murray, and David Peetz, *Women of the Coal Rushes* (Sydney: University of New South Wales Press, 2011), p. 288.

55 Jean Stead, *Never the Same Again: Women and the Miners' Strike* (London: The Women's Press, 1987).

56 Kuntala Lahiri-Dutt (ed.), *Gendering the Field: Towards Sustainable Livelihoods in Mining Areas* (Canberra: ANU E Press, 2011) Available from: http://epress.anu.edu.au/gendering_field_citation.html (accessed 9 June 2019).

57 Padmini Sathianadhan Sengupta, *Women Workers of India* (Mumbai: Asia Publishing House, 1960), p. 67.

58 Leela Fernandes, *Producing Workers: The Politics of Gender, Class, and Culture in the Calcutta Jute Mills* (Philadelphia, PE: University of Pennsylvania Press, 1997), p. 38.

59 Lahiri-Dutt, 'The shifting gender of coal'.

60 Jane L. Parpart, 'Class and gender on the copperbelt: Women in Northern Rhodesian copper mining communities, 1926–1964', in: Claire Robertson and Iris Berger (eds), *Women and Class in Africa* (New York and London: Africana Publishing Company, 1986), pp. 141–160.

61 William Hipwell, Katy Mamen, Viviane Weitzner, and Gail Whiteman, *Aboriginal People and Mining in Canada: Consultation, Participation and Prospects for Change* (Working Discussion Paper, North-South Institute, 2002).

62 Nesar Ahmad and Kuntala Lahiri-Dutt, 'Engendering mining communities: Examining the missing gender concerns in coal mining displacement and rehabilitation in India', *Gender, Technology and Development 10*(3) (2007), pp. 313–339.

63 Sharmistha Bose, 'Positioning women within the environmental justice framework: A case from the mining sector', *Gender, Technology and Development 8*(3) (2004), pp. 407–412.

64 K. Bhanumathi, 'Mines, minerals and PEOPLE, India', in: Ingrid MacDonald and Claire Rowland (eds), *Tunnel Vision: Women, Mining and Communities* (Victoria: Oxfam, 2002), pp. 20–24.

65 Robinson, *Stepchildren of Progress*, p. 12.

66 Indira Rothermund, 'Women in a coal mining area', *Indian Journal of Social Science 7*(3–4) (1994), pp. 251–264.

67 Kuntala Lahiri-Dutt, and Petra Mahy, *Impacts of Mining on Women and Youth in Two Locations in East Kalimantan, Indonesia* (2007). Report available from http://empoweringcommunities.anu.edu.au (accessed on 11 September, 2008).

68 *Ibid.*
69 Atu Emberson-Bain, 'Atu. De-romancing the stones: Gender, environment and mining in the Pacific', in: Atu Emberson Bain (ed.), *Sustainable Development of Malignant Growth? Perspectives of Pacific Island Women* (Suva: Marama Publications, 1994), pp. 91–110.
70 Macintyre, Martha, 'Women and mining projects in Papua New Guinea', pp. 26-30.
71 Lahiri-Dutt (ed.), *Gendering the Field: Towards Sustainable Livelihoods in Mining Areas.*
72 O'Faircheallaigh, 'Women's absence, women's power'.
73 Lahiri-Dutt, 'The shifting gender of coal'.
74 Christina Hill, and Kelly Newell. *Women, Communities and Mining: The Gender Impacts of mining and the role of gender impact assessment* (Melbourne: Oxfam Australia, 2009).

9 Accidents and mining

The problem of the risk of explosion in industrial coal mining in global perspective

Michael Farrenkopf

Introduction

In public perception, the mining industry is widely considered as one of the most dangerous industries. This generalised consideration is largely shaped by the public perception of outstanding mining accidents with a high death toll. If we only consider mining of raw materials from underground, there are some fundamental and specific factors which set mining apart from the rest of the economy. Within the earth's surface, mining creates a special work environment that is subject to constant changes. This is a fundamental difference to the processing industry. And the security requirements in mining are not only limited to the actual factors of production. In addition, the whole mining area affected by the degradation of the earth's crust must be controlled in favour of a safe working. Disturbances may occur and go back to geological phenomena that lie outside the immediate space of exploration. On the one hand, mine accidents are related to natural disasters that are generally defined as events with cosmic, biological, geological or meteorological reasons.[1] On the other hand, such events originate in the exploitation of minerals by using technical methods. So basically they are also technical disasters which per definition are caused by human action, based on technical procedure and relevant for technical and economic development.[2]

Mining history mainly in Great Britain and Germany some 40 years ago focused on structural social history and had a significant upswing. Later on it underwent some specific turns concerning for example cultural, environmental or economic topics but a specific accident history of mining has hardly been addressed. Analysing the state of research in Germany and Great Britain, Helmuth Trischler and John Benson argued in the late 1980s that although there was an overwhelming literature on mining history, huge deficits remained concerning questions of illness and death in the course of the mining industry. Regarding the German situation, Trischler demanded that accident history should not strive only for an analysis of the reasons for accidents in a narrow sense, but it should also explore the impact of economic and employment changes as well as the consequences of mechanisation and rationalisation.[3]

More than Trischler, John Benson scrutinised accident statistics in relation to mining safety in 19th century Britain. This work was presented as part of a lecture at the International Mining History Congress in Bochum, Germany, in 1989. He argued that due to absolute numbers, it was known for some time that in 1850–1914 an average of more than 1000 miners were killed annually, albeit with sharp regional differences. However, concerning the enormous growth of the British coal industry in that period, the reference of absolute numbers of deaths would disguise the fact that between 1850 and 1914 the likelihood of a miner being killed in an industrial accident declined to 30%. As one of the most serious problems of the history of British mining accidents up to that time, Benson recognised that the problem of mining safety was almost exclusively seen in terms of fatal accidents – and in particular of major disasters. 'Nonetheless it is easy to be misled. For we know that despite the horror and destructiveness of such disasters, they accounted in fact for only a small proportion of the lives lost in the British coal industry'.[4]

Benson's argument that a complete analysis of mining accidents should not just refer to fatal accidents, let alone to mine accidents with multiple deaths, points to a general problem for an accident history of the mining industry from a global perspective: it is hardly possible to use a correspondingly large statistical dataset for a variety of mining regions around the world as the basis of historical analysis.

Because of this, on the one hand, this chapter will deal primarily with large mining accidents or disasters, because in a global perspective it is only for these hazards that relatively accurate data are available for longer periods of time. On the other hand, there are also substantial arguments for such an approach. A key argument may be that especially larger mining accidents – even if they represent only a small portion of the real number of accidents – most likely not only cause the questioning of national regimes of technical, operational and administrative accident prevention but possibly also lead to relevant or even fundamental changes. My topic is limited in the sense that the chapter focuses mainly on the coal industry and the risk of explosion that in the course of industrialisation since the 18th century very often caused accidents with very high personal and property damage for certain reasons. Therefore we direct our attention now to the north of France at the beginning of the 20th century.

Science and strategies for reducing explosions prior to the 20th century

In the morning of the 10 March 1906, a coal dust explosion devasted the underground mine of the company of Courrières in the region of Nord-Pas-de-Calais. This turned out to be the largest ever catastrophe in European hard coal mining.[5] In the end 1099 miners were killed, many more injured and 13 miners escaped the drastically destroyed mine after a horrific ordeal of 20 days after the explosion. Like every disaster of a catastrophic size, this event was of a singular character. There must have been a complex of explosion triggering factors, which only occurred at this very place and in this very time.

At the same time, we can see the disaster of Courrières – to quote Charles Perrow – as a 'normal catastrophe' of European hard coal mining in its phase of high industrialisation.[6] European hard coal mining in this sense really means all European countries that have been engaged in this sort of mining ever since, although this chapter will concentrate on Great Britain, Germany and France, as those countries coal production reached the most greatest amounts. Besides all the specific circumstances, the explosion was part of a general risk phenomenon that all bituminous hard coal mining had been confronted with in the course of the 19th century. And it remained a great problem even after the case of Courrières. The term 'risk phenomenon' can be seen as an important theoretical approach to incorporate the catastrophe of Courrières into the history of industrialised hard coal mining.

One dimension of the risk phenomenon is primarily of a technical nature. With regard to Courrières, we first of all have to point out that we talk about a certain class of mining risks – namely the risk of explosion. This risk was determined by a large number of technical factors which were each specific to the explosion uncertainty in the period addressed. From this point of view, historical research has to analyse the selected and obviously failing protection drafts of the mine in question in relation to the general and structural level of explosion protection.

This is a methodically demanding approach because it requires interdisciplinary proceedings. Especially explosion protection in mining was and is not characterised by technical measures alone, but also integrates chemical and physical conditions as well as geological influences. The approach is moreover demanding because it has to consider current, advanced knowledge, which is very different to the level of understanding in the period of the catastrophe observed. It might be misleading therefore to judge historical actors in the light of today's knowledge.

Recent, mostly social history research on mining – especially in Germany – argues that the extremely catastrophic explosions on the eve of the First World War are evidence of an accident risk that had steadily increased in the course the 19th century. In these works, first and foremost the explosion disasters are seen as a result of an economically liberalised mining system, in which security requests were sacrificed to profit interests. This argument was based on the analysis of contemporary sources of the mining workforce, but the interpretation has to be more differentiated. The lack of historically critical reflection led to the fact that the contemporary perception of the accident risk by the miners, who were most directly concerned by accident consequences, is taken as an objective perception of the risk of explosions. This however is incorrect insofar as the explosion disasters of this time period do not reflect the trend of an unbroken intensified explosion risk. To explain this we have to deal with the development of the structural means of explosion protection in the 19th century.

Hard coal mining in Great Britain, Germany and France was aligned in giving priority to the prevention of firedamp ignition. However, it remained highly disputed among mining specialists that the coal dust, which deposited itself underground in the course of the coal-mining process and was responsible for the expansion of the explosion in Courrières, had at all been a part of the explosion risk. Mining explosion protection therefore was determined by two basic strategies: first, sufficient ventilation was to be guaranteed in order to keep methane gas in the pits below the critical ignition boundaries. The second strategy was based on the inspection of the ignition sources, especially the lamps of the miners and the shot firing used for drifting during building of the pits underground.

Both strategies entailed two security concepts. One pursued the establishment of a system of control, which was supposed to compensate individual breaches of safety measures by the introduction of specialised monitoring technology. This security concept aimed at the development of self-regulating technology that would be able the avoidance of the ignition of an explosion without human influence. After 1816, we find a well-known example of this strategy in the miners' safety lamp invented by Sir Humphrey Davy. The national firedamp commissions established in some European countries in the second half the 19th century concentrated particularly on this security concept.

In spite of evident progress during the 19th century, the safety lamps as well as safety explosives reached only a very limited constructive security level. Therefore it was necessary also to regulate the operation of these technical means by the miners. The leading idea of this second security concept – that we can call operational security – was the establishment of regulations. Although regulation practice differed throughout Europe in terms of legal traditions, this strategic duality for explosion protection was a common approach.

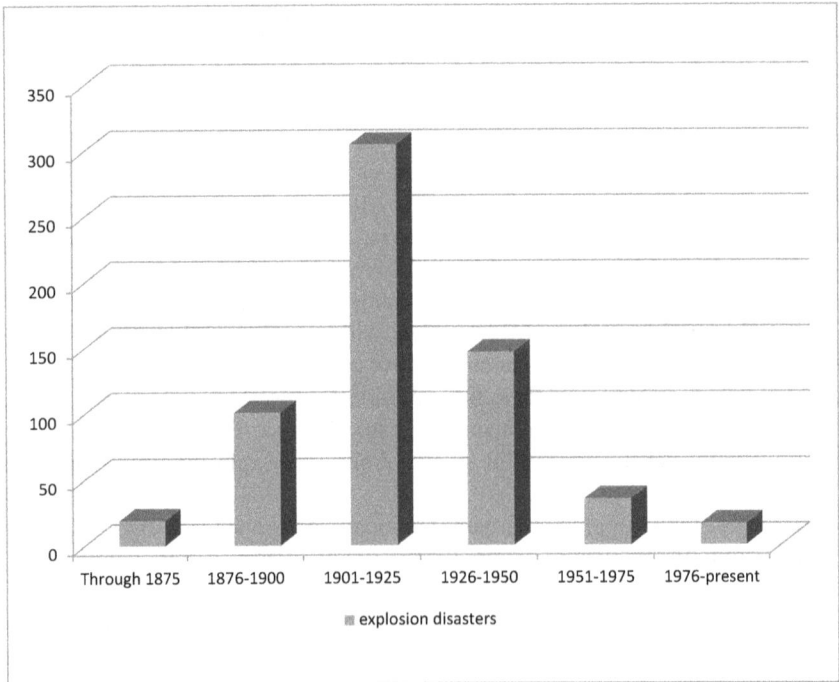

Figure 9.1 **Number of explosion disasters (where five or more deaths occured) in the USA, 1800–2013** (www.usmra.com/saxsewell/historical.htm)

As the industrial growth of hard coal mining gained momentum after the middle of the 19th century, the established concepts of explosion protection increasing failed. The decisive reason was that above all, the ventilation became critical because of the growth of pit expansion. As a consequence, the problem of the appearance of mine gas, which could not be dispersed by ventilation, massively intensified and could not be compensated by the highly unstable concept of ignition control. The final result was a massive increase of the explosion risk, both in the number of explosion events occurring per year as well as in the number of miners who were injured or killed in each explosion.

In principle, this development can also be shown for the US coal industry. According to statistics from the United States Mine Rescue Association, the number of explosions with a minimum of five miners killed developed as shown in Figure 9.1.[7] The statistics show, first, that the number of explosion disasters in the United States grew strongly until the beginning of the 20th century, only to decrease again after the 1920s. It is also striking that the number of explosion disasters in the period 1926 to 1950 was still higher than in the period 1876 to 1900. As shown in Figure 9.2, the development of explosion disasters in US hard coal mining correlates with the number of mining personal, which also was at its peak in the middle of the 1920s.

On the basis of available statistical data, it can be stated that a large part of the disasters in the phase of increasing risk of accidents in the United States correspond to mine gas explosions. Apart from the Avondale anthracite coal mine disaster of the Luzerne Corporation in Plymouth, Pennsylvania, where 110 miners died on

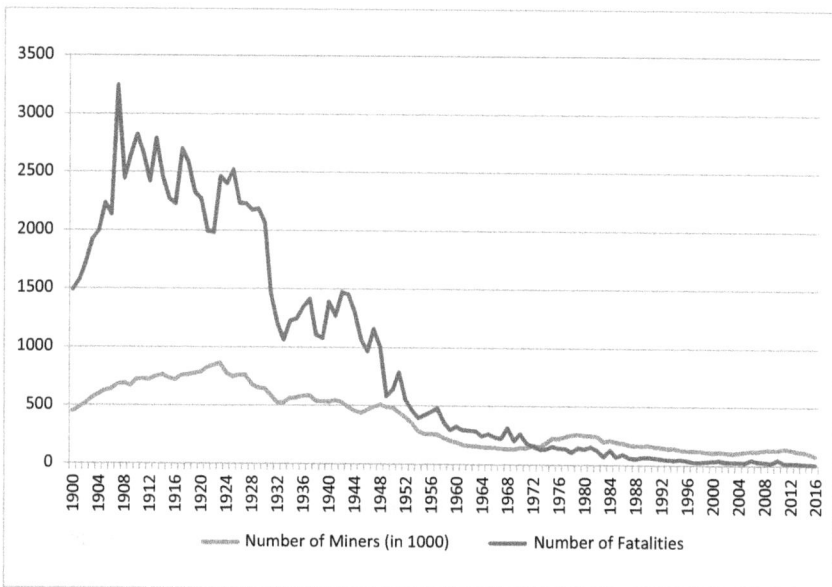

Figure 9.2 **Number of miners and fatalities in the USA, 1900–2016** (www.usmra.com/ saxsewell/historical.htm)

6 September 1869 because a fire broke out on the surface, the fatal accidents after 1890 were almost exclusively mine gas explosions. This was also true for the Hill Farm Mine in Dunbar, Pennsylvania, when 31 miners were killed on 16 June 1890, followed by 112 miners who lost their lives on 10 July 1902 in the Rolling Mill Mine in Johnstown, Cambria, Pennsylvania. Only two years later, on 25 January 1904, 179 miners died in a coal mine gas explosion, which was caused by shot firing at the Harwick mine in Cheswik, Pennsylvania.[8]

We may also recognise a trend in the increasing size of the disasters measured in the number of fatalities per single event that continued until 1907. With 3242 miners being killed altogether, 1907 turned out to be the year with the highest death rate in the history of the US mining industry (see Figure 9.2). In addition to a wide range of explosion accidents, the catastrophic death rate was a consequence also of the disaster in No. 6 and No. 8 pits at Monongah, West Virginia, in which a total of 362 miners were killed on 6 December 1907. This catastrophe was the worst ever mine disaster in the United States.

The period of major disasters in the 'old' mining countries in the early 20th century

If we compare the Monongah catastrophe with the hazards in British, German and especially French coal mining, the similarities to the worst disaster in Courrières that occurred only one year earlier are striking. In detail, however, some differences can be seen that feed the hypothesis that the level of explosion protection in Europe was already more advanced than in the United States.

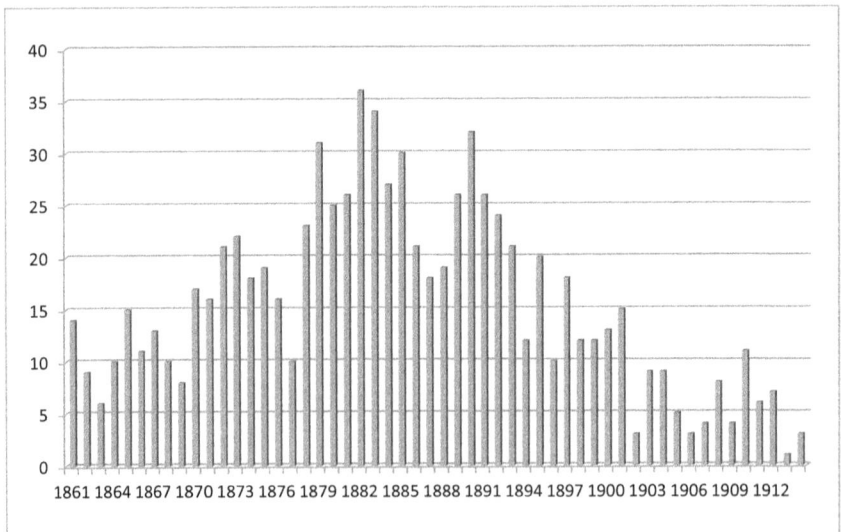

Figure 9.3 **Number of fatal explosions in Prussian hard coal mining, 1861–1914**
(from Michael Farrenkopf[9])

Again based on statistical data we can assess that the trend of an unrestricted intensification of the explosion risk was reversing in Germany already around the middle of the 1880s. This is certain with regard to Prussian hard coal mining (see Figure 9.3), and a similar trend can be ascertained for the other European countries. This was primarily a success of the national firedamp commissions being established in many European countries after the end of the 1870s.

These commissions, which were formed as a rule by technical experts of the governmental and private sectors, concentrated their investigations on the self-regulating technical means of explosion protection. As a central topic they searched for a better understanding of the chemical and physical characteristics of mine gas. Furthermore they aimed for rational criteria to measure the ventilation systems as well as to understand the constructive boundaries of the technical means in the case of danger. The extensive experimental programs that were carried out mostly in established laboratories helped to determine the degree of relative security according to the usual contemporary procedure of 'trial and error'. The increasing scientific knowledge and the constructive recommendations had a decisive impact. Nevertheless they were only able to optimise the relative degree of the self-regulating protection known at the time. Especially in the field of ignition inspection, explosion protection therefore remained an extremely unstable system for many decades.

It cannot be doubted that the extensive test programs of the European firedamp commissions after the 1880s led to substantial progress in the scientific understanding of the risk factors concerning the explosion problem. And their effectiveness was even more a result of imbedding this scientific progress into reformed security regulations. Only in this way was the optimised scientifically constructed knowledge implemented into operational regulations. As a result of the Prussian firedamp commission for example, the mining authority decreed a new and very complex so-called weather regulation in 1887/88, which was similar to the regulation of ventilation standards and mining safety of 1895 in France.[10] As a consequence of this considerable change in explosion protection efforts, the number of disasters per year started to diminish by the 1890s.

Compared with the achieved level of explosion protection in the most well-developed European hard coal mining countries at the beginning of the 20th century, the accident in the American mine in Monongah on 6 December 1907 reveals some significant shortcomings. Not surprisingly the persons that had been responsible for mine safety at Monongah stated in the aftermath of the disaster that the mine had been reviewed constantly with safety lamps, but critically gas content had not been assessed. This would have been the job of several fire bosses to fulfil this task responsibly. While they also argued that no appreciable coal dust existed, the miners, however, claimed the exact opposite, which seems to be more likely.[11]

From a systemic point of view it is especially important that in contrast to so-called firedamp mines for example in Germany, in Monongah, despite the general awareness of the existing explosion risk, all the miners were working with open-flame lights that gave them no protection at all when an explosive gas mixture occurred. In several European countries this practice was legally prohibited after the second half of the 19th century.

Also the scientific understanding of the physical, chemical and natural conditions of the explosion problem seemed to be little developed. In Europe it was known at least since the last quarter of the 19th century that the explosion potential of mine gas is only substantial if the proportion of methane in the mine air ranges between approximately 4% and 15%, and in particular strong ventilation of the mine reduces the risk of explosion. In Monongah, however, the persons responsible for mine safety believed that too much air in a mine would provide the oxygen needed by the methane gas to lead to a blow-up. Interestingly, there was a rumour that because the Monongah mine was idle the day before the explosion, the fans had also stopped working and thus providing a means of protection.[12]

In addition, the Monongah mine did not have any organisational structures for the rescue of the miners after the explosion. Following the explosion, the coal company employed a troop of doctors from Fairmont to report to Monongah. It is said that they 'stood around bonfires all night waiting to administer to survivors', but none ever came. Since the rescue work needed to be done by volunteers and not organised teams, the Fairmont Coal Company, as the owner of the mine, stationed a man at each pit entrance to escort every man who entered and in turn to check him as he came out.[13]

Although Monongah remains the worst catastrophe, in the following years there were frequent mining accidents with enormous fatal consequences. With 19 hazardous events in coal mines and six in metal and nonmetal mines, 1910 was the year of the largest number of major mine disasters in the USA overall. And very similar to Europe, Monongah also experienced forms of disaster tourism. The great public interest and an enormous reverberation in the media created huge public pressure, which finally led to a fundamental reform of the administrative control of security requirements in US mining. In 1910, Congress created the Bureau of Mines, 'whose primary roles were to investigate accidents, advise industry, conduct production and safety research and teach courses in accident prevention, first aid and mine rescue'.[14] By institutionalising such practice, the United States caught up with a century-old legislation tradition of mining that had been much more common in the old world. In Prussia and the German Empire, for example, this had led to a strong bureaucratic supervisory practice of the mining industry since the 18th century.

The development of an orderly rescue and education system concerning the risks of mining production in the newly formed US Bureau of Mines had clear parallels to Europe. Regarding the technical means, particularly in respiratory equipment, a transfer of technology began because such special technical means had been developed and tested mostly in Europe since the end of the 19th century.

In Germany it was also the consequence of a mine disaster that caused a reform of the administrative safety regulation as well as of the mine rescue service similar to that in the United States. On 12 November 1908 there had been an explosion in the Radbod colliery in the eastern Ruhr area, which led to the highest death toll the German coal industry had experienced until that date. A total of 350 miners died because of a firedamp and coal dust explosion that was probably sparked by an open-flame lamp.[15] All rejoicing that prevailed in Germany after the helpful intervention of German mine rescue teams in the explosion disaster in Courrières could not hide the fact that German coal mining was still vulnerable to explosions with even worse consequences.

Already under the influence of Courrières, a discussion about the participation of miners' representatives in the security control of mines had started in Germany. In collaboration with the supervising mining authority, mining entrepreneurs had quite successfully fought such claims for many years as an intrusion into their operational autonomy. In fact, the specific problems of the mining explosion risk in this phase turned out to be so complex that the mere participation of the miners would have had little effect on safety control. In the eyes of the workers and the labour movement, however, the very fact that the German entrepreneurs strictly avoided the possibility of miners' participation underlined their view of being neglected for ultimately political reasons.

In this respect, the explosion of Radbod came as a shock. At the political level in Germany, there had been intensive discussions after Courrières about the question of workers' participation in mine safety control. Facing the disaster at Radbod and with respect to the political pressure of the German workforce, the mining authorities got on the defensive. In the spring of 1909, the resistance against the miners' interest in security control could no longer be maintained. As a result, in summer 1909, a reform of the existing legal framework in Germany came into effect.[16] The Mining Act amendment conceded that so-called security men (*Sicherheitsmänner*) should be chosen from the workforce to be involved in safety control. At first their skills were limited so narrowly that they could hardly gain influence. Nevertheless, this was a deep cut in mining safety monitoring previously carried out strictly by entrepreneurs and state officials.[17]

Only a few months later – and again very similar to the situation in the US – selected representatives of the mine owners in the Ruhr area formed a commission that was supposed to provide a new basis for mine rescue in the major German coal district. It was the task of this commission to develop proposals for a systematic ordering of the hitherto uncoordinated rescue services in the Ruhr district in order to increase their efficiency and effectiveness. As a result, a central office for mine rescue was established as an organisation for the Ruhr mining industry in summer 1910. The main tasks of the office included the examination of existing and new equipment for the mine rescue service, training and monitoring of the mine rescue teams and the establishment of a standardised rescue plan for the district.

Comparing Germany and America, on the eve of the 20th century similar structures for technological and organisational accident prevention to reduce the risk of explosion are obvious, however, one significant difference cannot be overlooked. Whereas in Germany mining legislation gave state officials an extensive influence on the design of security matters, the US Bureau of Mines was given no inspection authority until 1941. Only then did Congress empower federal inspectors to enter mines, and in 1947, it authorised the formulation of the first code of federal regulations for mine safety.[18]

Improvements in the 'old' mining countries after World War II

Looking at the accident statistics of the coal mining industry in Western Europe and the United States one can undoubtedly state that the risk of accidents during the 20th century diminished constantly and significantly. Great progress has been made in the period after the Second World War. The reasons for this are complex and cannot be pursued in detail here. Thus, for example, the absolute number of fatal accidents decreased in the British coal industry from about 600 in 1946 to 63 in 1978. In the same period the number of serious, non-fatal accidents decreased from about 2500 to just below 500. So while the number of fatal accidents was reduced to about 10% within three decades after the Second World War, the number of non-fatal accidents declined at least to a level of about 20%.[19]

This significant reduction in the risk of accidents some authors attribute to the nationalisation of British mining industry in 1947,[20] but similar developments can be tracked on a statistical basis with respect to other mining regions in the Western world. Within the context of a general increase in labour safety, the risk of the occurrence of coal mine gas explosions in the former heartlands of industrialisation has greatly reduced, so that in this way the number of mine disasters also decreased. For example, this can be shown regarding the coal mining industry in the Federal Republic of Germany in the second half of the 20th century (see Figure 9.4).[21] Nevertheless, this does not mean that the risk of explosions in the highly mechanised coal mines of the old industrial countries can be excluded

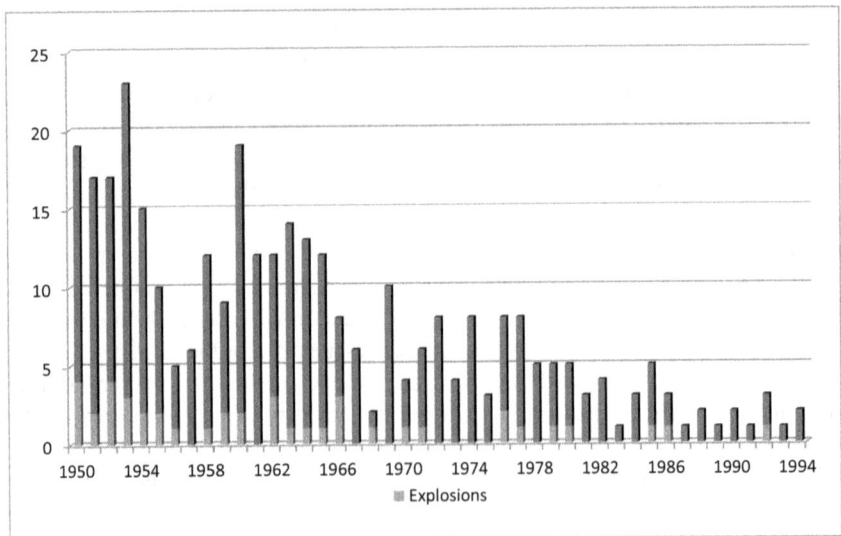

Figure 9.4 **Mining disasters (two or more miners killed) in the Federal Republic of Germany, 1950–1994** (from Michael Farrenkopf, *Schlagwetter und Kohlenstaub*, p. 98)

entirely even today. For the US, all accidents with five or more fatalities since 1970 amounts to 21 major accidents, of which 15 were explosions. The last disaster took place on 4 April 2010 at the Upper Big Branch mine of the Massey Energy Company of Montcoal, West Virginia, which resulted in 29 dead miners.[22]

Nevertheless, significant progress is shown statistically, even of the risk of explosion, probably because of the following reasons. First, the framework for an international solution to the problem of coal mine gas was significantly improved in Western Europe by the transition to the European Coal and Steel Community (ECSC). A major factor was a sustainable development of scientific understanding of the natural foundations and factors of mine gas problems, especially during the 1950s and 1960s. Only then, scientific methods led to a relatively reliable prediction of the natural gas load of a minefield.

A second factor was closely linked to the new scientific knowledge of coal mine gas. It was the development of a primarily technical process of so-called mine gas extraction. In this context, from the beginning the motives for explosion avoidance were linked with economic considerations. On the one hand, methane was to be considered as a gaseous raw material. This, on the other hand, offered a new rational to approach minefields that had been left aside previously due to high gas loading. Therefore these technical procedures are often summarised under the term of mine gas extraction. Such new ways of extraction of the gas before it could pass through degradation or rock into the underground ventilation corresponded to a profoundly new strategy of risk prevention. In contrast to the measures used for centuries of diluting mine gas below the critical limit of explosion by ventilation, the new approach aimed for a more general reduction of the possibilities of methane gas in the underground system.

A third factor can be seen in the close link between partial or full mechanisation of technical devices. These were supplemented by complex and increasingly automated test procedures which, despite mine gas extraction, helped to detect occurring methane concentrations very early. In case of danger, this resulted in an automatic shutdown of the production machines. All this was finally connected to a continuous adaptation of the technical regime of accident prevention to modern forms of organisation and management as well as increasing the qualifications of the workforce.

Mining explosions in the 'new' mining countries in the late 20th century

In recent decades, the risk of explosion disasters has mainly shifted to the coal mining industry in the former Soviet and now Commonwealth of Independent States and especially in China. The Asia-Pacific economy has become the main carbon market in the world, and the largest coal producers are now based in India, China, the United States, Australia and South Africa.[23]

For quite some time the world's largest producer of coal has been China, with 35% of global coal output produced in China in 2013. However, its safety record is devastating. About 80% of all fatal accidents in the global coal mining industry take place in China. According to a statistical comparison of coal mine fatality rates measured by number of deaths per million ton mines between 1990 and 2009, the number of deaths in China was far above India, South Africa and the United States (see Table 9.1).[24] We may assume that in the last few decades, both in South Africa and India, the accident rate in the mining industry has been more successfully reduced than in China. However, precise data for comparative studies over longer periods of time obviously do not exist.

South Africa and India

Concerning South Africa, we know that from 1981 to 1995, some 70 methane explosions resulted in the deaths of miners. South Africa's cheap coal was not only due to its relative abundance, but also a reflection of low wages, questionable attention to mine safety and the long-term health of mineworkers. Historically, black mine labour has been cheap in the South African coal mining industry. In one incident in 1926, the entire night shift of 125 men was killed. Nevertheless ongoing debates about mine accidents and high death rates in South African mines resulted in frequent public stand-offs, and the South African minister of minerals and energy promised to tighten up on health and safety regulation. From

Table 9.1 Coal mine fatality rates (measured by number of deaths per million tons), 1990–2009[25]

	China	*USA*	*South Africa*	*India*
1990	6.10	0.07	0.53	0.28
1991	5.21	0.07	0.48	0.24
1992	4.65	0.06	0.65	0.23
1993	4.78	0.06	1.57	0.25
1994	5.15	0.05	0.96	
1995	5.03	0.05	0.53	
1996	4.67	0.04	0.75	
1997	5.10	0.03	0.72	0.54
1998	5.02	0.03	0.73	0.47
1999	5.30	0.03	0.51	0.45
2000	5.86	0.03	0.54	0.43
2001	5.07	0.03	0.38	0.41
2002	4.94	0.02	0.44	0.27
2003	3.71	0.02	0.45	0.30
2004	3.08	0.02	0.53	0.24
2005	2.81	0.02		0.25
2006	2.04	0.04		0.28
2007	1.05	0.03		0.14
2008	1.18	0.02		0.15
2009	0.89	0.01		0.11

1984 to 1993, on average 69 coal miners died, and on average 508 experienced injuries in South African coal mines per year. Within 15 years, the rates improved significantly. According to official statistics, there were only 15 working fatalities reported in 2007–2008.[26] In the period 2006 to 2010, only 3.5% of underground hazards in South African coal mining were caused by explosions, probably due to the low gas content in coal in that region.[27]

For about a century, the growth of Indian coal mining that started in the 1770s remained sluggish for want of demand but the introduction of steam locomotives in 1853 gave a fillip to it. Within a short span, production rose to an annual average of 1 million tons and India could produce 6.12 million tons per year by 1900 and 18 million tons per year by 1920. Production got a sudden boost from the First World War but went through a slump in the early 1930s. The staggering increase in the late 1920s and 1930s in colliery roof collapses, coal dust and gas explosions, fire and inundation brought about a thorough examination of the problem and public demand for comprehensive statutory control of the industry in order to avert reckless methods and shortsighted production agendas in Indian mining. The Burrows Coal Mining Committee constituted by the Government of India in 1936 made a comprehensive investigation into the reasons and consequently a series of regulations and amendments to the Indian Mines Acts was stipulated in the late 1930s.[28] From that time, the fatality rate per 1000 persons, which had reached almost 1.4 in the mid-1930s, started to decrease consistently. Official statistics show that average fatality rates and the number of serious accidents have been coming down in India. Figures from the Directorate General of Mines Safety suggest that the annual fatality rate was 0.36 in 2012 and 0.21 in 2016, taking into account the total number of mines in India.[29]

With the advent of Independence, India embarked upon five-year development plans. On account of the growing needs of the steel industry, a boost had to be given to the systematic exploitation of coking coal reserves in Jharia Coalfield. Adequate capital investment to meet the burgeoning energy needs of the country was not forthcoming from the private coal mine owners. Unscientific mining practices adopted by some of them and poor working conditions of labour in some of the private coal mines became matters of concern for the government. On account of these reasons, the central government took a decision to nationalise the private coal mines.

Nationalisation was done in two phases, the first with the coking coal mines in 1971–1972 and then with the non-coking coal mines in 1973. The mines were brought under the management of Coalmines Authority Ltd and merged in the holding company Coal India Ltd in November 1975. According to statistics of the Ministry of Coal, five-yearly average fatal accidents have reduced from 157 in 1975–1979 to 59 in 2010–2013 (i.e. 62%) and the five-yearly serious injury rate per million tons of coal produced has even reduced from 14.24 to 0.51 (i.e. 96%) in the same period.[30] Nowadays, the safety status of coal mines is said to be continuously monitored at different levels. At the mine level the safety committees, where workmen are represented, review the safety status of every mine. Workmen's inspectors make inspections of the mines and the reports of the inspectors are rigorously acted upon. Bipartite committees of management and workmen's representatives review the safety status of each and every area. At the subsidiary company headquarters level, tripartite committees which include representatives of the Directorate General of Mines Safety (DGMS) also review the safety performance of the company and suggest measures for further improving safety standards.

The special case of China

A considerable number of accidents in China are explosion disasters. However, this cannot simply be explained by a general backwardness of the Chinese coal mining industry in security matters. At the end of this chapter, we will therefore try to formulate some hypotheses to explain this problem.

From the 1950s to the 1970s, the Chinese coal mining industry was shaped by state-owned enterprises (SOEs), because Mao's governmental strategy of industrialisation largely followed the model of the Soviet Union. In the late 1970s about 95% of China's, coal production originated in SOEs. Starting in the late 1970s, the Open Door Policy ushered in an era of privatisation, decentralisation and modernisation. Entering the 1980s, soon after Deng Xiaoping launched economic reforms, these had a profound impact on Chinese society, creating, on the one hand, a rising quality of life and incomes, but also contributing significantly to rising inequalities. In these years, the Chinese coal industry underwent a significant structural change that led into decentralisation. After the State Council had issued a series of directives that set policy guidance for expanding coal production by non-state enterprises, the government also offered some generous state support through credits, subsidies and investment in rural infrastructure.[31]

After the 1980s, the central government of China began to encourage coal production wherever possible and by whatever means. In a policy directive issued in 1985 it declared 'state, collective and private enterprises are equally welcome. Mines of all sizes, large, medium or small, can operate side by side'.[32] It is estimated that at the beginning of the 1980s approximately 10,000 town and village coal mines were in operation. By the end of the decade, this number had increased tenfold to 100,000.[33] By the end of 1996, China's operational coal mines had reduced to 84,000, of which 81,000 were small town and village enterprises with an annual output of less than 300,000 tons. Compared to the United States, total coal production was similar to China's in 1997, but the US had only 2200 operational coal mines.[34]

Since the implementation of modern safety standards in coal mines is expensive, it is hardly surprising that especially the mostly private Chinese town and village enterprises could not afford their application. Accident statistics for the 1990s show that the differential in fatality rates between non-state and state mines in China remained over 5 to 1 and briefly approached 10 to 1 in the early 2000s. In 2001, 90% of fatalities came from non-state controlled or illegal mines, which were contributing almost 50% to total coal output in China in 1996.[35] The serious problems concerning workers' safety and long-term health issues in small town and village mines were reinforced by the fact that their exploitation rate was approximately not more than 10 to 15%, compared to the larger state-owned coal mines, which reached generally around 50%.[36] Therefore, coal-mining licenses had created significant opportunities for corruption in the local economy, as well as a readiness to undergo a comprehensive, sophisticated and complex national legal regime dealing with coal mine safety from the 1990s onwards.[37]

Especially with regard to the risk of explosion in the Chinese coal industry, the enormous growth of mostly private small mines in the 1980s and 1990s had numerous adverse effects also for the larger SOEs. But in contrast to the small mines, explosion protection methods were applied there since the 1980s that largely corresponded to European and US standards. To address the explosion risk, ventilation systems had been modernised significantly in about half of the state-owned mines by the end of the 1980s. Until the mid-1990s, China had equipped 110 state-owned mines with central monitoring systems that handled, for example, mine gas concentrations, data on current ventilation, carbon monoxide levels and the operating status of relevant technical devices. Very similar to the industrialised countries of the West, these mines also focused on the introduction of preventive mine gas extraction.[38]

However, these costly measures put considerable pressure on the state-owned mines during the 1990s, because the prosperity of the small mines indirectly reduced the competitiveness of large state-owned mines. In contrast to the boom of the small mines, the state-owned mines withered from lack of funding. The increase in coal production from the small mines created a surplus on the market and depressed the price of coal. At that time, the state-owned mines desperately needed capital to explore new deposits, to update technologies and to improve safety infrastructure. Instead, they were told to be financially independent and not to count on the central government for help, even with regard to social welfare that once was shouldered by the state. This obviously had detrimental consequences for the safety regime of the state-owned mines during the 1990s.

Since the turn of the century, new aim of Chinese coal policy has been to strengthen the productivity of the large state-owned mines. To regulate the coal industry, the Chinese government has withdrawn coal-mining licenses for many small mines and has stopped issuing licenses to those market entrants since the second half of the 1990s.[39] From 1995 to 2010 more than 85% of town and villages mines were shut down; in 2009 they accounted for only 3% of Chinese national output.

In the meantime, high compensation standards for mining accidents have forced mine owners to invest more in safety facilities and technical training of employees. Compared to the revenue gained from mining, reparations before 2000 were affordable. Therefore there was no incentive for mine owners to improve production technology and safety or to provide vocational training of mine employees. Since 2001, China's mortality rate in coal mining has declined rapidly by 80%.[40] In 2014, the ratio of deaths per million metric tons of coal mined fell to 0.24, as calculated from official reports.[41]

Conclusion

Mining has always been a hazardous undertaking, no matter the period and no matter which form of extraction and raw material we might consider. Due to the problem of reliable statistical data for a history of mining from a global perspective this chapter has concentrated on the risk of firedamp and coal dust explosion, which very often led and still lead to disastrous mine accidents.

There is no doubt that by the beginning of the 20th century, a thoroughly differentiated system of explosion protection had been established in hard coal mining. Concerning the massive growth of European hard coal mining after about 1850, the strategies and concepts of explosion protection all over Europe were concentrated only on the prevention of the initialisation of explosions and not on their limitation underground. A phase of decreasing security to the middle of the 1880s was followed by a phase of optimisation of protection measures in Europe. In spite of the continuous growth of the mines, explosion protection had already attained an improved effectiveness, although explosion security as a whole remained a very unstable system. This was especially the case in the United States, where explosion protection caught up with the European standard only from the beginning of the 20th century onwards.

The lack of any means to avoid the expansion of explosions underground with the consequence of a further increasing number of victims per each disaster dogged the efforts and successes of prevention strategies. This was a common problem throughout European mines and even more so in global hard coal mining up to the beginning of the 20th century. This was shown by a number of explosion disasters that happened very shortly after the catastrophe of Courrières in France. Especially the firedamp and coal dust explosion of Radbod in Germany in 1908, which killed around 350 miners, revealed as wishful thinking German contemporary beliefs that the industry was well prepared in the field of explosion protection. Nevertheless, these catastrophic disasters not only in Europe, but also in the United States, triggered basic organisational and technological restructuring of mining rescue to avoid explosions of mine gas and coal dust in the Western world in the early 20th century.

While the risk of mine explosions in Europe and the United States was reduced substantially in hard coal mining after the Second World War, it has now shifted mainly to the less-developed countries. China faced this problem in recent decades mainly because of a widely uncontrolled and enormous growth of its coal mining industry. However, as argued, this cannot simply be explained by a general backwardness of the Chinese coal mining industry. China stopped its uncontrolled growth of hard coal mining in the late 1990s by reducing the number of small mines in order to strengthen the productivity of the large state-owned mines. In this context, new organisational structures and the appliance of modern standards of explosion protection in recent years have led to a significant decline of the mortality rate in Chinese coal mining.

Notes

1 Helmut Lackner, 'Technische Katastrophen und ihre Bedeutung für die technische Entwicklung. Ein Überblick', *Ferrum – Nachrichten aus der Eisenbibliothek 69* (1997), pp. 4–15, here p. 4; also Andreas Schmidt, *'Wolken krachen, Berge zittern, und die ganze Erde weint . . .'. Zur kulturellen Vermittlung von Naturkatastrophen in Deutschland 1755 bis 1855* (Münster et al.: Waxmann, 1999).

2 Lackner, 'Technische Katastrophen', pp. 4f.

3 Helmuth Trischler, 'Arbeitsunfälle und Berufskrankheiten im Bergbau 1861 bis 1945. Bergbehördliche Sozialpolitik im Spannungsfeld von Sicherheit und Produktionsinteressen', *Archiv für Sozialgeschichte 28* (1988), pp. 111–151.

4 John Benson, 'Mining Safety and Miners' Compensation in Great Britain 1850–1900', in: Klaus Tenfelde (ed.), *Sozialgeschichte des Bergbaus im 19. und 20. Jahrhundert/ Towards a Social History of Mining in the 19th and 20th Centuries. Beiträge des Internationalen Kongresses zur Bergbaugeschichte Bochum, Bundesrepublik Deutschland, 3.–7. September 1989/Papers presented to the International Mining History Congress Bochum, Federal Republic of Germany, September 3rd–7th, 1989* (Munich: C.H. Beck, 1992), pp. 1026–1038, here p. 1027.

5 Michael Farrenkopf and Peter Friedemann (eds), *Die Grubenkatastrophe von Courrières 1906. Aspekte transnationaler Geschichte* (Bochum: Deutsches Bergbau Museum, 2008); *10 mars 1906: la catastrophe des mines de Courrières . . . Et après? Actes du Colloque européen organisé par le Centre historique minier du Nord-Pas-de-Calais à Lewarde les 9, 10 et 11 octobre 2006* (Lewarde: Centre Minier, 2007).

6 Charles Perrow, *Normale Katastrophen. Die unvermeidbaren Risiken der Großtechnik* (Frankfurt am Main, New York: Campus, 2nd edition, 1992).

7 <www.usmra.com/saxsewell/historical.htm> (accessed 1 November 2013).

8 Gerald E. Sherard, *Pennsylvania Mine Accidents 1869–1972* (November 2011), <www. genealogy.com/users/s/h/e/Jerry-Sherard/> (accessed 1 November 2013).

9 Michael Farrenkopf, 'Le risque d'explosion dans les mines de houille à la fin du (long) XIXe siècle. Aspects d'un problème européen', in: *10 mars 1906: la catastrophe des mines de Courrières . . . Et après? Actes du Colloque européen organisé par le Centre historique minier du Nord-Pas-de-Calais à Lewarde les 9, 10 et 11 octobre 2006* (Lewarde: Centre Minier, 2007), p. 34.

10 *Ibid.*, pp. 30–40.

11 Lacy A. Dillon, *They Died in the Darkness* (Ravencliff, West Virginia: McClain Print Company, 1976), p. 73.

12 *Ibid.*, p. 73.

13 *Ibid.*, p. 78f.

14 Andrew W. Homer, 'Coalmine Safety Regulation in China and the USA', *Journal of Contemporary Asia 39*(8) (2009), pp. 424–439, here p. 433.

15 Friedrich Menneking, *Radbod 1908. Rückblick auf die große Explosion und andere Explosionen im Steinkohlenbergbau* (Dortmund: Bellmann 1984); Olaf Schmidt-Rutsch and Ingrid Telsemeyer (eds), *Die Radbod-Katastrophe. Berichte und Zeichnungen des Einfahrers Moritz Wilhelm* (Essen: Klartext, 2008); Michael Farrenkopf, 'Die Radbod-Katastrophe von 1908 – Dimensionen des Explosionsrisikos im Ruhrbergbau des Kaiserreichs', *DER ANSCHNITT 61* (2009), pp. 330–344.

16 'Gesetz, betreffend die Abänderung des Allgemeinen Berggesetzes vom 24. Juni 1865/1892 und 14. Juli 1905. Vom 28. Juli 1909', *Zeitschrift für das Berg-, Hütten-und Salinenwesen im Preußischen Staate (ZBHSW) 57* (1909), part A, pp. 107–116.

17 Trischler, 'Arbeitsunfälle und Berufskrankheiten', p. 120.

18 Friedrich-Wilhelm Wellmer, 'Bergbehörden und Grubensicherheit in den USA', *Glückauf 106* (1970), pp. 224ff; Fritz Schuermann and Leopold von Buch, 'Der amerikanische Steinkohlenbergbau im Aufschwung', *Glückauf 102* (1966), pp. 389–397.

19 K.-H. Rauch, 'Das Unfallgeschehen im britischen Steinkohlenbergbau 1978', *Glückauf* *116* (1980), p. 1175.

20 J.L. Collinson, 'Grubensicherheit im britischen Bergbau', *Glückauf 117* (1981), pp. 510ff.

21 Michael Farrenkopf, *Schlagwetter und Kohlenstaub. Das Explosionsrisiko im industriellen Ruhrbergbau (1850–1914)* (Bochum: Deutsches Bergbau Museum, 2003), p. 98; Evelyn Kroker and Michael Farrenkopf, *Grubenunglücke im deutschsprachigen Raum. Katalog der Bergwerke, Opfer, Ursachen und Quellen* (Bochum: Deutsches Bergbau Museum, 2nd Edition, 1999), pp. 442–509.

22 <www.usmra.com/saxsewell/historical.htm> (accessed 1 November 2013); <www.usmra.com/saxsewell/Upper_Big_Branch.htm> (accessed 1 November 2013).

23 Stefan Siemer and Jens Scholten, 'Weltmarkt Kohle – Produzenten und Verbraucher', in: Ulrike Stottrop (ed.), *Kohle.Global. Eine Reise in die Reviere der anderen. Katalog zur Ausstellung im Ruhr Museum vom 15. April bis 24. November 2013* (Essen: Klartext, 2013), pp. 156–159.

24 Yibing 'Bing' Wu, *'Blood Coal': A Flawed Development Strategy behind the Death Toll in Chinese Coalmines* (Jackson School of International Studies, University of Washington, May 2008), p. 2.

25 C. Chu, R. Jain†, N. Muradian†, G. Zhang, 'Statistical Analysis of Coal Mining Safety in China with Reference to the Impact of Technology', *The Journal of the Southern African Institute of Mining and Metallurgy 116* (2016), pp. 73–78; South Africa Mine Accident Statistics, www.sacollierymanagers.org.za/docs/oh&s04_107%20Key%20 OH&S%20issues%20June%2004_Att.pdf (accessed 26 March 2017).

26 Victor Munnik et al., *The Social and Environmental Consequences of Coal Mining in South Africa. A Case Study* (a joint initiative of Environmental Monitoring Group, Cape Town, South Africa and Both ENDs, Amsterdam, The Netherlands, January, 2010), pp. 13f.

27 Jill Harris et al., 'Comparative Analysis of Coal Fatalities in Australia, South Africa, India, China and USA, 2006-2010', *14th Coal Operators' Conference, University of Wollongong, The Australasian Institute of Mining and Metallurgy & Mine Managers Association of Australia* (2014), pp. 399–407, here p. 403ff.

28 Dhiraj Kumar Nite, 'Slaughter Mining and the "Yielding Collier": The Politics of Safety in the Jharia Coalfields 1895–1950, in: Kuntala Lahiri-Dutt (ed.), *The Coal Nation: Histories, Ecologies and Politics of Coal in India* (Farnham, Surrey: Ashgate, 2014), pp. 105–128.

29 Anil Sasi, 'After Jharkhand Toll, 2016 One of Deadliest Years for Mine Workers', (http://indianexpress.com/article/india/after-jharkhand-toll-2016-one-of-deadliest-years-for-mine-workers-4456477 (accessed 27 March 2017).

30 Government of India, Ministry of Coal, *Annual Report 2013–14*, pp. 83–88, www.coal.nic.in/welcome.html (accessed 29 December 2014).

31 F. Su, 'The Political Economy of Industrial Restructuring in China's Coal Industry, 1992–1999', in: B. Naughton and D. Yang (eds), *Holding China Together* (Cambridge/ UK, 2004).

32 Wu, 'Blood Coal', p.11f.

33 *Ibid.*, p.12.

34 K. Pan, J. Pu and T. Xiang, 'Comparative Research of the Coal Market Concentration Degree between China and America', *Management World 12* (2002), pp. 77–88 (in Chinese).

35 Tim Wright, 'Small Mines in the Chinese Coal Industry', *Working Paper No. 80* (June 1998), National Library of Australia, p. 4.

36 F. Wang, 'Development Problems and Policy Analysis of Small Coalmines in China', *Journal of China University Geosciences (Social Science Edition) 6* (2006), pp. 61–67 (in Chinese).

37 Homer, 'Coalmine Safety Regulation in China and the USA', p. 429.
38 Chen Minghe, 'Grubensicherheitswesen im Steinkohlenbergbau Chinas', *Glückauf 132* (1996), pp. 157–160; Li Dingjian, Zhang Shecan and Helmut Schönfeld, 'Wandel im Steinkohlenbergbau Chinas', *Glückauf 124* (1988), pp. 931–935.
39 Stefan Siemer, 'Der Mensch als Arbeitsmaschine', in Ulrike Stottrop (ed.), *Kohle. Global*, pp. 204–207, here p. 206.
40 Hangtian Xu and Kentaro Nakajima, 'The Role of Coalmine Regulation in Regional Development', *PRIMCED Discussion Paper Series 45* (2013), p. 20.
41 Michael Lelyveld, 'China Cites Rise in Coal Deaths Despite Drive to Cut Output', www.rfa.org/english/commentaries/energy_watch/china-coal-08152016150446.html (accessed 25 March 2017).

10 On fatalities, accidents and accident prevention in coalmines

Colliers' safety discourse in oral testimony from the Ruhr in Germany and the Witbank collieries in South Africa

Paul Stewart and Dagmar Kift

Introduction

All coalfields share certain basic conditions. In the 1980s, Klaus Tenfelde noted these conditions as the primeval nature of production (i.e. the dependency on geology), the need for large investments, similar work processes and a 'close relationship between working and living'.[1] While Tenfelde perhaps overestimated the extent of the similarities of work processes when viewed over time and across locations, the high risk and frequent occurrence of mine disasters resulting in fatalities and accidents could well be added as a further condition applying to all coalfields. The question of worker safety in mining, generally measured in terms of the frequency of fatalities and accidents in collieries, however, 'was and is a very complex and difficult one' as the historian of the Natal coalfields in South Africa, Ruth Edgecombe, well appreciated.[2]

As part of its legislative and regulatory functions regarding oversight of their respective mining industries, states have long recorded both fatalities and accidents in mining as much as the coalmining industry globally has made continued attempts to improve its safety statistics – in both Germany and South Africa as elsewhere. For the colliers in the industry, however, fatalities, accidents and safety at work are an underlying condition of life and work. Their perceptions, views and attitudes regarding recorded statistics and the attempts of state, industry and trade unions at making mining safer feature prominently in two sets of recollections of colliery life and work, which is the subject of this chapter.

What consequently follows here compares and contrasts two sets of chronologically overlapping oral testimonies as told by colliery workers on two different continents. By gleaning from what colliers had to say about mining accidents and safety underground on the Ruhr and around Witbank and setting these accounts alongside one another, we sought to explore whether they would tell us something of the global experience of coalmining and if so, in what respect they might do so.

Coalmining at Zollern on the Ruhr in the heart of Europe and the Witbank collieries on the southern tip of Africa overlap by nearly a century. Deep-shaft coalmining in the Ruhr Valley goes back to the beginning of the 19th century and picked up speed in the 1850s. Coalmining in the Witbank area goes back to the last decade of the 19th century, established itself in the first few years of the 20th century and continues to be the most prominent coalfield in South Africa. While coalmining finally completely ceased in the Ruhr Valley by 2018, around Witbank opencast coalmining in particular, as well as a slew of smaller companies operating in the wake of new legislation, continues apace.

The oral evidence from the Zollern colliery on the Ruhr was collected in the 1980s as part of a broader project by the LWL-Industriemuseum in Dortmund to preserve something of the legacy of the members of an industry that had powerfully contributed to the economic development and consequent wealth of the Ruhr and beyond.[3] Workers' accounts from the Witbank collieries were collected in this century to give individual voice to black African workers in a new South African democracy, recently liberated from the travails of a racially divided society reintegrating itself into the global economy.[4] The two sites epitomise the stark contrasts between developed and developing societies, which today dominates the globe. Which basic conditions, we asked, do they share in terms of accidents, fatal and otherwise, accident prevention and safety? And how were these conditions reflected and perceived by the miners as told in their stories?

Regarding both the Ruhr and Witbank, attention is given to state legislation and interventions by the coalmining companies, which frame the experience of the miners themselves. Accidents and safety are the specific focus used to explore the similarities and contrasts between coalmining and workers' experience at these two sites. Despite the contrasting nature of the two coalfields, the similarities are suggestive of continuity and a shared experience between colliers worldwide and over time. The contrasts, however, strongly suggest that the extent to which colliers are compelled to face the risks and dangers which attend coalmining and which often result in fatalities and accidents, are powerfully shaped by geological and technical factors as well as by social context, state legislation and the interventions around safety by coalmining companies, unions and workers' representatives. What emerges from comparing the collective experience and attitudes of colliers towards both danger and safety under different conditions is that the raising of colliers' voices speaks strongly to the broader institutional narrative of how worker safety in coalmining globally can be enhanced.

Coalmining in the Ruhr area

Major pit disasters did not happen in the Zollern colliery. When the mine came into being at the turn of the 20th century, it was built according to the latest state of the art. It boasted one of the first electrically driven main shaft winding engines, a social policy of control and care and the insistence on adhering to safety precautions. As Theo Rittterswürden, head of the rescue team between 1945 and 1966, emphasised, Zollern colliery had a perfectly working ventilation system which, among other things, eliminated the risk of methane gas explosions. In fact, his team was rescuing miners in other collieries. Their first mission was the Grimberg pit disaster (a firedamp and coal dust explosion) in Bergkamen near Dortmund in 1946, which cost the lives of 406 miners. Rittterswürden, who was born in 1912, was a miner from 1931, a deputy (*Steiger*) from 1943 and was made head of the Zollern rescue team in this year. In his testimony, he describes the conditions in which he and his men had to work in the Grimberg mine:

> It was so draughty – all the ventilation doors were destroyed – that we had to crawl on our stomachs and hang on to the rails. Had we tried to stand up we would have been flung through the air like pieces of paper. At the stable shaft, we saw the first dead men lying on the timbers.[5]

Ritterswürden's narrative is unsentimental and displays the mental condition he and his men had to adopt in order to be able to do their jobs. Emotion only comes in when he mentions that he did not lose a single man during his 20 years in office.

The number of pit disasters, mostly firedamp and/or coal dust explosions, had risen during the second half of the 19th century when shafts were sunk deeper and the production of coal and methane gas increased. In order to make mining safer, the Prussian Firedamp Commission was set up in 1881, consisting of technical experts from the government and mining companies. As Michael Farrenkopf points out in this volume, prevention eventually was directed towards ventilation (to keep the methane below its ignition point) and against potential ignition sources (such as lamps and explosives). In both cases, Farrenkopf argues, more safety was hoped to be achieved by constructive and operational measures, i.e. by improving the equipment and by issuing rules and regulations how to handle it properly. The number of firedamp explosions eventually fell, but the number of fatalities per disaster rose since mines and workforces grew bigger. More had to be done – and was done – especially when disasters revealed specific problems which would then subsequently be addressed. This pattern also occurred in the South African mining industry, as will be seen.

After the firedamp and coal dust explosion in the Radbod colliery in 1908 (350 casualties), which was probably sparked off by a lamp, lamps fuelled by petrol were prohibited at Radbod and consequently in other coalmines as well. The disaster also initiated the setup of a testing site where explosions and explosives were studied, after which the latter were made safer. After a firedamp explosion in Dortmund's Minister Stein colliery in 1925 (136 casualties), a second testing site was set up to test explosions and fires under realistic conditions underground, analyse them and develop remedies.[6] In the Grimberg pit disaster in 1946 most miners had not died in the explosion itself but suffocated in the carbon monoxide fumes which developed afterwards. Consequently, CO-self-rescuers were introduced and became part of the miners' equipment.

While, as noted above, the Zollern colliery was not hit by any major disaster, 'standard' accidents happened frequently. It is them, above all, which are recollected in the testimonies of the Zollern workforce. How did miners and deputies from the Zollern colliery recall the accidents and the fatal disasters of their time? What incidents and developments do their narratives refer to? What do these narratives and developments tell us about the history of accident prevention in German coalmining during the 20th century? These are the questions the following paragraphs will try to answer. Longer passages from an interview with a former safety officer will illustrate how accident prevention actually worked – or was made to work – in the Zollern colliery during the 1950s.

Some of the testimonies emphasise how quickly accidents could happen – and end fatally. Erich Adam, for example, recounts the death of his brother in an accident in 1937. Adam's brother was trying to help some colleagues solve a technical problem when the iron props collapsed on top of him.[7] Adam emphasised that you often did not see or hear 'it coming', for example, when coal wagons were approaching. 'You do not hear anything because the axle bearings are so well greased'.[8] At that time protective clothing had not been introduced and made obligatory. Wilhelm Brand suffered a basila skull fracture in 1944 when he got hit by a loose part of the props: 'There were no helmets at the time. We were wearing just an old cap or an old hat. I passed out straight away and was taken to Langendreer by horse and cart', i.e. to a hospital which continues to specialise on miners' accidents and diseases, about 8 km from the Zollern colliery. Brand was on sick-leave for five and a half months.[9]

Faulty equipment could also be dangerous, as Friedrich Steinhoff recalls. During his time, a faulty miner's lamp exploded underground and the lamp's alkaline injured two men and a deputy. Under these circumstances, it was important to be vigilant, familiar with one's environment and careful. Ignorance and carelessness could quickly lead to the loss of a finger or more serious injuries by rock fall, which were often fatal.[10]

The need for constant vigilance is why educating the miner in safety matters became an important element of the general safety package. Miners had to know the rules and regulations, be aware of potentially dangerous situations – and know what to do and what not to do. Experience alone was not considered sufficient in the 20th century, which is why accident prevention became part of vocational training when the latter was systematised from the 1920s onwards.[11] Together with their miner's certificate, hewers were given a copy of the book *Der Bergmannsfreund* (The Miner's Friend). It consisted of numerous before-and-after-accident pictures, the accidents themselves caused by miners being too ignorant, too daring or too careless. This was intended to raise the miners' awareness of potentially dangerous situations. Various kinds of awareness training were continued in the colliery and even in the miners' homes. Posters reminded the workforce of potential dangers and how to avoid them and articles in the company's magazine made accident prevention a family issue. The magazine was supposed to be read by the miner's wife as well and therefore also contained articles tailored towards her supposed interests, such as cooking recipes, educational and gardening advice. It also made her part of the safety package. In fact, one sometimes cannot help but get the impression that it was, above all, the miner's wife on whom safety underground depended. In an open letter, published in 1961 in the company's magazine, the mine's doctor argues that accident prevention begins at home. Miners' wives should provide a happy home so that their husbands were not distracted by anger and frustration during their shifts, that they should not drive their husbands to overwork themselves so that they could spend more money – and possibly more than their neighbours – and that they should make sure that their husbands got enough sleep during the night.[12] Some of these issues were certainly raised with a reason. But the magazine – being a company magazine – reflects above all the management's conviction that accidents were not only or not so much caused by technical or organisational risk factors (which were managerial responsibilities), but by 'human failure' on the part of miners.[13] This refrain will be seen to recur time and again in the testimonies of miners in our two contexts. The immediate difference of course is that the wives of indigenous African migrant mine workers in South Africa were not privy to, let alone appealed to by the companies in the interests of safer mining.

German miners (and their families) did, of course, have an interest in a safe work environment and had tried for decades to have a say in safety matters and accident prevention.[14] After the Radbod disaster of 1908, the question of including miners' representatives in safety matters – a question already raised after the mining disaster of Courrière two years earlier – came back on the agenda. Despite objections by the entrepreneurs, the Mining Act Amendment of 1909 introduced safety representatives (*Sicherheitsmänner*) from within the workforce who were to be involved in accident prevention in their respective deputy's underground department (*Revier*).[15] This did not go down too well with the deputies since safety matters were their responsibility. It was, in fact, the deputy who was accountable to the state authorities in safety matters. The amendment thus caused internal rivalries as well as creating new structural problems. First, the deputy was the superior of the safety representative who was a member of his team. Second, each of the two people now responsible for accident prevention served a different master and were torn between conflicting loyalties and objectives. Deputies where not only responsible for safety, but also had to meet certain production goals. Safety representatives were also workers' representatives. Demanding improvement in safety matters from the management was one thing; reporting fellow miners for violation or neglect of safety rules another. This did not advance the development of a safer work environment. Perhaps not surprisingly, the safety representative system was abolished in 1920.[16]

After the Second World War the mining companies began to introduce a different model and created the so-called safety officer (*Sicherheitsbeauftragter*). In 1953 the new Mining Police Bye-Law (*Bergpolizeiverordnung*) demanded that the mines employ full-time officers responsible for accident prevention and health protection above and below ground.[17] The new model avoided the problem of conflicting loyalties. The safety officers were chosen from the ranks of the deputies because of their technical experience, knowledge of safety regulations and authority over miners and acceptance by other deputies, but being full-time safety officers, they were not responsible for an underground district and its production quota as well. Furthermore (and unlike the former safety representatives), safety officers had the power to act and intervene straight away if anything or anybody undermined safety underground.[18] As will be seen, while white production supervisors in South African mines were responsible for safety from 1911, it would take until 1996 for black worker safety representatives to be able to command the equivalent degree of authority regarding safety underground.

Probably the first safety officer in the Zollern mine was Friedrich Nordmeier, who was born in 1908, was a hewer from 1939, a deputy from 1942 and safety officer between 1952 and 1958. Nordmeier left us a detailed account of his duties regarding accident prevention and accident documentation, his routines – and his tricks. His testimony is one of the few sources which gives us an insight into the practical implementation of the mine's safety regime.

Having been one of the lads himself, as well as acting as a person of authority, helped Nordmeier to communicate safety measures to miners and deputies who were both aware of the dangers underground but who, according to his testimony, were not always terribly cautious. For them Nordmeier developed a rhetorical strategy which was as gentle as it was overpowering – and which brought him the nickname 'itinerant preacher'. Nordmeier relates that:

> I did not shout, I did it quietly, from around the back. I can remember one hewer whose soul I really kneaded. He had taken a whole pile of coal out without setting provisional props. So I would ask him: 'Everything taken out already?' – 'Yes.' I would say: 'Why don't you sit down for a bit?' So we would sit down and I would say: 'How are things at home?' – 'Fine.' – 'Family?' – 'Yes.' – 'Children as well?' – 'Yes.' – 'Well, it's Christmas soon, isn't it?' – 'Yes.' – 'And your wife? Knitting jumpers for the children?' – 'Yes.' I would say: 'Look, you are on your shift now and your family is at home. Can you imagine me ringing at your doorbell and asking your wife: 'Are you Mrs. So-and-so? My name is Nordmeier, I am the safety officer of the Zollern colliery. I am sorry to tell you that your husband has had a heavy accident.' You are well on your way to make me do this job.' He never neglected his props again.[19]

This approach seems to have been very efficient. But if he had to, Nordmeier could also read the riot act:

> Smoking underground was forbidden. Had I caught somebody smoking there I would have gone straight to the telephone and informed the management. Then he would have had to go . . . instant dismissal, criminal charge! Just think what could happen: a firedamp explosion followed by a coaldust explosion. We had to control those who went underground. That's what the state authorities prescribed. Then we were standing by the shaft: 'Well, any tobacco products?' That's what we did every three months.[20]

Another one of Nordmeier's regular tasks was to check each district underground in turn – a task which brought the deputies to his attention. Again he did his job quietly. To avoid unpleasant situations for both parties, Nordmeier rang the respective deputy the day before he was coming.

> I did not want to grab him by the neck and he did not want me to find too many faults since I had to report them all to the management. If despite all that something was not quite as it should be, I noted it down in my little book and told the deputy . . . and checked it again eight days later. In most cases it was done straight away.[21]

In the end, Nordmeier got done what needed to be done without having to involve himself and others in arguments and paperwork. But he does give the impression that he was perhaps a bit lenient towards his fellow deputies and by doing so also prevented the mine and its management to get into the bad books of the state authorities. A representative of the miners might not have been as lenient.

Miners' representatives did try to get involved again in accident prevention during the 1950s and based their claim on the new Law on Co-determination of 1952. The law demanded, among other things, the creation of a director for social and personal affairs in the mining companies. The new director was to be provided by the workforce and put on the same footing as the director for technical affairs and the director for sales and marketing. Since accident prevention was structurally connected both to social and to technical affairs, it took quite some time until the new director got his foot into safety matters as well. What helped in the end was that the new directors also put quite some emphasis on awareness training and developed a range of successful incentives such as exhibitions, prize competitions and bonuses which went down much better than the previous long lists of 'don'ts'. At the same time the new directors insisted that accident prevention also needed structural improvements and better equipment such as helmets and safety boots.[22]

With regard to firedamp and coal dust explosions, Grimberg with 406 fatalities was the last major firedamp disaster on the Ruhr. Here, the number of firedamp-related deaths fell to 76 in 1950, 41 in 1955, 31 in 1962, 17 in 1968 and 7 in 1979 respectively, while output more than quadrupled from 1,208 kg per person and shift in 1946 to 5,254 kg per person and shift in 1994.[23] As Figure 10.1 shows, in terms of fatalities and injuries, mining in West Germany was made safer in the late 1920s and permanently from the late 1950s onwards. Firedamp explosions kept occurring, but declined in frequency, as Figure 10.3 demonstrates, indicating that the constructive and operational measures developed since the late 1900s seem to have worked in the long run. Figure 10.2 confirms this long-term development by showing that the number of accidents related to output also went down. However, Figure 10.2 also demonstrates that the number of accidents related to the number of mines went up, and Figure 10.3 illustrates the reason: while firedamp accidents more or less disappeared, rockfall and spillage accidents increased in numbers during the 1960s and even more so in the 1970s, when mechanisation took off and longwall mining with coal ploughs and shearers replaced getting the coal by pneumatic pick-hammers. In 1957, 25% of the coal was won by machines, in 1990, 100%.[24] Longwall mining with machines meant that the props at the coal face disappeared, potentially destabilising the roof.[25] Mechanisation thus posed a new challenge to mining safety in the second half of the 20th century, but did not

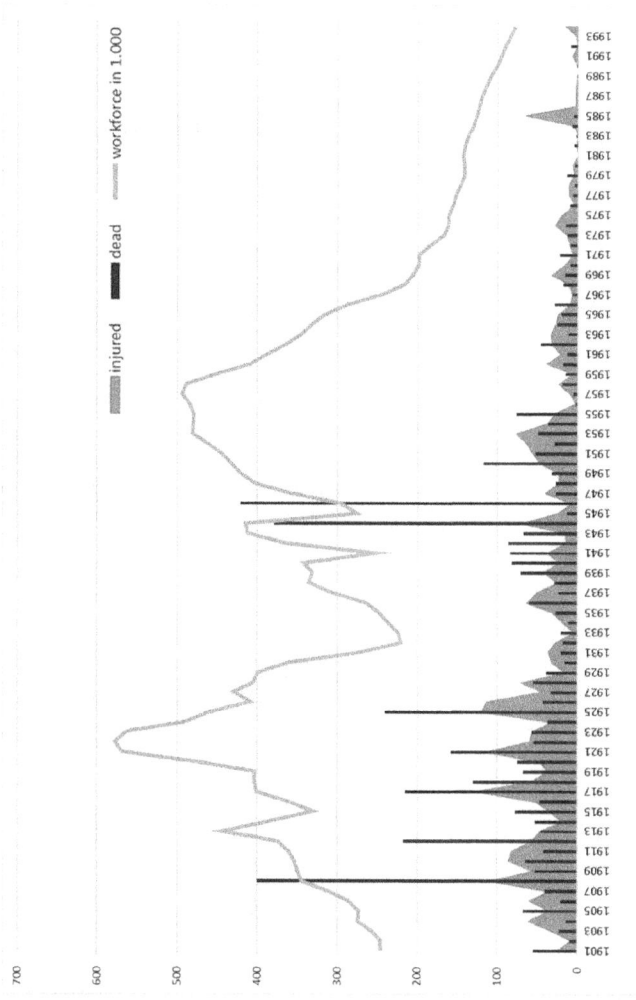

Figure 10.1 Workforce, dead and injured in accidents, Ruhr 1901–1994 (The underlying figures for the number of mines and workforce in Figures 10.1 to 10.3 are from Joachim Huske, *Die Steinkohlenzechen im Ruhrrevier: Daten und Fakten von den Anfängen bis 2005* (Bochum: Deutsches Bergbau-Museum Bochum, 2006, 3rd ed.). I would like to thank Michael Farrenkopf and Stefan Przigoda from the German Mining-Museum Bochum for providing a very helpful list and summary. The figures for the dead and injured from 1900 to 1994 are from Kroker and Farrenkopf's compilation of mine accidents (Evelyn Kroker and Michael Farrenkopf, *Grubenunglücke im deutschsprachigen Raum. Katalog der Bergwerke, Opfer, Ursachen und Quellen* (Bochum: Deutsches Bergbau-Museum Bochum, 1999), pp. 93–509), their definition of a mine accident being an event resulting in at least two people dead or three people injured which is in accordance with the definition of the German mining authorities (*ibid.*, p. 12))

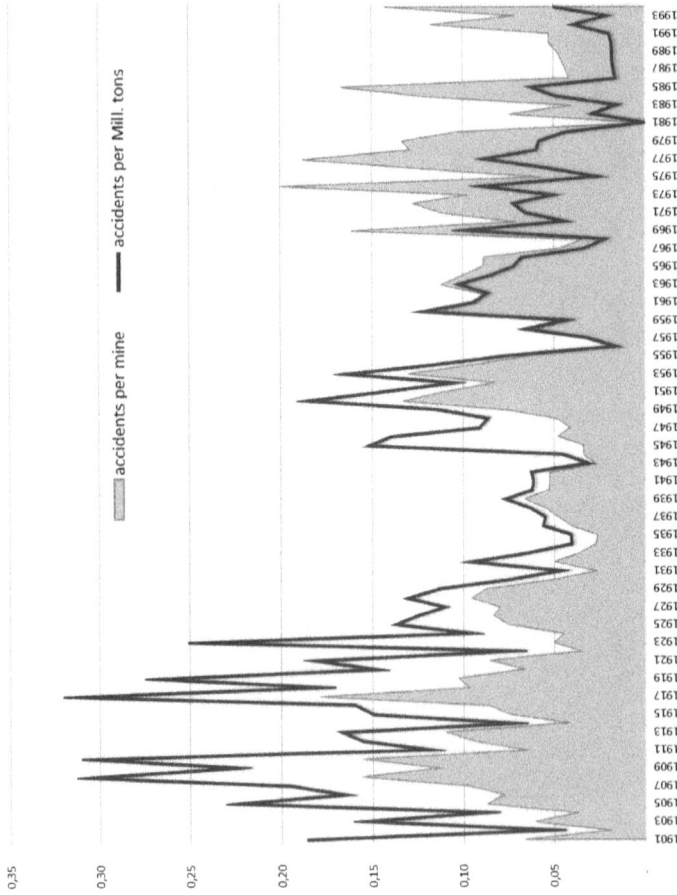

Figure 10.2 Accidents related to output and to number of mines, Ruhr 1901–1994

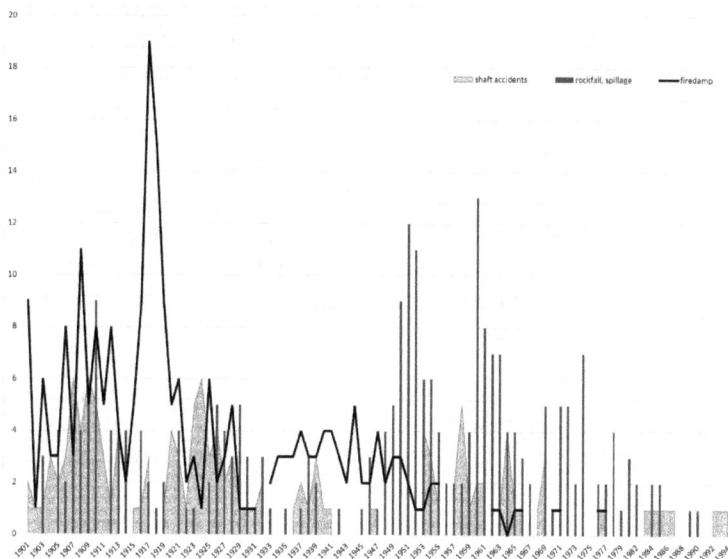

Figure 10.3 Type of accident, Ruhr 1901–1994

result in an increase of the fatality and injury rates: rockfall and spilling accidents were local and happened in places where machines had replaced many men and the men still working there were better protected in term of equipment (helmets, safety boots) as well as thoroughly trained and supervised in accident prevention. With the introduction of shield-support-systems and their implementation during the 1990s, both the accident rate per mine and the number of rockfall and spillage accidents finally went down as well.[26]

Coalmining in the Witbank area

If miners interviewed on the Ruhr had to cast their minds back to a working life they once knew, for those on the collieries around Witbank mining remains central to everyday life. While coal was first discovered in the British colony of Natal in 1838 and noted in 1868 in the areas around Witbank, coalmining in South Africa had to await the discovery of gold in 1886 to find a viable commercial market. The Natal Mines Law (No. 34 of 1888) accompanied the birth of the coal industry but had 'clearly become inadequate' a decade later with the first explosion occurring in 1891, with nearly 40 further explosions in collieries by 1909 causing 135 deaths.[27]

With coalmining established by 1895 around the town of Witbank, founded in 1903 by the Neumann's Witbank Colliery,[28] the surrounding coalfields emerged within a few years as South Africa's premier source of coal.[29] Somewhat like the Ruhr, at this colliery modern technology was quick to be installed. The compressed-air Sullivan 'coal-puncher' machines reduced the number of indigenous black African coalmining labourers on the Witbank Colliery from 1100 to 800.[30] Yet, as on the Ruhr, modern technologies were introduced alongside traditional forms of work as pit ponies only eventually retired from the modern Zollern colliery in the 1950s and from Coalbrook Collieries and one other colliery in the Transvaal in 1960.[31] Even candles, for a long time the main source of light in mining, were still in use[32] in the 1950s, even though the carbide lamp was introduced into some collieries as early as 1907. It was in this year, however, that the Mining Regulations Commission began to sit and which became the basis of the Mines and Works Act (No 12 of 1911)[33] and provided the legislative framework for mining safety until the contemporary era in 1996.

After the Union of South Africa in 1910, the government-appointed Coal Dust Commission of 1912 required all mines in the Witbank area – 'a decade ahead of the United Kingdom'[34] – to apply either stone dust or water down working areas to prevent firedamp explosions, but this was reversed after a court challenge in the 1940s by Douglas Colliery management as the mines in the then province of Transvaal were considered to be 'non-fiery'.[35] By 1920 ventilation fans would be installed in almost all collieries and shot firing by electricity was universal.[36] By the 1940s, black attendants of coal cutter machines were wearing rubber boots which they were required to pay for themselves.[37] In racially segregated South Africa in 1943, black African workers had to appeal to their immediate white supervisor in terms of the Mines and Works Act if they found 'a place that looks dangerous'. Seeing this limited right was being denied on the gold mines, as the president of the African National Congress (ANC), Dr A.B. Xuma, averred to the Native Mine Wages Commission; this would certainly also have been the case on the collieries.[38] If the Natal coal owners were 'conservative' regarding safety until the 1950s,[39] so were those in the Transvaal as the modern coal industry was 'taking shape'.[40] Regarding awakening the awareness of safety specifically, while initiated on the South African gold mines in 1915, an Inter-Mines Prize Scheme awarded to the mine with the lowest accident fatality rate, only appears to have been instituted on coalmines in the 1950s.[41]

An anecdote from the literature speaks to the largely informal character of safety considerations on the coalmines in these years. John Buti Mashiame, who started work on the collieries in 1954, tells how workers were not permitted to kill rats, but instead were required to take evidence that a rat had eaten what was a favourite bread (*mponyani*) taken underground and be paid out its price on the grounds that: 'If you see rats running away, you must run away as well, because the roof might collapse and you might die underground'.[42] By this time, in a space of 50 years, the coalmining work process, defining the nature of work and its attendant conditions regarding safety, appears to have been transformed. According to a stern supporter of the industry, in the bord and pillar method of mining which dominated the South African coalfields '"hand-got" mining with picks had given way to electric cutters, and hand drilling to electric drills; hand tramming had given way to animal power, in turn replaced by electric power and other methods of transport', as well as the mechanisation of loading.[43]

Whatever the unevenness of technologies employed on the South African coal-fields, there is no doubt that since the mid-1950s, when our informants Philemon Mamuhjele, Sechaba Matiase and John Masilela refer to accidents resulting in dressing cases (minor injuries), 'lost time' incidents (serious injuries) or fatalities, safety continued to improve significantly over the next quarter of a century. It was only after the largest ever disaster in South African coalmining at Coalbrook in 1960, resulting in 437 deaths, that more formal and systematic research was begun into safety on the country's collieries.[44] The Coalbrook disaster directly led the state and the industry to establish the Coalmining Research Controlling Council (CMRCC) in May 1961, but on which no colliery employees were rep-resented, and nor was the continuity of research maintained.[45] The Collieries Research Laboratory (CRL) was established in 1966.[46] It conducted important research around methane monitoring and other mine safety-related issues, with this period also being marked by a 'remarkable improvement in living and work-ing conditions' for all colliers.[47] The decade saw further technological changes in underground coalmining such as continuous miners and longwall shearers. While longwalling did not take root in the collieries as elsewhere internationally,[48] the continuous miner revolutionised coalmining, being capable of 'cutting, drilling, blasting, loading and even roof supporting', and was first installed at Coalbrook in 1957.[49] Certainly by this time, according to the first chief inspector of mines under democracy, Dirk Bakker, most of the then used personal protective cloth-ing (helmet, boots, overalls, gloves, safety glasses and ear protection) was being issued by the mines. Coalmines nevertheless remained hazardous working places, especially for disenfranchised and unorganised workers.

Closely mirroring workers' experience of safety on the Ruhr in at least four respects, Philemon Mamuhjele, who started at Greenside colliery in 1961 and worked in mine hospitals until 2004, tells how:

> There were accidents underground. I used to undertake this safety thing that 'You need to be careful and check everywhere in order to be safe. Yes, that you must close here and there and wear suitable clothes'. Some even died. Some died, truly. Management blamed the person who was injured. They also wanted to protect themselves. At the end of the day, they presented the [evi-dence for the] accident and injury and it looked like you were just not clever, that is why you got injured. They presented it to look like it was the fault of workers, whereas there was nothing that workers could have done. When workers got injured and management blamed them, as workers we did not say anything. Those were difficult times. The union did not exist then.

With only white supervisors being able to decide whether a working place was safe or not, Sechaba Matiase, a migrant worker, oscillating not only between home and mine but also between asbestos mines from 1972 and coalmines from 1976 to 2012, tells how at Douglas Colliery near Witbank:

> Yes, management blamed workers for injuries borne. They thought we were stupid if we injured ourselves. They would say: 'Why would you not see and prevent yourself from falling in an incident that might kill you?' If we reported something they would say 'You are not a supervisor. There is a supervisor for this place and he will see it'. But there were few safety arrangements made by management. There was only a supervisor of the place who was in charge and who could report problems such as the roof not being stable or if machinery was not working.

For Matiase, mining has always been dangerous and his narrating of witnessing accidents underground on both asbestos and coalmines is graphic and disturbing. On one occasion after having been helpless to stop a moving winch from crushing to death an already injured miner in the late 1970s and having been told to say nothing, 'within plus one or two hours after that he was put away and I was soon compelled to go on pulling production with that winch'.

Despite the early introduction of advanced technologies in the 1920s and South African coalmines being the most globally productive by the 1940s,[50] in 1970 half of South African coal was still 'essentially won by hand' by being 'cut, drilled and blasted', while in fully mechanised mines, the 'coal was virtually untouched by hand'.[51] On the Ruhr, the pneumatic pick-hammer was widely introduced in the 1920s, and in 1928 nearly 90% of the coal was won by it.[52] In 1967 80% of the coal was won by machines.[53] In South Africa, industry safety research continued, but seemingly in episodic bursts, with the Fuel Research Unit, for instance, conducting explosibility tests based on equipment developed by the Westfälische Berggewerkschaftskasse (WBK) in Germany.[54] By 1971 opencast mining, featuring massive draglines weighing 3,600 tons would strip the soil – or 'overburden' – to expose the coalseams and would finally begin to completely revolutionise coalmining.[55] Yet in the continued absence of systematic safety training, African workers relied on their own council. Matiase's father, who worked and died on the collieries, as well as older colleagues, continually admonished younger workers to always be watchful of the roof, the sides (of the working faces), the moving (winch) ropes in the tunnels (the gullies and travelling ways), the gases and the heat in the *skwere* (the working stopes).

In the 1970s, John Masilela, who was compelled to work at a Witbank colliery from 1976 due to racialised legislation and who was still working there in 2012, tells how 'everyone had to stand for himself' by virtue not only of the racial divide between black and white, but also its corollary of enforced tribal divisions:

> There were *indunas* (head clerks) representing every region and tribal community. A Xhosa *induna* guy represented the persons coming from the Eastern Cape, an Ndebele *induna* for persons from Ndebele and a Shangani for people from Mozambique. Even at hostel rooms, we were grouped like that, according to the language and the place of origin.

This tribal representation system clearly did not work except to entrench existing differences of language and origin between black workers and certainly appears to have failed grossly in communicating the occurrence of fatalities to families in far-flung rural villages. Echoing Matiase's testimony, Masilela tells how when a fatality occurred, emblematic of prevailing racialised social attitudes and practices predating apartheid:

> we just continued with the job after the fatality. We waited until the manager arrived to check the situation and the body was removed. We continued with what we were doing as if nothing happened over there. The mines had their grave yards. Once someone died they would bury them there. They would take his rubber boots and hat. At home, the family members assumed that the guy forgot home and was not visiting. Management did not inform the family members.

Clearly neither did the *induna* system relay the news, for it was very largely confined to the mines and without resources to fulfill its communicative role.

As far as colliery workers interviewed were concerned, however, attitudes to safety only improved significantly after the arrival of the National Union of Mineworkers (NUM) in 1982 when black workers started to become involved in safety on the South African mines and collieries.[56] The NUM would also take on the *induna* system and replace it with formal trade union representation,[57] but had safety as its first mobilising issue and commissioned sympathetic university scholars to pursue research on safety.[58] This led directly to the institutionalisation of new safety regimes on the South African mines, but which would nevertheless take over a decade before modern safety legislation was promulgated and then another decade before legislation took any real effect. It was, however, for disenfranchised black mine and colliery workers, the right to have a say over safety which arguably constituted the most significant change wrought by unionisation from the early 1980s. As an aside, were it not for the power of the organised strength of the mineworkers, initially mobilised around safety by the NUM, national liberation might have taken longer to achieve.

On the coalmines, the NUM's safety-oriented mobilising drive coincided with the introduction in July 1985 of the concept of 'alternative or separate intake airways' to provide a safe route for escape from coalmines in the event of fire,[59] obligatory in the Ruhr since the 1880s. This crucial development accompanied the enforcement in October 1986, 40 years after the Ruhr, of carrying self-contained, self-rescue packs even though they were almost immediately shown to be inadequate.[60] While the environmental problems associated with the developments in mechanisation underground were only addressed within the CRL in the 1980s, industry research had as their particular focus 'dust control on continuous miners, face ventilation and more recently the protection of mine workers in the event of fires and explosions'.[61] At this point in 1986, the fatality rate was cited as a then all-time low of 0.36 deaths per 1,000 workers,[62] however official state and industry measurement methods and figures have been contested.[63]

With the advent of democracy in 1994, the mid-1990s saw colliery management beginning to move away from the old tendency of accusing production
workers for injury and fatality. This new 'safety culture' is, however, a relatively recent development. When John Booi started working on the coalmines
in 1992 'there was no training center. The training center came into operation
in 1994–5'. This was around the achievement of democracy in 1994 and after
the exertion of 'relentless pressure' on the industry by the NUM, the Leon
Commission of Inquiry into Health and Safety in the South African mining
industry sat in 1995.[64] The following year saw the passing of the Mine Health
and Safety Act (Act 29 of 1996) (MHSA) which introduced the contemporary
era in safety on South African mines. Employers were now required to formally assess the risks to the health and safety of mineworkers with a range of
preventative measures and modern risk assessment and management systems
implemented in major coalmining operations.[65] Institutional and legal formality
had replaced much of what had gone before,[66] regarding which colliers on the
Witbank mines had much to say.

Reminiscent of the *Der Bergmannsfreund* issued to colliery workers on the
Ruhr over 70 years before, but in a modern media, John Booi tells how:

> Videos of hazards and safety for new recruits became available in 1997. Until
> 1994–95, we were just told: 'You should be careful and respect your co-
> workers and team leaders. Those are the people who will guide you and will
> let you know about the environment underground'.

Hence like Matiase and Masilela, Booi had to rely on 'Old colleagues and other
co-workers [who] gave us some guidance and instructions like: "You should not
walk on the rail line and you should be careful of the gas that does not smell
ordinary"'. The well-meaning advice strikes a poignant note because methane
and carbon monoxide have no smell, but such was the extent to which coalminers
were left to their own devices even in the first few years of the democratic era.

After a fall of ground in 2001, Booi recalls how: 'Others ran away for their
safety. I remained because I was the person who had to know what was going on',
but Booi openly admits that: 'When I saw a piece of body there, that was when I
started to run away'. Yet a few years later, being now a more experienced miner
and echoing the dispassionate attitude of Ritterswürden on the Ruhr in 1946, Booi
is unsentimental when he notes the continuing occurrence of accidents: 'A few
deaths are owed to the release of gases out of mining activity. When I went down,
I found them lying there in 2008'.

Among workers, old superstitious attitudes, such that the presence of a woman
in a mine would result in the roof falling, does 'not happen anymore', says one
of Booi's friends, London Mkhomqo. Employed on a colliery since 1994 and a
miner since 2009, Mkhomqo is one of the new generation of coalmine workers.
After being called to an accident and contacting the emergency services, the manager, the mine captain and doctor:

The clinic delayed in treatment. They waited for another doctor to come while Mr John was bleeding a lot. He died along the way. I was so sad because I tried my best. I was very sad when I saw his family who asked me: 'London, why, what happened? I observed his kids crying 'where is our father?' They were from Lesotho. Even now when I see his picture in the control room I shiver remembering that accident.

Yet where work was reported, by a number of informants, to have simply continued after an accident, the underground section implicated or even the whole mine is now often closed down in terms of Section 54 of the MHSA, yet not without controversy and opposition within the industry and even among the miners themselves due to losing bonuses when the state enforces this section of the Act.[67] In the past especially, the attitude to accidents was, it appears, often fatalistically accepted as not under human control. As Martha Maseko, a coalmine worker since 1986 at Khuthala mine and head of the NUM Women's Desk, put it: 'Coal and methane work in their own way'. Many of the new generation instead identify causes, provide reasons and make suggestions as to preventing the occurrence of accidents.

Refusal to work in an unsafe environment, for John Booi and others, is no longer synonymous with dismissal and a more collaborative worker/management relationship regarding safety has emerged, though later this was to amount to uncomfortable complicity in tolerating unsafe acts in the interests of production and bonuses. One of the current major safety issues, however, is attempting to ensure that older workers, acculturated in a previous, less safety-conscious era, adhere to safety regulations. This new generation of trained NUM shaft stewards and safety officials, such as John Booi and London Mkhomqo, are pitted against the older workers who continue to manifest blasé and macho attitudes towards safety at work. These younger miners are aware of the hazards, know the rules, are articulate defenders of safety procedures and take proactive steps to prevent accidents from happening in the first place. The testimony of younger, more educated and better-trained workers 20 years into democracy clearly expresses a considerably greater awareness and positive attitude to safety and working safely which is often, but not always, missing in the stories older workers tell. John Booi implies that management on the collieries manipulates the older generation of workers by giving them gifts, which both lends to the ignoring of safety procedures and diminishes the authority of the NUM safety and shaft stewards. Simphiwe Litchfield, who started work as an electrician on a coalmine in 2003, attributes accidents to a lack of vigilance once 'people get used to the mines' and who 'then do not follow safety procedures'. Currently, managerial initiatives, which emphasise that everyone should be responsible for their own safety, are being articulated by workers themselves. This is a far cry from the past. While there was always the practice of inspecting a working face and 'making safe' prior to entry, safety has become a serious and regular talking point in the mining industry today.

At an institutional level, there have also been developments. In 2006 the measures for safety statistics were aligned with international standards to enable comparison with international benchmarks, and the South African Mine Health and Safety inspectorate (MHSI) set milestones based on the actual accident statistics of Australia, the USA and Canada.[68] While recognised as a problem due to no statutory requirement that industry records actual hours worked, the fatality rates were 0.13% in 2005 and 0.16% in 2006.[69] The highest fatality rate on the Ruhr was 0.14% in the Grimberg disaster. In all other years between 1901 and 1994 it was considerably lower (see Figure 10.1).

Along with the gradual improvement in safety over the decades, the voice of workers is now being heard as they identify safety issues featuring in old practices and must confront new issues arising. Regarding the old issue of working *zama zama* (overtime) shifts on the weekends, John Booi decries the fact that safety again plays second fiddle to meeting production targets. Yet despite unionisation, the advent of democracy and even the best efforts of especially current senior management and safety stewards, workers still fear to testify at official accident inquires, potentially compromising establishing the causes for accidents once they have occurred, as an old refrain continues to assert itself: 'Management usually blames workers for any accident', says London Mkhomqo. Other old issues remain: safety is subordinated to production, the structure of the bonus system, and as Simpiwhe Litchfiled says, safety is ignored as workers 'want to finish early'.

As ever, new issues arise. A recent phenomenon militating against safer mining is the entrant of a new union, the Association of Mineworkers and Construction Union (AMCU), which, while rising to prominence on the platinum mines in 2012, had its origin at Douglas Colliery in 2001.[70] Recent research shows how management has indeed manipulated the inexperience of the new AMCU incumbents on health and safety committees, resulting in health and safety issues being compromised.[71] But then AMCU has made new demands for greater recognition and payment of health and safety committees, not just full-time health and safety representatives as the NUM negotiated. Issues pertaining to health and safety, as Ruth Edgecombe noted over a generation ago, are indeed 'very complex and difficult', remain part of what it means to mine and are some of the ongoing unresolved problems in the quest for safer mining occurring much more broadly than in the collieries around Witbank today.

Similarities and contrasts

In comparing the two regions discussed in this chapter, state legislation and mining company initiatives regarding safety clearly frame colliery workers' safety discourse and practice. In both cases disasters prompted both the state and industry to intensify safety procedures. Testing sites were set up in Dortmund after 1908 (on the surface) and after 1925 (underground) and while not noted in the text, in South Africa similar initiatives happened later. Yet whereas worker safety representatives were appointed on the Ruhr in 1909, only to be abolished in 1920 due a conflict of interests with the safety officials' system but with full-time safety officers appointed in the early 1950s as a result of state regulation, this would only occur after 1996 in the South African mining industry. Awareness training appears to have assumed different forms, with a series of initiatives occurring on the Ruhr from the 1920s, but competitions for the lowest fatality rates (and others not noted) only initiated on the South African collieries in the 1950s.

Against the backdrop of these contrasts in how the state and industry framed the health and safety environment, the experience of safety for workers on the Ruhr was more of an enabling one when compared to particularly black workers around Witbank. A safety policy of control and care and insistence that safety precautions be taken seriously on the Ruhr from the earliest days in the 20th century empowered the safety deputies, while safety stewards at Witbank (and elsewhere) had to await unionisation, democratic enfranchisement and safety legislation before enjoying a more enabling safety environment. In both cases, however, miners took risks in the interests of earning production bonuses, wanting to get their job done or for other reasons and they were not always terribly cautious.

Whereas safety on the Ruhr was exercised in the context of a developed society, although ethnically heterogeneous due to immigration from other German provinces or from outside Germany, the discourse around safety for African colliers in South Africa was embedded in a developing society, marked by disempowering racialised and ethnic heterogeneity. Workers on the Ruhr were dedicated skilled coalminers. On the collieries around Witbank, while most mineworkers were oscillating migrants, some of who had also worked on asbestos and gold mines, they were for the most part of the long period under review here, generally accorded unskilled labourer and mineworker status. While around a quarter of these workers settled around the collieries, the majority were oscillating migrant workers and despite the somewhat more stable world of the coalmines in the mining sector as a whole, their voices were muted and their lives were considered to be of inferior or little value. Being on the receiving end of racial discrimination and tribally divided on the mines, the testimony of many of the Witbank colliery workers attests to this clearly being a prominent aspect of underground colliery work and which clearly had the implications of not always being able to work safely.

Despite very different work processes in the two sites, the safety experience of workers displays many similarities. Taken from the late 1940s in Germany and from the late 2000s in South Africa, the experience of Ritterswürden on the Ruhr and Booi in Witbank reveals similar matter-of-fact attitudes towards accidents underground. In another two cases, also 60 years apart, both Nordmeier as safety officer and Litchfield as union shop steward, assumed similarly subtle strategies to convince colliery workers on two continents, separated by time and space as well as linguistic and cultural traditions, to look after their own safety. That black mineworkers, prior to enfranchisement, were compelled to look out for one another in a racialised context and in one where regulatory oversight was clearly poor, asserts itself powerfully. In the period after unionisation in the mid-1980s, the voices of disenfranchised South African colliery workers assumed a different, stronger tone. Indeed, in the late 1990s it became a confident one after democracy and the legislation, which framed the conditions in which they worked. By that time miners on the Ruhr could look back on nearly 40 years of co-determination and an even longer history of trying to have a say in safety matters. The number of fatalities on South African coalmines, however, dropped from 90 in 1993 to 5 in 2015.[72] More broadly across mining in South Africa, in 2014 the fatality rates of all mines was 0.09% in comparison with the USA (0.08%), Canada (Ontario) (0.1%) and Australia (0.03%).[73]

Conclusion

Long after mining had ceased at Zollern, the more educated current generation of South African colliery workers and coalminers under democracy firmly hold that accidents can be avoided, though complaints and laments about race continue to inform many of the accounts. But it needed the introduction of democracy and unionisation to improve safety conditions, awareness and prevention.

What clearly does unite these two sets of testimonies is the risk of accidents attending work in underground coalmines. In their telling of these dangers, on both the Ruhr and around Witbank, what the colliers have to say both reflects the character of their different societies, but also their shared experience of injury and fatalities in the collieries. In both instances, how they protected themselves at work underground comes to be seen as framed by the work they performed and state legislation, which sought to protect mining personnel. In the interests of safety in mines, it is hard not to conclude that mineworkers' voices, reflections and experience of safety need to be more systematically documented and reflected in legislative regimes, in mining companies' policy and practice and integrated into the working practices of the coalminers themselves.

Notes

1 Klaus Tenfelde, 'Comparative Research in the History of Mining Workers: Some Problems and Perspectives', in: Gustav Schmidt (ed.), *Bergbau in Großbritannien und im Ruhrgebiet. Studien zur vergleichenden Geschichte des Bergbaus 1850–1930*, (Bochum: Studienverlag Brockmeyer, 1985), p. 26.
2 Ruth Edgecombe, *A Brief History of the Natal Coal Industry 1888–1987* (Dundee: Talana Museum), pp. 5–6.
3 See Dagmar Kift, 'Die Zolleraner-Interviews von Martin Rosswog', in: Martin Rosswog (ed.), *Schichtaufnahmen. Porträtfotografien und Erinnerungen an die Zeche Zollern II/IV* (Essen: Klartext, 1994), pp. 10–14; Dagmar Kift, 'Heritage and History: Germany's Industrial Museums and the (Re-)presentation of Labour', *International Journal of Heritage Studies 17*(4) (2011), pp. 380–389.
4 See Dhiraj Nite and Paul Stewart, *Mining Faces: An Oral History of Work in Gold and Coal Mines in South Africa, 1951 –1911* (Johannesburg: Jacana Media (Pty) Ltd, 2012).
5 Martin Rosswog, *Schichtaufnahmen. Porträtfotografien und Erinnerungen an die Zeche Zollern II/IV* (Essen: Klartext, 1994), p. 124, translated by Dagmar Kift.
6 *75 Jahre Berggewerkschaftliche Versuchsstrecke in Dortmund-Derne der Westfälischen Berggewerkschaftskasse 1894–1969* (Westfälische Berggewerkschaftskasse, Herne, 1969);

M. Hildebrandt, Werner Foit and Bernd Margenburg 'Die Versuchgrube Tremonia in Dortmund – Sicherheit im Bergbau', *Bergbau 48*(12) (1997), pp. 557–562.
7 Rosswog, *Schichtaufnahmen*, p. 86.
8 *Ibid.*
9 *Ibid.*, p. 47.
10 *Ibid.*, p. 221.
11 See Olge Dommer and Dagmar Kift, *Keine Herrenjahre. Jugend im Ruhrbergbau 1898–1961. Das Beispiel Zeche Zollern II/IV* (Essen: Klartext, 1998).
12 Anon., 'Brief an eine Bergmannsfrau', *Werks-Nachrichten der Dortmunder Bergbau-AG 11*(5) (1961), p. 105.
13 See also Norbert Ranft, *Vom Objekt zum Subjekt. Montanmitbestimmung, Sozialklima und Strukturwandel im Bergbau seit 1945* (Köln: Bund-Verlag, 1988), pp. 342–343.
14 See Heinrich Imbusch, *Arbeitsverhältnis und Arbeiterorganisationen im deutschen Bergbau. Eine geschichtliche Darstellung* (Essen: Verlag des Gewerkvereins Christlicher Bergarbeiter); (reprint: Berlin: J.H.W. Dietz Verlag, 1980), pp. 133–134, 625.
15 Gerhard Adelmann, *Die soziale Betriebsverfassung des Ruhrbergbaus vom Anfang des 19. Jahrhunderts bis zum Ersten Weltkrieg unter besonderer Berücksichtigung des Industrie- und Handelskammerbezirks Essen* (Bonn: Ludwig Röhrscheid), 1962, p. 145.
16 Ibid., pp. 145–149.
17 Ranft, *Vom Objekt zum Subjekt*, p. 343.
18 See Christa Steinberg, *Der Unfallgefährdete und die Unfallverhütung im Ruhrbergbau* (Berlin: Duncker & Humblot, 1957), p. 29.
19 Rosswog, *Schichtaufnahmen*, p. 99.
20 *Ibid.*, p. 102.
21 *Ibid.*, p. 100.
22 Ranft, *Vom Objekt zum Subjekt*, p. 347–358.
23 See Evelyn Kroker and Michael Farrenkopf, *Grubenunglücke im deutschsprachigen Raum. Katalog der Bergwerke, Opfer, Ursachen und Quellen* (Bochum: Deutsches Bergbau-Museum Bochum, 1999), pp. 435–509; and Michael Farrenkopf et al. (eds), *Glück auf! Ruhrgebiet. Der Steinkohlenbergbau nach 1954. Katalog der Ausstellung des Deutschen Bergbau-Museums Bochum vom 6.12.2009 bis 2.5.2010*, (Bochum: Deutsches Bergbau-Museum Bochum, 2009), pp. 534–535.
24 Günter Hegermann and Wolfhard Weber, 'Bergbautechnik nach 1945', in: *ibid.*, pp. 329–341.
25 Dietmar Bleidick, 'Bergtechnik im 20. Jahrhundert: Mechanisierung in Abbau und Förderung', in: Dieter Ziegler (ed.), *Geschichte des deutschen Bergbaus, Bd. 4: Rohstoffgewinnung im Strukturwandel* (Münster: Aschendorff Verlag, 2013), pp. 355–411.
26 *Ibid.*, p. 391.
27 Ruth Edgecombe, 'Dannhauser (1926) and Wankie (1972) – Two Mining Disasters: Some Safety Implications in Historical Perspective', *Journal of Natal and Zulu History 13*(1990/91), pp. 72–73.
28 John Lang, *Power Base: Coalmining in the Life of South Africa* (Johannesburg: Jonathan Ball Publishers, 1995), p. 51.
29 Lang, *Power Base*; Peter Alexander, 'Challenging Cheap-Labour Theory: Natal and Transvaal Coal Miners ca 1890–1950', *Labor History 49*(1) (2008), pp. 47–70.
30 R.W. Grout and Richard L. Lechmere-Oertel, 'A Brief History of the Progress of Efficiency in South African Collieries in the Last Seventy Years', *Journal of the South African Institute of Mining and Metallurgy* May (1958), p. 500.
31 Lang, *Power Base*, p. 139.
32 Grout and Lechmere-Oertel, 'A Brief History of the Progress of Efficiency in South African Collieries in the Last Seventy Years', pp. 485–509.
33 W.G. Pyne Mercier, 'The Mines, Works and Machinery Regulations and their effect on Accident Prevention and the Health of Workmen', *Journal of the Chemical Metallurgical and Mining Society of South Africa 54*(2) (1953), p. 31.

34 Lang, *Power Base*, p. 55.
35 Edgecombe, 'Dannhauser (1926) and Wankie (1972)', p. 85.
36 Lang, *Power Base*, p. 56.
37 Edgecombe, 'Dannhauser (1926) and Wankie (1972)', p. 81.
38 Native Mine Wages Commission 13 July 1943, AH 646 Box 127 SATLC Disputes 1942–1946.
39 Ruth Edgecombe and Bill Guest, '"The Coal Miners' Way of Death": Safety in the Natal Collieries, 1910–1953', *Journal of Natal and Zulu History 8* (1985), pp. 65.
40 Lang, *Power Base Africa*, p. 128.
41 J.P. Harding, 'Opening Address: Safety in Mining Symposium', *Journal of the Chemical Metallurgical and Mining Society of South Africa 54*(2) (1954), pp. 70–72.
42 Alexander, 'Challenging Cheap-Labour Theory', p. 9.
43 Lang, *Power Base*, p. 138.
44 H. Wagner, 'Coal Mine Safety Research in South Africa – Achievements and Developments', Symposium on Safety in Coalmining, Pretoria, *Council for Scientific and Industrial Research 1*(4), 5–8 October (1987), pp. 1–9.
45 *Ibid.*
46 *Ibid.*
47 Edgecombe and Guest, '"The Coal Miners' Way of Death"', pp. 62–63.
48 J.P. Leger, 'Coalmining: Past Profits, Current Crisis?', in: Stephen Gelb (ed.), *South Africa's Economic Crisis* (Cape Town: David Phillip and London: Zed Books, 1991).
49 Lang, *Power Base*, p. 139.
50 Alexander, 'Challenging Cheap-Labour Theory'.
51 Lang, *Power Base*, p. 178.
52 Uwe Burghardt, *Die Mechanisierung des Ruhrbergbaus 1890–1930,* (München: C.H. Beck Verlag, 1995), p. 290.
53 Hegermann and Weber, 'Bergbautechnik', p. 335.
54 Wagner, 'Coal Mine Safety Research in South Africa'.
55 Leger, 'Coalmining: Past Profits, Current Crisis?'.
56 J.P. Leger, 'Towards Safer Underground Gold Mining: An investigation Commissioned by the National Union of Mineworkers', *Labour Studies Research Report 1, Sociology of Work Programme* (University of the Witwatersrand, 1985).
57 D. T. Moodie (with Ndatshe V) *Going for Gold: Mines, Men and Migration* (Berkely: University of California Press, 1994).
58 Leger, 'Towards Safer Underground Gold Mining'.
59 R.M. Stroh and G.H.J. Kruger, 'A Prevention and Rescue Strategy for Fires and Explosions in Collieries', Symposium on Safety in Coalmining, Pretoria, *Council for Scientific and Industrial Research 8*(3), 5–8 October (1987), p. 7.
60 A.J. Kielblock and James P. van Rensburg, 'Self-contained Self-rescuer Performance as a Function of Human Mechanical Efficiency and of Escape Route Terrain', Symposium on Safety in Coalmining, Pretoria, *Council for Scientific and Industrial Research 18*(2) 5–8 October (1987), p. 6.
61 Wagner, 'Coal Mine Safety Research in South Africa', p. 3.
62 Lang, *Power Base*, p. 187.
63 Herbert P. Eisner and J.P. Leger, 'Safety in South African Mines: An Analysis of Safety Statistics', *Journal of the South African Institute of Mining and Metallurgy 88*(1) (1988), pp. 1–7.
64 Timothy S. Pakhathi, 'Worker Agency in Colonial, Apartheid and Post-apartheid Gold Mining Workplace Regimes', *Review of African Political Economy 39*(132) (2012), pp. 288.
65 Patrick J. Foster, H.J.M. Rose and C.F.Talbot, 'Risk Assessment: An Opportunity to Change Health and Safety Performance', *The Journal of The South African Institute of Mining and Metallurgy* (Nov/Dec 1998), pp. 333–338.

66 *Ibid.*, p. 334.
67 May Hermanus, Nancy Coulson, Dirk Bakker, N. Pillay and Huw Phillips, *Enforcement Mechanisms Study: Stakeholder Perspectives on the Enforcement of the Mine Health and Safety Act and Consolidates Recommendations,* Mine Health and Safety Council (MHSC), Circular Q1 03-MRAC-2013-4, Centre for Sustainability in Mining and Industry (CSMI) (University of the Witwatersrand, 2014).
68 Mine Health and Safety Inspectorate 2006, *Annual Report* 2006/07, p. 4.
69 *Ibid.*, p. 16.
70 Peter Alexander, Thapelo Lekgowa, Botsang Mmope, Luke Sinwell and Bongani Xezwi, *Marikana: A View from the Mountain and a Case to Answer* (Jacana: Auckland Park, South Africa, 2012).
71 Bezuidenhout, Christine Bischoff, Nathaniel ka Dlamini, Boitumelo Malope, John Mashayamombe, Sandla Nomvete and Paul Stewart *Organise or Die: Findings of the NUM Health and Safety Audit – Summary Report*, Final Report, Consortium from the University of Pretoria and the University of the Witwatersrand, Business Enterprises at the University of Pretoria (BE@UP), 2015.
72 Department of Minerals and Energy, '2015 mine Health and Safety Statistics', Media Statement, 28 January 2016.
73 *Ibid.*

11   The state, labour conflicts and coal mining

Chris Wrigley

For various periods historians have argued for German, French, US or British exceptionalism when writing labour history. Such notions have been contested, but in this chapter, I argue that in some respects British trade union history, at least before the First World War, differed from much of continental Europe; although I also believe that comparisons bring out more similarities than might be expected.[1] In the case of industrial relations in coal mining, there were very different political cultures between Britain, France, Germany, Austria-Hungary and Russia before the First World War as well as very substantial differences in outlook and strength between employers, whether formally organised or not. There were also contrasts of degree. For instance, trade unionism was not as deeply divided in Britain as elsewhere, with no significant confessional unions and weaker communist bodies, though Scottish breakaway mining unions were substantial.[2] Other differences stemmed from the length of time mining had taken place and the timing of industrialisation in countries.[3] Later, as in international economic history, there was a degree of convergence and similarities became more marked, with the more recent industrialised economies catching up the older ones, albeit in decline. The chapter reflects on some aspects of the state and industrial disputes in mining, primarily in coal mining.

The problems of comparative history, and of comparative mining history in particular, have often been considered over the years.[4] The issues for coal have been discussed among others by the late, great historian of coal mining, Klaus Tenfelde, in several of his many important publications, as well as by Stefan Berger, Andrew Taylor and others in *Towards A Comparative History of Coalfield Societies*.[5] There have been several books making comparative studies of coalfield communities and it is not the intention here to repeat what has been done well in such detailed studies.[6]

A major problem of comparative history is to acknowledge variations in circumstances between the places being compared without retreating into considerable detail that is likely to accentuate differences and makes generalisations seem foolish. Yet generalisations in comparative studies can be very fruitful, not only in bringing to the fore similarities but also differences that either require explanation or fresh research in one of the countries' or places' history to check whether the missing feature has simply been overlooked by previous historians.

Comparative studies add depth to the study of one country or place, bringing in to sharper focus distinctive features.[7]

The British state was very reluctant to intervene in industrial disputes, other than in a 'law and order' capacity, until the late nineteenth century. Britain was unusual in being committed to free trade from 1846 to 1932, and Gladstonian economic orthodoxy favoured free market economics and a small state role in the economy. However, there was a long-established role for the state in mining for health and safety matters.[8]

The structures and organisation of coal industries varied considerably. Britain and Germany both saw massive increases in production of coal from the mid-nineteenth century to the First World War. In Britain coal output rose from 64.7 million tons in 1854 to 117.4 million in 1871, to 225.2 million tons in 1900 and 287.4 million tons in 1913.[9] In Germany coal output rose from 5.8 million tons in 1851 to 29.4 million tons in 1871 to 60 million tons in 1900 and 114.5 million tons in 1913.[10] In Britain the state regulated working conditions but did not control private enterprise. After a Royal Commission on the employment of children in mines, there was legislation in 1842 banning women and children under the age of ten underground in mines. For several decades, a few women evaded this legislation and worked underground. Women and girls continued to work on the surface until 1952.

There was less extensive legislation in Belgium, which excluded young children from working underground, but was amended in 1878 to limit underground work to boys of 12 or more and girls of 13 or more. In the late nineteenth century at Charleroi Mine and others, women continued to haul coal trucks. In contrast to labour restrictions in Britain, in British controlled India women and children continued working in mines. In 1924 there were 42,000 women and 471 children under the age of 12 working with 65,000 men in India's mines.[11] Restricting child and female labour in mines was a state matter before German unification, with Prussia limiting child labour in 1839. After the mid-nineteenth century, the state largely withdrew, leaving private firms to expand coal production. In the late nineteenth century the German state took some measures to curb the private sector's powers, including limiting child and female labour in mines, but, overall, the state and big business shared a common outlook on issues of efficiency and on disciplining labour.[12]

In both Britain and Germany, the biggest mines got bigger before 1914, but in both countries there were many small mines as well. In Britain, finance for fixed capital came predominantly from a range of local sources, notably from landowners, other coal owners, businessmen and merchants. Later, many British mines became limited companies, which secured additional funds from the stock market. Some mines were financed by other industries, most notably in iron and steel. Short-term capital came from local banks. Unlike Germany, however, it was rare for banks to control mines, and in most cases where they did, it was due to closure on bad debts. In 1873 57 per cent of collieries were owned by partnerships or individuals but by 1894 this proportion had fallen to 16 per cent and by 1905 to 5 per cent. Other coal companies owned 26 per cent of collieries in 1873 and 66 per cent in 1894 and

77 per cent in 1913.[13] In Germany there was a development of cartels dominated by joint stock companies, especially with the adoption of tariffs in 1879. Some smaller mines remained outside the cartel system, and about 10 per cent of output was in state-owned mines, but generally, the industry was controlled by big business. Most companies were members of the Rhenish-Westphalian Coal Syndicate from 1893.

In both Britain and Germany, as in other countries, the state generally reinforced employers through enforcing law and order against discontented miners. In Britain, as elsewhere, there were notable paternalist employers who provided housing and other benefits to attract experienced and skilled labour and to thereafter bind it to their company. In some famous cases, such as Krupp in Essen and Schneider in Le Creusot, they provided their own welfare for their workers and opposed the coming of state welfare as it weakened their control of their labour force.[14]

Another unusual feature of Britain was that governments and employers in the late nineteenth century did not fear working class insurrection, in contrast to their attitudes earlier, notably on several occasions in 1812–48. This was also a contrast to the concerns of autocratic or semi-autocratic governments in Germany, Austria, Hungary and Russia as well as democratic governments in France (where memories of 1789, 1830, 1848 and 1871 remained strong),

Indeed, by the third quarter of the nineteenth century some British employers were willing to deal with trade unions. This was partly due to politics. After the defection of the Whigs and many middle-class Radicals led by Joseph Chamberlain over Ireland in the mid-1880s, skilled working-class men, especially nonconformists, were much more important supporters of the Liberal Party, with two coal miners elected to Parliament in 1874.

The willingness of some employers and most trade unionists to come to terms with each other did not prevent frequent and sometimes bitter strikes. But it did lessen conflict in British coal mining, especially in the period of falling coal prices (roughly, 1875–88) after the peak prices of 1871–5. In this downturn of the trade cycle several areas of the Miners' Federation of Great Britain (MFGB) made sliding scale agreements whereby wages automatically followed the movement of prices. For the miners this avoided repeated strikes and moderated the pace of decline in wages. However, miners were angered by such agreements when coal prices recovered in an economic upturn in the late 1880s. Nevertheless, such understandings as the sliding scale agreements marked a different style of coalfield industrial relations to most of continental Europe, though in France there was trade union recognition of the relationship of wages to changing coal prices.[15] By 1910, Clegg Fox and Thompson estimated that 900,000 miners working in all coalfields were covered by collective bargaining. In contrast, only 82 German miners were covered by collective bargaining.[16] In France the first collective bargaining agreement was in the Pas-de-Calais coalfield in 1891, and there were later regional negotiations. However, pre-First World War industrial relations were very regional and trade unionism in French mining was weak.

The first time a British government intervened in an industrial dispute, other than for the maintenance of law and order, was during the great coal lockout of

1893, which lasted over 16 weeks, involved some 300,000 miners across many British coalfields and resulted in some 24.4 million working days lost.[17] With coal supplies falling and the price of coal going up, with winter approaching and other attempts at conciliation failing, the government intervened. That Gladstone should be prime minister when this occurred, was symbolic of changed times as well as Gladstone's own more radical attitudes towards labour over his younger days when he had been hostile to the 1832 Reform Act and volunteered to be a special constable at the time of the great London Chartist demonstration of 1848.[18] It is difficult to think of another European former and future head of a government who would have applauded trade union victories as Gladstone did in his later years. He heralded the dockers' victory in the great London dock strike of 1889 as a 'real social advance' that 'tends to a fair principle of division of the fruits of industry' or reflected in a public speech (in 1890) on past industrial disputes that 'where it has gone to sharp issues . . . I believe that in the main and as a general rule, the labouring man has been in the right'.[19] In the US, President Theodore Roosevelt also intervened in a coal strike, the major dispute of 1902, settling it with higher wages and shorter hours for the miners and higher prices for coal and no trade union recognition for the owners.

Even the great lock-out of 1893 was not a full British dispute. It did not apply to Scotland. The MFGB was very much a federated body, and this was a source of weakness in 1926. The same was true of the National Union of Mineworkers (NUM), the successor of the MFGB, with the degree of local autonomy causing problems in 1984–85, and not only with the breakaway Union of Democratic Mineworkers (UDM). Divisions in coal mining trade unionism in continental European countries helped the state and employers to win industrial disputes. This was famously so in Germany in 1905, when the Christian coal union did not participate. There were also problems due to divisions between the labour of the host country and the migrant workers, notably Polish in Germany and France and also Belgian in France.

Coal was a crucial commodity for industrialising countries, as well as being important as a fuel for navies and for domestic heat. It was also valuable for chemical industries as a raw material, as in the manufacture of TNT. Its importance was reflected in the size of the workforce employed in coal mining. In Britain in 1892 there were some 517,100 coal miners, in 1901 the total rose to 751,700, by 1911 to 1,016,000, and peaked at 1,298,300 in 1921. So in 1911 coal miners constituted roughly 1 in 16 of the potential workforce or 1 in 10 of the potential male workforce. In Germany in 1889 some 90,000 coal miners on strike in the Ruhr forced the Kaiser to intervene and to abandon an intransigent attitude in favour of flexibility when dealing with the situation. The dispute led to the creation of the Alter Verband, the first effective mining union, the same year as in Britain the MFGB was formed. This was when German trade unionism was relatively small (343,300 in 1891) in comparison to Britain (1,468,000 in 1892).[20]

The changing conditions which influenced the beginning of British state intervention in industrial disputes also included increased competition from other industrialising nations, notably Germany and the US, as well as the inter-dependence of much of the British industrial economy. Strikes in coal, railways or docks had big impacts on each other, and coal disputes seriously affected steam-driven factories and workshops. The length of the 1893 dispute effectively forced Gladstone's intervention. In writing to the secretaries of the MFGB and the Coal Owners' Federation, he referred to 'the national importance of a speedy termination of the dispute'.[21]

It is not surprising that the next major interventions by government ministers were in coal, rail and docks, with other Liberal ministers, most notably David Lloyd George and Winston Churchill, being involved. They were both more active users of the powers to intervene in industrial disputes under the Conciliation (Trade Disputes) Act, 1896, than their Conservative predecessors had been.[22] In the massive 1912 coal strike when over a million miners were on strike, conciliation, which worked in 1893, failed to resolve the dispute and the government took the unprecedented step of legislating to arrange the means to fix a minimum wage in the coal industry (the Minimum Wages Act, 1912). At the time Lloyd George commented that this initiative 'sounded the death-knell of the Liberal Party in its old form'.[23]

There was much bitterness among organised labour in Britain over the use of the forces of law and order. During the 1893 coal lock-out at Featherstone, West Yorkshire, troops fired at a large crowd demonstrating against strike-breakers loading coal. Eight people were injured, with two passive by-standers dying of their wounds. In 1910–11 in South Wales, troops were deployed in the Rhondda valleys. Rioting occurred in Tonypandy but the only death occurred in Pontypridd, a miner dying from head injuries most probably inflicted by a police baton. Two men were shot in August 1911 by troops at Llanelli, where a train was attacked during the first national railway strike, and a further four people were killed due to an explosion during a subsequent riot. Miners supported the striking railwaymen, but none were killed. The sending in of troops, whether in 1893 or 1911–12, was deeply resented, as an invasion of coal mining communities' space. Thomas Ellis, a junior government whip and the Liberal MP for Merionethshire, was informed that a meeting at Pentre had protested about the sending of troops to the district and called for them to be withdrawn. Ellis was told that the troops were 'not wanted in the Rhondda. They always cause disturbance'.[24]

The shooting of miners, or of other protesters, was unusual in Britain and 'Featherstone' (with two dead) and 'Tonypandy' (with one dead) became part of the martyrology of the British labour movement, both long remembered and held against the home secretaries of the time, H.H. Asquith and Winston Churchill, and the place names were used to taunt them for the rest of their careers. However, in Britain, governments normally opted for conciliation and by the second half of the nineteenth century repression was unusual.

Elsewhere in Europe before 1914, the numbers of fatalities were often far worse. In Russia, 250 people were killed and 150 injured (or 150 killed and 100 injured according to the employers' papers) at the strike at the Lena goldfields in 1912, the horror of which, like the massacre of Father Gapon's followers in St Petersburg in 1905, was followed by a strike wave.[25] In France, middle-class

Republican governments remained hostile to the labour movement after the Commune, 1871. Then some 300,000 Communards had died, very large numbers of Parisians being shot by the army as it forced its way through Paris. In 1891, the army shot dead nine people, including several women and children, who were demonstrating on May Day at Fourmies, a woollen textile town.[26] In Germany the state was also ready to use the military against strikers, not least against miners during the 1889, 1905 and 1912 coal disputes. In 1889, troops fired at Bochum and Gladbeck, where eight were killed or wounded and subsequently the area had a large membership of the free miners' union (the BAV).[27]

There were high levels of violence in the United States. In big disputes, employers often hired armed men and used local militias to deal with strikers. In Tennessee in 1891, coal miners in Briceville, Anderson County, struggled against employers who used convict labour, thereby keeping wages low and undercutting trade union solidarity. From 1894 to 1914, there were many incidents in which miners and even their families were killed. In April 1894, during the eight-week Bituminous Coal Miners' strike, mine guards at Uniontown, Pennsylvania, opened fire killing eight miners and wounding five. On 10 September 1897 at the Lattimer Mine near Hazelton, Pennsylvania, a sheriff's posse shot 17 miners dead, their wounds being in the back and in some cases multiple. At Virden, Illinois, on 12 October 1898, a gun battle between striking miners and detective agency men protecting a train load of strike-breakers, left four detectives and seven miners dead. At the Ludlow Mine, Colorado on 20 April 1914, National Guards and company guards fired on 1,200 miners and their families in a tent colony during a long and bitter mining dispute in which in all between 69 and 199 people were killed.[28] There were many more deaths in 1920–22 in West Virginia at Matewan Mine and Blair Hill and Herrin Mine, Illinois.

After the 1905 strike, the German state brought in compulsory worker committees. Such committees were not popular with employers or strong trade unions in Britain. The most notable exception was worker committees which were seen as complementing wages boards as part of a constructive bid for urban votes by some Conservative and Unionist tariff reformers, who were eager to follow the example of Germany in protectionist economic policies.[29] The British system of industrial relations was marked by voluntary agreements and by the state abstaining as far as it could from intervention.[30]

The First World War strengthened the bargaining position of labour in the belligerent countries. The war was like a magnified upturn of the trade cycle, marked by immense demand and reduced amounts of available labour, given the huge numbers of men going into the armed services. In Britain, by 1918 the armed forces accounted for roughly 39 per cent of the total potential male labour force. The voluntary recruiting of 1914–16 took key industrial workers. In the case of mining, by July 1915 the total male labour force fell by 21.8 per cent. Output fell nearly as much as the fall in labour. By the October 1917 revolution in Russia, 15.3 million men had been called up (of a pre-war total population of 139.3 million). By the end of the war some 11.1 million men had been enlisted in Germany (total pre-war population of 64.9 million), France 8.3 million (45.2 million),

Austria-Hungary 7.8 million (50.8 million), France 8.3 million (39.1 million), Britain 5.7 million (45.2 million) and Italy 5 million (34.7 million).[31]

Labour was usually unwilling to push its economic advantage very far in the belligerent countries, given patriotic war-time concerns. In France, the Germans had occupied half the Pas-de-Calais coalfield. Elsewhere in French mines, miners worked very hard to make up the shortfall, there were few strikes and the unions were weak (not least because the strong trade unionism in the German-occupied areas was eliminated). Women undertook surface work elsewhere in France, as at the coal mines of Gard in southern France.[32] In Britain, the MFGB was strong enough to prevent the use of female labour or younger boys underground but did not succeed in preventing an increase in the number of women working on the surface. There was an increase in their number from 5,762 at the end of 1913 to 8,312 at the end of 1918.[33]

Nevertheless, even with restraint, labour in the belligerent countries was powerful, and deteriorating social conditions (scarcities at home, greater pressures at work) encouraged it to clash with the government in all countries. In Britain, the demands of the South Wales miners led in July 1915 to major concessions from David Lloyd George, then minister of munitions. The Cabinet had intended him to be tough with the miners, but the realities of the miners' strong position led Lloyd George to concede nearly all their demands.[34] The South Wales coalfield was taken under state control in December 1916, according to Redmayne, a major government adviser on coal control, 'in order to eliminate war profits, to avoid industrial disputes during the war, and to secure the best results from the labour of the mines'.[35] In fact, owners received guaranteed profits from coal and made huge amounts of money from subsidiary undertakings such as coke ovens which were excluded from government financial control, while the state did very well out of the boom conditions until the sharp economic downturn from January 1921. The rest of the coalfields were taken into state control from 1 March 1917, one of Lloyd George's promises to Labour when he secured the premiership in December 1916. While control was not nationalisation, it was soon accompanied by trade union membership being a condition of employment in the mines during the war and the MFGB having equal representation with the employers on the Coal Controller's Advisory Board, as well as representatives in some mines on pit committees. State control resulted in miners' industrial pressure being focused on the state, not the coal owners. As Lloyd George spelled out in February 1919, 'if there is a conflict it will not be a conflict between the mine owners and the miners, it will be a conflict between one industry and the whole of the state'.[36]

The strong position of labour in the British coal industry during the war and the post-war boom, 1915–20, was marked by coal accounting for a higher proportion of all strikes than in the pre-war period, other than exceptional highs of 1893, 1898 and 1912. In 1915, mining accounted for 56.1 per cent of all days lost through industrial disputes. With state control, mining's proportion of strikes dropped to 13.3 per cent, before going up to 21–22 per cent during 1917 to 1919. However, the MFGB's bid to move from state control to nationalisation, a demand carried at its conference and at the Trades Union Congress (TUC) in the 1890s, as well as to protect the newly achieved seven hours working day and higher pay, led to major clashes with the government. In 1920, 65.9 per cent of all strikes were in mining.

When international coal prices fell steeply at the start of 1921, the British miners could not prevent the government decontrolling the industry three months early and so leaving the miners to fight the coal owners over wages and conditions. The huge mining dispute of 1921 ensured that mining accounted for 85.0 per cent of all days lost through disputes. In France, the coal employers successfully went on the counter-offensive, secured wage cuts and blacklisted militants. In Britain in 1925–6, as the owners threatened worse wages and hours and an end to collective bargaining, the MFGB forced the government to provide subsidies for nine months in 1925 but failed to secure further government intervention in May 1926. The TUC's nine day's solidarity strike, the General Strike, failed to extract help for the miners and the MFGB was weakened by a long coal lock-out which lasted to nearly the end of the year. Not surprisingly, mining that year accounted for 90.3 per cent of all industrial disputes.[37]

The Second World War saw similar labour market conditions to those of the First World War. In all belligerent countries the state intervened to maximise war production and minimise the huge withdrawal of labour for the armed forces. In Britain, as early as December 1940, the men in the armed forces amounted to 20 per cent of the pre-war male labour force below retirement age. By June 1942 this percentage had risen to 32. At their peak, there were 4,635,000 men in the armed forces and 467,500 women in the Women's Auxiliary services. The worst industrial relations in Britain during the war were in the coal industry, where there was a legacy of bitter feelings from 1926 and the tougher managerial attitudes in the decade after. In 1941, a dispute at the Betteshanger colliery in Kent led to the government prosecuting strikers but, as in South Wales in 1915, it was not practical to prosecute a thousand determined miners and the government backed down. In 1940–4 the coal industry accounted for over half the days lost through stoppages of work and over half the workers involved in stoppages in Britain.[38]

In several countries during the Second World War, women were substitute labour in coal production. In Germany, women and girls over 18 worked as hewers at opencast mines, but not underground.[39] In Britain, as in the First World War, the MFGB was sufficiently strong for governments not to ignore its hostility to women working underground. However, a few thousand women continued to work on the surface in Lancashire and parts of Scotland. Ernest Bevin, the most powerful trade unionist of his generation and minister of labour and national service, conscripted substitute boy labour through a ballot of those aged 18–23 who were due to be called-up for the Armed Forces in 1943–5. There were nearly 21,800 Bevin boys drawn by ballot from all classes between 1943 and the end of the war in 1945.[40] While women continued to be excluded from underground work in Britain, they were readmitted underground in India during the war and until 1 February 1946 (having been banned in the 1930s) and worked for very low rates of pay.[41]

In occupied parts of Europe under the Axis Powers, the substitute labour was often forced labour. In Germany itself it has been estimated that in 1944 7,615,970 people were unfree labour, with most from the Soviet Union, Poland and France and lesser numbers from Italy, Rumania, Croatia and Slovakia. Of these people, 1.9 million were prisoners of war and 5.7 million were civilians, including Jews, communists, socialists, trade unionists, homosexuals and others hated by the Nazis.[42] There was forced labour elsewhere. For instance, in Czecho-Slovakia in 1942 some 1,200 Jews, including women and girls, were sent to work in mines in the Moravska Ostrava and Karvina Basin areas.[43]

The experience of the Second World War ensured changes in coal industries and their industrial relations. In Britain and France, coal was nationalised. In Britain, this was carried out on grounds of rationalisation, with beliefs that the state could make the industry more efficient and could fund the scale of investment required to facilitate the post-war recovery. In France, many owners were discredited by their cooperation with the German forces of occupation, even in some cases assisting in the German round-ups of union militants. The collieries of the Nord-Pas-de-Calais were nationalised in December 1944. However, in both countries the miners were soon disappointed by state ownership. This was partly because in Britain and France there was a great push for increased output. In France, there was a return to pre-war Bedaux systems and to wartime tough measures. In 1947, a national coal strike was followed by another in November 1948, marked by six miners being killed, 2,000 imprisoned and 6,000 sacked. Such actions turned miners against the state as employer[44] In Britain there was disillusionment with the old style of management continuing (and even managers staying in post after nationalisation). In both countries, the ultimate authority was the state, and so serious industrial clashes were likely to end up with confrontations with the government. This was so with the big British strikes of 1972, 1974 and 1984–5.

The post-Second World War years in the West were marked by Cold War fears of communism. In Australia, in 1949, Ben Chifley's Labour government confronted the miners in a bitter seven weeks strike. The government froze the mining trade union funds, had trade union leaders arrested and used 2,500 troops to work in opencast mines to maintain electricity supplies. Chifley exaggerated the power of communists in the mines, just as in Britain at the same time, Clement Attlee (and later Harold Wilson in 1966), over-reacted to communists in the British docks.[45]

In Britain, the state managed the run-down of the coal industry. With the closures of the 1950s onwards, the moderate leaders of the coal trade unions managed decline. In the 1960s and 1970s, large numbers of pits were closed in Britain by agreement. In France, at Decazeville in 1961, 2,000 miners occupied the pits, but the matter was settled by offers of transfers and retraining, with the same outcome occurring at Faulquemont in 1973, when 15,000 miners went on strike. The closures often had major effects on communities, and resistance to closures was marked by substantial action by women.

In Germany, the government managed the massive run-down of the coal industry without incurring a massive strike of the scale of the 1984–5 British dispute. As in other countries, German coal production reduced in scale in the face of alternative energy, notably oil and natural gas (while electricity generation used much coal). The number of mines fell from 148 in 1960 to 39 in 1980 to 12 in 2000, while output dropped from 150 to 20.7 million tonnes between 1957 and 2006 and the labour force dropped from 610,000 to under 50,000 in 2006. A particular problem for the state was the restructuring of the industry in the 1990s following German reunification. Another issue has been the impact of coal burning on the environment, with a strong Green movement. In 2007, Germany was a pioneer in 'clean' coal technology, beginning to build a non-polluting coal power plant. With the planned ending of state subsidies for deep mines in 2018, it is very likely that that part of the industry will follow the British deep mines into history. However, Germany has huge lignite mines, notably the Garzweiler (output 35–40 million tonnes per annum) and the Hambach (output 40–45 million tonnes per annum) opencast mines, which provide fuel for the power stations, and remain crucial for the foreseeable future as Germany phases out its nuclear power stations by 2022. The Swedish state energy company Vattenfell announced in October 2014 that it was selling off its German lignite interests in order to meet its climate change goals. By 2012, Germany, with an output of 196.2 million tonnes, had fallen to being the eighth largest coal producer in the world.[46]

Poland was Europe's biggest coal producer in the 1970s, with an annual output of around 150 million tonnes. The regime became increasingly repressive with the growth of the trade union Solidarity. On 16 December 1981, armed police and the army shot dead nine striking miners at the Wujeck coal mine, Katowice, Silesia, who were protesting at the introduction of martial law three days earlier. In the summer of 1988, miners were part of a strike wave against the communist government, with strikes in the July Manifesto mine and nine other mines. With the collapse of communism and the industry facing the same alternative fuels as other countries, coal contracted. As in Germany, opencast mining of lignite contributes a substantial share of output. In 2012 output was still at 79.2 million tonnes, with a workforce of 113,000. The industry remained entirely state run until 2009, but then a policy of privatisation began, with large European energy companies, including Vattenfall, RWE and EDF, buying parts of the industry. In early 2015, the Polish government faced strikes when it announced the closure of four mines of the state-owned Kompania Weglowa. Some 20,000 miners and their families held an underground protest in mid-January, which succeeded in securing a deal which, at least for a while, kept the mines in production.[47]

In the Czech Republic, the coal mines were state run for a decade after the fall of communism. Coal output fell from 26.4 million tonnes to 14.9 million tonnes between 1985 and 2004. The mines were privatised when the country joined the European Union. All the black coal mines of the Ostrava area were bought by OKD. However, with the fall in oil prices and the subsequent fall in international coal prices, OKD demanded state subsidies if it was to keep the Paskov mine open. As in Germany and Poland, opencast lignite mining has remained profitable, and two private companies mine in the Bohemian Basin, Czech Coal (50 per cent owned by the financier Pavel Tykac) and Severoceske doly.

By the late twentieth century, the largest producers of coal were no longer in Western or Central Europe, with Russia the remaining European huge producer of coal. The United States remains at the forefront of coal production, but it has shown signs of following Germany and other European countries away from heavy reliance on deep mines and towards greater dependence on opencast mining. In 2012, the United States was the second greatest producer of coal at 922 million tonnes. In the 1960s, there was a change in the structure of the industry with a series of mergers, with coal production more in the hands of large companies and less with small independents. The United Mine Workers declined from 160,000 members to 16,000 in 2005, partly because of increased mechanisation and a failure to unionise miners in new western coalfields. By 2010, the biggest coal output was in Wyoming, West Virginia, Kentucky and Pennsylvania and 31 of the biggest 52 mines were opencast.

From the last quarter of the twentieth century, China was the greatest producer of coal. In 2012 it produced 3.6 billion tonnes of coal, over 47 per cent of the world's total, from some 12,000 mines. Nevertheless, it also faced problems arising from the international economic slowdown, including less demand for coal. Wang Xianzheng, chairman of the China Coal industry, was quoted as saying over half the coal mining enterprises were making a loss and even more were struggling to pay their workers. So many coal disputes were focused on wage arrears and pit closures. In January 2015, hundreds of Shandong mine workers went on strike over unpaid wages and social welfare contributions by the Jinda Mine, controlled by the Chenlong Energy Group set up by the Communist Party. There was also conflict in several industries over the privatisation of state-owned enterprises. In Hunan in August 2009, the state-owned Jinzhushan Mining Industry of Hunan Coal, which was seeking to be privatised and to have shares quoted on the stock exchange, sought to force workers to give up their rights to a month's pay for every year worked. Not surprisingly, there was a strike, involving some 5,000 workers at four coal mines. According to the *China Labour Bulletin*, there were 382 strikes and protests in 2012, 656 in 2013 and 1378 in 2014 (the increase being partly due to strikers publicising their disputes more).[48]

In India, most of the coal has been mined for over 40 years by Coal India, which in 2012 employed some 3.7 million coal workers. India's output of 602 million tonnes of coal made it the world's third largest producer in 2012. In India, as in China, Poland, the Czech Republic, Britain and elsewhere, moves to privatise coal have led to strikes and demonstrations. In January 2015, government moves to open up mines to private enterprise led to a million miners going on strike. It had been intended to strike for five days, but on the second day the action was called off when the government agreed to reconsider its proposals.[49]

Australia is the fourth largest coal producer, with some 100 private mines producing 413 million tonnes of coal in 2012. Over half Australia's output is exported, most of it to Asia (40 per cent going to Japan and 16 per cent to South Korea). Three quarters of Australia's coal comes from opencast mines. Lignite is mined in Victoria and South Australia, while black coal predominantly comes from Queensland and New South Wales. In recent years, the mines have suffered from 'a critical skills shortage' (as stated in advertisements for more women mine workers), hence the active recruitment of women. In 2014 women already accounted for 18 per cent of the labour and worked not only in clerical and cleaning jobs but also in drilling and truck driving.[50]

Indonesia, with a coal output of 386 million tonnes in 2012, was then the world's fifth largest coal producer and the biggest exporter of coal. Its coal is produced in three regions – South Sumatra, South Kalimanta and East Kalimanta – by a few big producers and many small producers. The Indonesian government tries by such means as export taxes to boost domestic use of coal to meet more of the country's fuel needs. There has been friction over overseas companies, often Australian, taking large profits and putting not much money back into the local economy. Also, Indonesian security forces have beaten up and arrested striking miners, as at a KPC mine with an Australian contractor where the refusal to renew a collective bargaining agreement led to a strike from 10 November to 27 December 2011, with a further strike in January 2012. Worse still, at the Grasberg Mine, a copper, gold and silver mine on Papua owned by Freeport-McMoRan of Australia, some 10,000 miners and subsidiary workers went on strike following 28 deaths in 2013 and four in 2014 as work became less safe as the mine was turned from opencast to underground mining.[51]

Russia, with an output of 354.8 million tonnes in 2012, was the sixth largest producer of coal. It has been the third biggest coal exporter, with 20 per cent going abroad. Coal strikes in 1989 and 1991 helped to undermine President Gorbachev and the Soviet system. After the fall of communism, the Russian coal industry was restructured in order to make it attractive to private investors Much of the industry was privatised between 1993 and 2001. With privatisation there were more closures: 180 deep mines and 15 opencast mines were shut between 1992 and 2010. During privatisation there was substantial unrest in coal mining. In 1995, 215 of 235 mines were closed by strikes called over the failure to pay wages. In 1996, 75 per cent of the country's coal mines were affected by another big coal strike, again over US$200 million was owed to miners in back pay. There were other strikes against the cutting of government subsidies and mine closures. With mergers, two huge private companies came to dominate much of the market: Krasnoyarskugol and the Siberian Power Company, with two state companies – Vorkuta Coal and Inta Coal – being important producers.[52]

South Africa, with an output of 280 million tonnes of coal in 2012, was the seventh largest producer of coal. Most of the coal was mined in the Highveld (31 per cent), Witbank (30 per cent) and Ermelo coalfields (14 per cent) of eastern South Africa. There were 64 coal mines in 2004, ranging from the huge to small-scale ventures, and 85 per cent of the coal mines were owned by five companies: Anglo-American Thermal Coal, BHP Billiton's Ingwe Collieries, Sasol Mining, Eyesizwe and Kumba Resources. In the Mpumalonga Highveld area, a workforce of some 55,000 produced 220 million tonnes, of which 29 per cent was exported.[53] Conditions in the coal mines, as in the gold and platinum mines, have often been bad, with 125 fatal accidents in 2011. There have been numerous strikes over very poor living conditions in compounds, pay and redundancy terms. In 2012, a wave of wildcat strikes culminated in police shooting dead 34 striking platinum miners at Lonmin's Marikana mine on 16 August 2012.[54]

In Britain, strikes in mining were frequently a large proportion of all strikes. Yet, as with other strike-prone industries, careful studies such as that by Roy Church and Quentin Outram have shown that a high proportion of strikes occur at relatively few collieries.[55] This is also clear from the National Coal Board records for the period after the 1984–5 strike, when managers of pits communicated daily even in short disputes. These reports showed that certain pits (often in South Yorkshire), certain coal faces and even certain shifts (such as the night shift) had repeated strikes.

While miners on strike have sometimes been violent, notably at Blair Hill and Herrin in the US in 1921–2, states have also been violent. This has been the case in democratic governments, whether in France, the USA, Britain (policing in the 1984–5 strike) or South Africa (Marikana), as well as in autocratic regimes, such as Tsarist Russia.

The European, US and Australian coal mining industries do not display exceptionalism in the long run. There has been convergence as economies have moved into deindustrialisation. China, like European and North American countries, has increasing numbers of derelict or recycled industrial buildings. Western coal production has been especially susceptible to international economic pressures. Such pressures include prices of alternative sources of energy, especially oil prices and cheap imported coal. In the case of other coal industries, such as those in China, India and Indonesia, there has been rapid growth, just as was experienced by the British, German and other coal industries as the countries industrialised. In the case of China, it may be that its consumption of coal peaked in 2013 and that, like the Western economies, coal production may fall, especially after 2020. The move from coal in China and elsewhere may be further encouraged by the Paris Agreement on Climate Change of December 2015. However, India and some smaller economies, such as Viet Nam and the Philippines, are demanding more coal.[56]

Notes

1 Ira Katznelson and Aristide Zolberg, *Working Class Formation: Nineteenth-Century Patterns in Western Europe and the United States* (Princeton: Princeton University Press, 1986); James Cronin, 'Neither exceptional nor peculiar: towards the comparative study of labor in advanced society', *International Review of Social History* 38(1) (1993), pp. 59–75. For a comparative approach in arguing exceptionalism, see: H. Chapman, M. Kesselman and M.A. Schain, 'Introduction: The Exceptional Trajectory of the French Labor Movement' in: idem (eds), *A Century of Organized Labor in France* (New York: Palgrave Macmillan, 1998), pp. 1–22. On some links between British and other European labour movements, see Chris Wrigley, 'The European Context: Aspects of British Labour and Continental Socialism before 1920' in: M. Worley (ed.) *The Foundations of the British Labour Party: Identities, Cultures, Perspectives, 1900–1939* (Aldershot: Ashgate, 2009), pp.77–93.

2 Alan Campbell, *The Scottish Miners*, Volume 2 (Aldershot: Ashgate, 2000), chapters 6 and 7.

3 For the classic statement, see Alexander Gerschenkron, *Economic Backwardness in Historical Perspective* (Harvard: Harvard University Press, 1962). For an earlier view of the benefits of later industrialisation, see: Thorstein Veblen, *Imperial Germany and the Industrial Revolution* (New York: Macmillan, 1915).

4 Gaston V. Rimlinger, 'International differences in the strike propensity of coal miners: experience in four countries', *Industrial and Labour Relations Review 12*(3) (1959), pp. 389–405.

5 Klaus Tenfelde, 'Comparative Research in the History of Mining Workers: Some Problems and Perspectives' in: Gustav Schmidt (ed.), *Berghau in Grossbritannien und im Ruhrgehiet* (Bochum: Universitätsverlag Brockmeyer, 1985), Gustav Schmidt (ed.), *Towards A Social History of Mining in the Nineteenth and Twentieth Centuries* (Munich: Beck, 1992) and Gerald Feldman and Klaus Tenfelde (eds), *Workers, Owners and Politics in Coal Mining: An International Comparison of Industrial Relations* (Oxford: Bloomsbury Academic, 1990). Stefan Berger, 'Introduction', pp. 1–11, and Andrew Taylor, 'So Many Cases but so Little Comparison: Problems of Comparing Mineworkers', pp. 12–28, and essays by Stefan Berger and Neil Evans, Sean Patrick Adams, Stephen Catterall and Keith Geldart, Brian McCook and Leighton James in Stefan Berger, Andy Croll and Norman LaPorte (eds), *Towards A Comparative History of Coalfield Societies* (Aldershot: Ashgate, 2005). See also Alan Campbell, Nina Fishman and David Howell (eds), *Miners, Unions and British Politics, 1910–1947* (Aldershot: Ashgate, 1996). More generally, see Christine Eisenberg, 'Comparative view in labour history', *International Review of Social History 34*(3) (1989); Ira Katzenelson, 'What now for labour history?', *International Review of Labour History* (Special Issue) (1994); and Stefan Berger, 'Comparative History' in: S. Berger, H. Feldner and K. Passmore (eds), *Writing History: Theory and Practice* (London: Arnold, 2003).

6 Examples in English include David Gilbert, *Class, Community and Collective Action: Social Change in Two British Coalfields 1850–1926* (Oxford: Oxford University Press, 1992); Roger Fagge, *Power, Culture and Conflict in the Coalfields: West Virginia and South Wales 1900–1922* (Manchester: Manchester University Press, 1996); John H.M. Laslett, *Colliers Across the Sea: A Comparative Study of Class Formation in Scotland and the American Midwest, 1830–1924* (Urbana: University of Illinois, 2000); Leighton James, *The Politics of Identity and Civil Society: The Miners in the Ruhr and South Wales, 1890–1926* (Manchester: Manchester University Press, 2008).

7 Dick Geary, *European Labour Protest 1848–1939* (New York: St. Martin's Press, 1981); Dick Geary, *European Labour Politics from 1900 to the Depression* (Basingstoke: Macmillan, 1971) and 'The Myth of the Radical Miners' in: Berger, Croll and LaPorte, *Towards a Comparative History of Coalfield Societies*, pp. 43–64. Stefan Berger, *The British Labour Party and the German Social Democrats, 1930–1931* (Oxford: Oxford University Press, 1994), especially pp. 1–17 on some issues of comparative history. Marcel van der Linden, *Transnational Labour History* (Aldershot: Ashgate, 2003); Jan Lucassen (ed.), *Global Labour History: A State of the Art* (Bern: Peter Lang, 2006).

8 Catherine Mills, *Regulating Health and Safety in the British Mining Industries, 1800–1914* (Aldershot: Ashgate, 2010).

9 B.R. Mitchell, *British Historical Statistics* (Cambridge, Cambridge University Press, 1988), p. 247.

10 Klaus Tenfelde, 'Comparative Research in the History of Mining Workers'.

11 Official quoted in the House of Commons on May 1924, in: *Dundee Courier*, 15 May 1924.

12 Martin F. Parnell, *The German Tradition of Organized Capitalism: Self-Government in the Coal Industry* (Oxford: Clarendon Press, 1994), pp. 29–33.

13 B.R. Mitchell, *Economic Development of the British Coal Industry 1800–1914* (Cambridge: Cambridge University Press, 1984), pp. 54–60.

14 Dick Geary, 'Rapport/Bericht' in: Tenfelde, *Sozialgeschichte des Bergbaus*, p. 561.

15 Joel Michel, 'Industrial Relations in French Coal Mining from the late Nineteenth Century to the 1970s' in: Feldman and Tenfelde (eds.), *Workers, Owners and Politics*, p. 278.

16 H. Clegg, A. Fox and A.F. Thompson, *A History of British Trade Unionism Since 1889, Volume 1: 1889–1910* (Oxford: Oxford University Press, 1964), p. 362 and p. 471; Peter Stearns, *Lives of Labour* (London: Holmes & Meier, 1975), p. 165 and 180. Dick Geary (ed.), *Labour and Socialist Movements in Europe before 1914* (Oxford: Berg Publishers, 1989), p. 123. More generally, see Keith Burgess, 'Coal-Mining' in: *idem, The Origins of British Industrial Relations* (London: Taylor & Francis Books Ltd., 1975), pp. 151–230.

17 There are detailed older accounts: Thomas Ashton, *Three Big Strikes in the Coal Industry*, 3 volumes (Manchester: Ashton, 1926–27) and Robin Page Arnot, *The Miners: A History of the Miners' Federation of Great Britain* (London: Allen and Unwin, 1949). For briefer, more recent accounts, see Clegg, Fox and Thompson, pp. 106–111; Roy Church, *The History of the British Coal Industry*, Volume 3: *1830–1913 Victorian Pre-Eminence* (Oxford: Oxford University Press, 1986), pp. 736–738, and Chris Wrigley, 'The Coal Lock-out of 1893', in: A. Charlesworth, D. Gilbert, A. Randall, H. Southall and C. Wrigley, *An Atlas of Industrial Protest in Britain 1750–1990* (London: Red Globe Press, 1996), pp. 116–121.

18 Chris Wrigley, 'The Government and Industrial Relations' pp. 135–158 and John Benson, 'Coalmining' pp. 187–208 in Chris Wrigley (ed.), *A History of British Industrial Relations 1875–1914* (Hassocks: Harvester Press, 1982). Chris Wrigley, 'Gladstone and Labour' in: R. Quinault, R. Swift and R.C. Windscheffel (eds), *William Gladstone: New Studies and Perspectives* (Hassocks: Harvester Press, 2012), pp. 51–70.

19 Michael Barker, *Gladstone and Radicalism* (Brighton: Harvester Press, 1975), pp. 92–94.

20 G.S. Bain and R. Price, *Profiles of Union Growth: A Comparative Statistical Portrait of Eight Counties (Warwick Studies in Industrial Relations)* (Oxford: Blackwell Pub, 1980), p. 39, 45 and 133.

21 Printed in full in H.C.G. Matthew (ed.), *The Gladstone Diaries, Vol. 13: 1892–1896* (Oxford: Clarendon Press, 1994), pp. 324–325.

22 Roger Davidson, 'Government Administration' in: Wrigley (ed.) *A History of British Industrial Relations 1875–1914*, pp. 159–183, and *Whitehall and the Labour Problem in Late Victorian and Edwardian Britain* (London: Routledge, 1985). E.H. Phelps Brown, *The Growth of British Industrial Relations* (London: Macmillan, 1959).

23 Chris Wrigley, *David Lloyd George and the British Labour Movement* (Hassocks: Harvester Press 1976), pp. 50–77, and 'Churchill and the trade unions', *Transactions of the Royal Historical Society 11* (2001), pp. 273–293. R. Page Arnot, *A History of the Scottish Miners* (London: Allen & Unwin, 1953), pp. 107–111.

24 Richard Morris to Tom Ellis, 19 August 1893; Tom Ellis Papers, National Library of Wales, 1524.

25 Leopold Haimson and Ronald Petrusha, 'Two Strike Waves in Imperial Russia, 1905–1907, 1912–1914', in: Leopold Haimson and Charles Tilly (eds), *Strikes, Wars and Revolutions in an International Perspective* (Cambridge: Cambridge University Press, 1989), pp. 101–166.

26 Madeleine Reberioux (ed.), *Fourmies et les Premier Mai* (Paris: de l'atelier, 1994), especially pp. 7–20 and 23–38.

27 Karl Ditt and Dagmar Kift, *1889, Bergabeiterstreik und Wilhelminische Gesellschaft* (Hagen: Berg Verlag, 1989).

28 Jeremy Brecher, *Strike!*, second edition (Cambridge: PM Press, 1997), p. 84. Michael Novak, *The Guns of Lattimer* (New York: Transaction Publishers, 1978). Rosemary Feuer, 'Remember Virden! The Coal Mines War of 1898–1900', *Illinois History*

Teacher 13(2) (2006), pp. 10–22 (also online). Thomas G. Andrews, *Killing for Coal: America's Deadliest Labor War* (Harvard: Harvard University Press, 2008).

29 For the ideas of the tariff reformers before the First World War, see Alan Sykes, *Tariff Reform in British Politics, 1903–1913* (Oxford: Clarendon Press, 1979).

30 The classic statement is A. Flanders and H.A. Clegg (eds), *The System of Industrial Relations in Great Britain* (Oxford: B. Blackwell, 1954). For a recent overview, see Chris Wrigley, 'Industrial Relations', in: N. Crafts, I. Gazely and A. Newell (eds), *Work and Pay in Twentieth Century Britain* (Oxford: Oxford University Press, 2007), pp. 203–224.

31 For a survey and the sources used, see Chris Wrigley, 'Introduction' in: Chris Wrigley (ed.), *Challenges of Labour: Central and Western Europe, 1917–1920* (London: Routledge, 1993), pp. 1–23. On coal output in Britain early in the war, see Sir R.A.S. Redmayne, *The British Coal Mining Industry during the War* (Oxford: Clarendon Press, 1923), pp. 12–15. For a more general survey, see Chris Wrigley, 'The First World War and State Intervention in Industrial Relations' in: Chris Wrigley (ed.), *A History of British Industrial Relations 1914–39* (Brighton: Prentice Hall, 1987), pp. 23–70, and 'Organised labour and the International economy', in: Chris Wrigley (ed.), *The First World War and the International Economy* (Cheltenham: Edward Elgar Publishing, 2000), pp. 201–215.

32 'The burden France has borne', *National Geographic Magazine 31* (April 1917), p. 332.

33 Coal Mining Organisation Committee, *First Report*, 7 May 1915, summary of conclusions printed in G.D.H. Cole, *Labour in the Coal Mining Industry 1914–21* (Oxford, 1923), pp. 17–19. Sir R.A.S. Redmayne, *The British Coal Mining Industry During the War* (Oxford, 1923), p. 19.

34 Wrigley, *David Lloyd George and the British Labour Movement*, chapter 7.

35 Redmayne, *British Coal Mining Industry*, p. 27.

36 G.D.H. Cole, *Labour in the Coal Mining Industry, 1914–1921* (Oxford, 1923), pp. 40–41, 68–70. For Lloyd George's comments, see Chris Wrigley, *Lloyd George and the Challenge of Labour: The Post War Coalition, 1918–22* (Brighton: Harvester-Wheatsheaf, 1990), pp. 149–150. For a famous version of this, see Aneurin Bevan, *In Place of Fear* (London: Simon & Schuster, 1952), pp. 20–21.

37 Figures calculated from Labour Force, Table 17, B.R. Mitchell, *British Historical Statistics* (Cambridge: Cambridge University Press, 1988), p. 145. For the major involvement of the government, see Wrigley, *Lloyd George and the Challenge of Labour.* For the coal industry, see in particular, Barry Supple, *The History of the British Coal Industry, Volume 4: 1913–46* (Oxford: Oxford University Press, 1987). For an excellent reassessment of the 1926 lock-out, see John McIlroy, Alan Campbell and Keith Gildart, *Industrial Politics and the 1926 Mining Lock-Out* (Cardiff: University of Wales Press, 2004).

38 Chris Wrigley, 'The Second World War and State Intervention in Industrial Relations 1939–45', in: Chris Wrigley (ed.), *A History of British Industrial Relations 1939–1979* (Cheltenham: Edward Elgar Publishing Ltd., 1996), pp. 12–43. Supple, *History of the British Coal Industry*, pp. 525–526.

39 *The Cornishman* (7 August 1941).

40 Alan Bullock, *The Life and Times of Ernest Bevin, Volume 2: Minister of Labour 1940–1945* (London: Heinemann, 1967), pp. 160–161. Supple, *History of the British Coal Industry*, p. 475 and pp. 558–559. W.H.B. Court, *Coal* (London: HMSO and Longman Green, 1951).

41 Reginald Sorensen, a Labour MP, complained in the House of Commons, that Indian women were paid only 'a penny an hour upward' and was told by Leo Amery, secretary of state for India and Burma, that while pay was low, the rates 'compare favourably with the rates of other industries in that part of India', *Gloucester Citizen* (8 February 1945) and *Sunderland Daily Echo and Shipping Gazette* (1 November 1945).

42 The Jewish Virtual Library, 'Forced (Slave) Labour', www.jewishvirtuallibrary.org/forced-labor (accessed online on 22 January 2015).

43 According to reports from the Jewish News Agency (JTA) in Zurich (7 August 1942).
44 Michel, 'Industrial Relations in French Coal Mining', pp. 305–310.
45 Phillip Deerry, 'Chifley, the Army and the 1949 coal strike', *Labour History 68* (1995), pp. 80–97. Phillip Deery (ed.), *Labour in Conflict: The 1949 Coal Strike* (Sydney: Hale & Iremonger, 1978). Neville Kirk, *Labour and the Politics of Empire* (Manchester: Manchester University Press, 2011), pp. 183–184. Justin Davis Smith, *The Attlee, and Churchill Administrations and Industrial Unrest, 1945–1955* (London: Pinter Publishers Ltd., 1990).
46 Deutsche Welle staff, 'The Rise and Fall of Germany's Coal Mining Industry', online, 31 January 2007, www.com/en/the-rise-and-fall-of-germany's-coal-mining-industry (accessed 19 November 2014). *Guardian* (30 October 2014).
47 *Wall Street Journal* (5 January 2015). *Reuters' news release* (17 January 2015). *IndustriALL Global Union*, www.industriall-union.org/search?regions-181 (8 February 2015).
48 Radio Free Asia, 'Congyue Dai on strikes in China', www.rfa.org/English/news/china/strikes (5 January 2015). 'In Defence of Marxism', www.marxist.com/in-defence-of-marxism.htm (27 January 2015), www.com/en/the-rise-and-fall-of-germany's-coal-mining-industry (accessed 30 January 2015). Mining-technology Market and Insight, 'Coal giants: the world's biggest coal producing countries'www.mining-technology.com/ . . ./feature-coal-giants-the-worlds-biggest-coal-producers, (accessed, 20 January 2015).
49 BBC News India website (8 January 2015). More generally on Indian authorities being brutal to villagers, see the work of Pavritha Naryanan.
50 Australian Mining, www.pc.gov.au/inquiries/completed/. . ./subdr060-resource-exploration.pdf (29 October 2014 and 10 February 2015). AMMA, mining, oil and gas website on jobs for women in coal mining (accessed, 14 May 2013). In 1984 in the US women were 5.4 per cent of the mining labour force. The Department of Labor, Women's Bureau, *The Coal Employment Project: How Women Make Breakthroughs into Non-traditional Industries* (1985). Women have long worked in mines in China.
51 IndustriALL, www.amma.org.au/. . ./miners-call-this-international-women-s-day-we-need-you (1 April 2012). Australian Mining website (29 October 2014).
52 Ignatov and Company, 'Coal mining in Russia', https://d11p9je03a4iqr.cloudfront.net/preview/. . .coal/. . ./WorldCoal.2015-Preview.pdf (accessed 24 January 2015). Dmitry Butin, 'Coal industry, 1991–2001', *Kommersant* (11 February 2015).
53 African Mining IQ, www.projectsiq.co.2a/what-is-mining-iq.htm (accessed 24 January 2015). Energy Department, Republic of South Africa, 'Coal', www.energy.gov.za/files/coal-frame.html (accessed 24 January 2015). Mining Technology, Market and Customer Insight, www.intelligence.mining-technology.com (21 October 2014). https://oecd-library.org/medium-term-coal-report-june-2015-ntrcoal-2
54 Jade Davenport, *Digging Deep: A History of Mining in South Africa* (Johannesburg: Jonathan Ball, 2013), pp. 431–432 and 461–462. Peter Alexander, Thapelo Lekgowa, Botsang Mnope, Luke Sinwell and Bongani Xezwi (eds), *Marikana: The View from the Mountain and a Case to Answer* (Johannesburg: Bookmarks, 2012).
55 Roy Church and Quentin Outram, *Strikes and Solidarity: Coalfield Conflict in Britain 1889–1966* (Cambridge: Cambridge University Press, 2002). For an earlier influential detailed study, see the government funded study by C. Smith, R. Clifton, P. Makeham, S. Creagh and R. Burns (eds), *Strikes in Britain* (London: HMSO, 1978)
56 International Energy Agency, *Medium Term Coal Market Report 2015* (Singapore, December 2015).

12 This land is my land

Global indigenous struggles and the Adivasi resistance in Muthanga (Kerala, India)

Pavithra Narayanan

If we ever owned the land we own it still, for we never sold it. In the treaty councils, the commissioners have claimed that our country had been sold to the Government. Suppose a white man should come to me and say, "Joseph, I like your horses, and I want to buy them." I say to him, "No, my horses suit me, I will not sell them." Then he goes to my neighbor, and says to him: "Joseph has some good horses. I want to buy them, but he refuses to sell." My neighbor answers, "Pay me the money, and I will sell you Joseph's horses." The white man returns to me, and says, "Joseph, I have bought your horses, and you must let me have them." If we sold our lands to the Government, this is the way they were bought.

(Heinmot Tooyalakekt (Chief Joseph), leader of the
Wal-lam-wat-kain band of Nimi'ipuu, 1879)

Chief Joseph's narrative about the methods employed to usurp indigenous lands is located in a long and brutal history of colonial expansions, conquests, and arbitrary notions of ownership which disregarded pre-existing indigenous rights and land tenure systems. The creation of a rule of law was imperative to legitimate colonial authority, claim and control land and resources, and deny indigenous populations occupancy and sovereignty. In some cases, colonizers negotiated treaties with tribal communities, but they also used the formal legal arrangements to dispossess and displace indigenous peoples. In the second half of the twentieth century, indigenous rights have garnered greater international and domestic attention. While attempts have been made to address past injustices, as in the past, legislation continues to play a significant role in recognizing and at the same time, limiting indigenous rights.

The struggle for indigenous rights is inextricably linked to mining interests and history. As Subhabrata Banerjee points out, "Since the earliest times of European invasion, the two industries that led to the greatest dispossession of Aboriginal peoples from their land were the pastoral and mining industries."[1] In an era when "the discourse of development" is used to promote private and economic interests, Banerjee goes on to say that mining extractions on indigenous lands have only increased.[2] Given the abundant presence of precious mineral natural resources on indigenous territories, these lands have become the grazing grounds for capitalist expansions. Since the 1970s, governments have consciously enforced policies that favor mining and other commercial undertakings over indigenous rights. Within the context of contemporary neoliberal reforms that have destroyed and displaced millions of indigenous populations, this chapter examines indigenous opposition to government and corporate incursions on their lands. I specifically discuss the collective action by the Adivasis[3] in Muthanga in India's Wayanad district of Kerala and the government's failure to uphold the constitutional rights of its citizens. I argue that the resistance movements by various indigenous groups around the world against capitalist interests offer the last ray of hope for future generations.

Global indigenous movements

Over the last several decades, indigenous movements have gained momentum and increasing visibility in countries including Australia, Bolivia, Brazil, Canada, Ecuador, India, New Zealand, and the United States. From Wayanad to the Amazon, indigenous communities organize rallies and sit-ins and march the city streets with slogans that say "From colonies to farmlands," "Take back the land," "Land and Liberty," "For everyone, everything. For us, nothing," "No to land privatization," and "We shall give up our lives but not land." Restoration of alienated lands, protection of constitutional rights, recognition of indigenous territorial lands, and rights to self-representation are some of the demands of indigenous populations. While issues such as displacement, discrimination, and land rights have always been a central focus of indigenous movements, earlier, these struggles were against European colonial rule and expansions. With the rise of modern states, the movements developed to counter internal colonization and racism. Since the late 1960s, the emergence of indigenous coalitions such as the American Indian movement (AIM), the World Council of Indigenous People (WCIP), the Zapatista movement, the Indigenous Peoples' Network (IPN), La Via Campesina, Regional Indigenous Council of Cauca (CRIC), Indigenous Missionary Council (CIMI), Coordination of the Indigenous Organizations of the Brazilian Amazon (COIAB), Idle No More, the Jharkhand movement, and the Adivasi Gothre Mahasabha (AGMS) is linked to inequalities, injustices, and ecological crises engendered by economic liberalization. As Tom Hayden, the National Coordinator of the Mexico Solidarity Network, noted, "The globalization of trade and capital, also caused a globalization of conscience and resistance."[4] Most of "the world's remaining natural resources" are on terrains inhabited by indigenous people,[5] and the uprisings and demonstrations of indigenous groups in the twenty-first century are largely a response to the destruction and exploitation of their lands, lives, and livelihoods by corporations and governments.

Massive agricultural and industrial undertakings by corporations in countries including Australia, Brazil, Ghana, Guatemala, Honduras, India, Indonesia, Papua New Guinea, Peru, and Tanzania, that followed the implementation of neoliberal economic policies, have compounded problems of displacement, inequity, loss of livelihoods, food and water insecurity, deforestation, and loss of species. Philippines' San Roque dam project and its Mining Act of 1975, which allowed foreign companies to purchase nearly half of Cordillera's land area, have displaced and affected the livelihoods of thousands of indigenous Ibaloi people.[6] In Labrador and Quebec, the Lower Churchill Hydroelectric Project threatens to submerge ancestral lands of the Innu indigenous people.[7] Quebec's La Grande complex has already drowned approximately "10,000 caribou and countless other animals, along with 4,400 square miles of indigenous lands."[8] The Belo Monte Dam is projected to displace more than 20,000 indigenous people and destroy "over 1,500 square kilometers of Brazilian rainforest."[9] Other large scale dam projects such as Chixoy (Guatemala), Inambari (Peru), Narmada (India), Myitsone (Burma), and Bakun (Malaysia) have uprooted thousands of indigenous populations from their homes and flooded large acres of their land. The Chad-Cameroon Petroleum Development and Pipeline Project, financed by the World Bank, Exxon-Mobil, and Chevron Texaco, which runs through the lands of the Bagyeli, Brendan Schwartz and Valery Nodem have noted,[10] directly impacted 248 villages. Oil spills and security measures by the Cameroonian coast guard and Exxon private security have prevented the Kribians from fishing. "The lust for black gold and the actions of their corporations," the authors point out, have also adversely impacted the environment and communities in countries including Nigeria, Equatorial Guinea, Gabon, Congo-Brazzaville, Angola, and Sudan.[11] Logging in Indonesia and Peru has destroyed millions of hectares of forests; a large portion of these forest regions belongs to indigenous communities.[12] In Guatemala, 3,000 Q'eqchi Maya Indians were forcefully evicted in 2011, when an agribusiness firm purchased their land for a sugar plantation.[13] The World Development Movement (WDM) reported that Columbia's Cerrejón coalmine located in the middle of the Wayuu people's territory has destroyed several indigenous communities including Roche, Chancleta, Tamaquitos, Manantial, Tabaco, Palmarito, El Descanso, Caracoli, Zarahita, and Patilla.[14] Sixty-five of Columbia's 102 indigenous groups are now on the verge of extinction, yet, mining operations on their lands have not abated.[15]

In India, development projects have displaced over 40% of the country's 8% Adivasi population.[16] In the central Indian state of Chhattisgarh, where Adivasis constitute 32% of the population, over 50% of tribal populations have lost their lands.[17] With 55% of its rural population living below the poverty line, low literary rates, high malnutrition and mortality rates, and the lack of access to basic amenities, Chhattisgarh is categorized as one of the "most backward" states.[18] However, the attention that the state receives today has little to do with attempts to address its social, infrastructural, and economic problems. Governments and corporations, Sudha Bharadwaj[19] has observed, are scrambling for land in Chhattisgarh because it contains more than 28 precious minerals, including coal, iron ore, diamonds, gold, bauxite, uranium, and copper. Jharkhand is another government classified "backward" state that is a hotbed for foreign investment. This region, Mattew Areeparampil has noted, contains some of the world's largest deposits of coal, iron-ore, copper, uranium, and high-grade kyanite and "produces 48 per cent of the country's coal, 45 per cent of its mica, 48 per cent of its bauxite, 90 per cent of its apatite and all of its kyanite."[20] The colossal number of industrial and mining activities undertaken in Jharkhand has adversely impacted its landscape and communities, particularly the Adivasis, who comprise 28% of Jharkhand's population.[21] In 2005, the Jharkhand government and companies including Mittal, Tata, Jindal Steel, and Power Company Limited signed 42 memorandums of understanding (MOUs) to construct steel plants. It is estimated that this agreement will displace 10,000 families and cause further deforestation of 57.15 square kilometres of land.[22] Other Indian states with vast reserves of mineral resources that are caught in profit-driven land-grab schemes of corporations include Andhra Pradesh, Bihar, Gujarat, Karnataka, Orissa, Rajasthan, Tamil Nadu, and West Bengal. The human and environmental costs outlined above, which are direct results of unceasing quests to expand empires, represent far from an exhaustive inventory of the global reach of capitalist expansions and their impact on indigenous populations. In the era of neoliberalism, economic interests of corporations and governments override the rights of indigenous peoples; oil pipelines, big dams, and mining, alongside privatization of land, forests, and water, and billion-dollar acquisitions and investments by corporations have become landmarks of development.

Dominant narratives about the neoliberal trajectories of many Latin American, African, and Asian countries represent organizations such as the World Bank and International Monetary Fund as having mandated pro-market economic policies in return for financial aid. What is often conveniently omitted from such narratives is that the ruling and elite classes of countries that were forced to accept such deals shared the same interests and goals of the international financial institutions mandating liberalization. The financial crises of the 1980s and 1990s merely provided an opportune moment for different power players to further their capitalist interests.[23] Once trade agreements were in place, proponents of capitalism promoted the idea that economic liberalization was a prerequisite for development, and local governments instituted their own market-oriented national measures to transfer ownership of industries and lands to corporations and private investors. Argentina's economic policy changes in the 1990s primarily benefitted private investors, and today, the country has a wave of new landlords such as the Benetton brothers, Sylvester Stallone, George Soros, Ted Turner, and Joe Lewis. In Patagonia, 900,000 hectares of land, which includes Mapuche tribal land, is now owned by Benetton.[24] The Fujimori government began implementing neoliberal policies in 1991 with the aim of attracting foreign investors. In 1996, Peru implemented the National Mining Cadastre Law, granting "national and transnational mining firms exclusive control of the necessary land resources to implement their operations."[25] Within a decade, private industries from Australia, Canada, US, Japan, Mexico, South Africa, and Switzerland owned a majority of Peru's mining operations.[26]

India's move to liberalize its economy and open its borders to foreign markets began during Narasimha Rao's tenure as prime minister in 1991. His successor, Manmohan Singh, who was Rao's finance minister, zealously implemented market-based initiatives, removed rules and regulations for foreign investment, and shifted control of many state-owned enterprises to private entities.[27] Historian Ramachandra Guha notes:

Adivasis were displaced from their lands and villages when the state occupied the commanding heights of the economy. And they continue to be displaced under the auspices of liberalization and globalization. The opening of the Indian economy has had benign outcomes in parts of the country where the availability of an educated workforce allows for the export of high-end products such as software. On the other hand, where it has led to an increasing exploitation of unprocessed raw materials, globalization has presented a more brutal face. Such is the case with the tribal districts of Orissa, where the largely non-tribal leadership of the state has signed a series of leases with mining companies, both Indian and foreign. These leases permit, in fact encourage, these companies to dispossess tribals of the land they own or cultivate, but under which lie rich veins of iron ore or bauxite.[28]

In 2008, the Singh government implemented yet another market-driven national policy to encourage foreign companies to invest in mining operations in India. While the policy acknowledges that "mining activity . . . has the potential to disturb the ecological balance of an area" and specifies that "special care will be taken to protect the interest of host and indigenous (tribal) populations," it also categorically states that "the needs of economic development make the extraction of the nation's mineral resources an important priority."[29] Such national policies are part of a growing list of laws and acts that are inherently contradictory; although land rights of indigenous people seem to be recognized, there is no clause granting them the right to refuse or veto "development" activities on their land. They are offered one option – trade land rights for compensation. Many indigenous groups, however, have refused this offer, and remind the political and corporate class that they have at least one other option – resistance.

Across the world, indigenous movements are calling attention to systemic discriminations, violence, racism, and human rights violations. Asserting a politics of self-determination, of ethics, and justice, these movements challenge the notion that neoliberal policies are liberatory. Some of the recent industrial projects that indigenous groups protested against include Malaysia's Murum dam which will displace Penan and Kenyah native communities;[30] mining extractions by the British company, Beowulf Mines, on Sami land in the Arctic circle;[31] British Columbia's Arctos Anthracite Project, a coalmining operation on Tahltan First Nation land;[32] a proposed mining law that would affect the Panama's Ngöbe people;[33] the TransCanada Keystone XL pipeline that runs across Native treaty lands;[34] and uranium extractions on Ardoch Algonquin First Nation land in Canada.[35] Legal rights have done little to grant land ownership to the Ogiek of Kenya and the Bushmen of Botswana and Namibia, and to prevent expansions of ski resorts on Secwepemc first nation land, and the incursion of loggers and settlers in the eastern Amazon area of Brazil.[36] The Zapatistas rebellion of 1994 is one of the earliest uprisings against neoliberalism. The Zapatistas declaration that "NAFTA is death"[37] captures the violence of market-driven policies on communities and the environment. Murder, rape, torture, intimidation, interrogation, displacement, and destruction of livelihoods, forests, land, and water are among the many global manifestations of market-sponsored violence.

Political responses to organized indigenous resistance against neoliberalism have been primarily reactionary. Combining a variety of tactics, including the use of extrajudicial actions, anti-terrorist laws, and military, para-military, and private security forces, governments have made every attempt to destabilize and delegitimize indigenous movements. Protestors have been arrested, illegally detained, tortured, and killed. There is little regard for due process, and even the most peaceful demonstrations and blockades often end in violent state action. In every instance of collective resistance, conflicts have arisen because of human rights and constitutional violations, forceful evictions, rampant police and military violence, and government and corporate incursions in indigenous territories. There is a widespread condemnation of the armed resistance by indigenous groups, but the crimes committed by the state against indigenous people have largely been ignored and often perceived as necessary counterterrorist measures. Moreover, the "economic violence" on indigenous communities, Balakrishnan Rajagopal rightly argues, is "treated as out of bounds of human rights law,"[38] and "while the march of the market is celebrated unreservedly,"[39] marches by indigenous peoples are treated as criminal, anti-state and anti-progress activities. To rephrase the words of Mike Dodson, member of the United Nations Permanent Forum on Indigenous Issues and Australian of the Year, 2008, it is not the resistance by indigenous peoples or their demands, but the dispossession, the denial of constitutional rights, the human rights violations, and the crimes committed by the state that are "barriers to progress."[40]

Most of the literature addressing the conflicts between the state and indigenous peoples suggests that solutions can be found through legislation by dissolving colonial laws, upholding land acts and treaties, including indigenous people in decision-making processes and negotiations, sharing profits among stakeholders, and providing adequate compensation and rehabilitation. The underlying message here is that indigenous lands are a saleable commodity and indigenous communities have to accept that "development" operations will continue on their land. The idea that the stakeholders include governments and corporations,[41] and that monetary compensation or relocation adequately address problems of dispossession and displacement, further suggest that the land does not belong exclusively to indigenous people, and that governments have the right to offer petty concessions in return for land rights. Furthermore, the proposed solutions fail to consider that many indigenous communities do not support industrial activities on their land, they are unwilling to extinguish their land rights, and they refuse to trade their livelihoods and lands in return for any form of compensation. Ecological consequences that result from industrial projects also go routinely unaddressed. Instead of decentralizing power, the focus on legislation as the only avenue for redress, reinstates the political class at the center as the authority figure. As Howard Zinn has observed, "The rule of law does not do away with the unequal distribution of wealth and power but reinforces that inequality with the authority of law."[42]

Having faced never-ending legal battles, indigenous communities are fully aware that legislation can be and is manipulated and existing laws do not always benefit and protect them. For instance, market-oriented legislation undermines "traditional communal or tribal holdings and open the way to their dispossession by third parties or other private or corporate interests"; and some environmental and conservation policies work against indigenous communities, preventing them from accessing or using traditional lands, forests, and resources for subsistence activities.[43] Even Native Title Acts and constitutionally mandated legislation, which recognize the land rights of indigenous communities as pre-existing rights, place the burden of proof on indigenous people. In order for their land rights to be recognized, Aboriginal and Torres Strait Islander peoples have to successfully prove that the areas claimed are in accordance with "a system of traditional laws and customs which has its roots in a society that preceded the date on which the Crown asserted sovereignty (between 1788 and 1879 depending on where in Australia the claim is made)."[44] Similarly, in India, several courts ordered the restoration of alienated land to Adivasis,[45] however, the onus is on tribal communities to demonstrate that the land originally belonged to them. In many cases, alienation has been impossible to prove and the Adivasis have been forced to settle for alternate land. The lessons that indigenous people learned from their interactions with the state and the judiciary have forced them to take collective action.

Political powers might imagine that large-scale development projects and billion-dollar agreements are necessary to build stronger economies and nations, but the struggles of indigenous peoples reflect a brutal reality, the reality that nation formation has and continues to be extremely violent. In Australia, Aboriginal groups refer to the day that marks the anniversary of the arrival of the British in Sydney as "Invasion Day." Reflecting on the importance given to such anniversaries, at a protest organized in August 2013, Jay McDonald, an activist with the Tasmanian Aboriginal Centre, said:

> True reconciliation cannot be achieved and a just society cannot be built if we continue to celebrate the gains of one race at the expense of another. Invasion Day is a day to remember the wrongs that were committed against Aborigines, a day to remember the injustices forced upon one race of human beings by another. This is no day for celebrating; it's a day for mourning, a time to reflect, and a time to steel ourselves for the ongoing battle for a better society.[46]

On October 12, 2013, in Chile, the Mapuches also reminded the world that Columbus Day is not a day of celebration: 15,000 people participated in a march demanding a return of ancestral lands and the right to self-determination.[47] It is within this larger global history of indigenous struggles against imperialism, violence, dispossession, and disenfranchisement that the resistance of the Adivasis in Muthanga is situated.

The Muthanga struggle

On January 4, 2003, 820 Adivasi families moved to a section of a eucalyptus plantation in Muthanga in the Wayanad district of India's southern state of Kerala.

Making a conscious decision to avoid natural forests, the group chose to live in a degraded part of the plantation. The government received advance notification about the occupation.[48] The families built huts, dug wells, raised cattle, and built playgrounds for children. For 45 days, they encountered no problems. On the 46th day, someone set fire to the forests and the Adivasis were accused of the act. Alleging that the community was destroying the forest, on February 19, 2003, the Kerala police launched a brutal attack against the Adivasis.[49] Several television stations aired video footage of police opening fire and beating the Adivasis mercilessly, yet, Kerala's forest minister, K. Sudhakaran, claimed that the police launched a restrained attack. It was noteworthy, he said, that in spite of the police firing 18 rounds, only one person was killed, "that too by a plastic bullet." Asserting that women and children were not beaten, and those who were beaten, received blows only below their knees, the minister declared that the police deserved "praise for applying so little force."[50]

The government claims that only two people (an Adivasi and a policeman) died during the stand-off between the police and the Adivasis, however, eye witness accounts tell a different story. *Kairali* television's cameraman, Shaji Pattanam, who filmed the attack, recounted:

> The tribal people were shouting slogans but standing far away. The police attacked first. The women started screaming and tried to run. The police began shooting. They used .303s not rubber bullets. Those who were not shooting were lashing out with their *lathis*. They dragged the women by the hair and hit them viciously. One small child had his head split open. A pregnant woman fell down but still they hit her. They hit even dead bodies. Maybe they were unconscious. But at the scene of an accident you know instinctively by the posture, a person is dead. I counted four or five.[51]

Madhyamam reporter Mohammed Sharif's recollection of the events corroborates reports that the death toll was far higher than the government was willing to admit:

> When the police began firing, the Adivasis did not know what hit them. They began screaming. Jogi, the man who was standing with the lighted torch over Vinod, was the first to be hit. Without waiting, the man next to Vinod hacked at him. He was immediately arrested. Then we heard non-stop gunfire; pitiful screams; people were running. The police were in hot pursuit, firing relentlessly at even women and children. Others followed, hitting them with *lathis* and rifles. They surrounded me and said: 'We will bury you also along with these people here.' It was a face-to-face encounter with death. I cannot describe how I ran and managed to escape. At least 15 people fell to the bullets. That is what I saw. The police were chasing and shooting women and tiny children. There is no count of those who fell and died there. Though the police and the Chief Minister initially announced that five Adivasis had died, they later changed it to one.[52]

As Sharif says, there is no way of knowing how many people died, given the government's repeated refusal to conduct a judicial inquiry. Many Adivasis, adults and children, are still missing.[53] To date, no action has been taken against the police, and the events that transpired at Muthanga in 2003 continue to remain under investigation.

The occupation of Muthanga was a symbol of resistance and a collective demand for justice; it reflected a community's unshakable determination to carve out its own future. Muthanga became a home for 2,500 people, 500 of them children, when the state did not follow through on a 2001 agreement between Kerala's A.K. Antony's government and the Adivasi Gothre Mahasabha (The Grand Assembly of Adivasis, AGMS) to provide five acres of land to Adivasis who were landless or owned less than an acre of land.[54] Additional terms of the agreement included commencing land distribution on January 1, 2002 and completing the task on December 31, 2002; the agreement ceded Adivasis areas in Kerala under Schedule V, thereby granting Adivasis the right to self-governance; and the government agreed to abide by Supreme Court decisions regarding land transfers to Adivasis in Kerala.[55] However, at the end of 2002, of the 53,472 Adivasi families identified to receive land,[56] only 843 beneficiaries were allotted about three acres each.[57] It was after the government stopped allotting land and the deadline of the agreement expired that various groups of Adivasis across Kerala collectively decided the next course of action – to claim their rights by occupying state-owned lands.[58] A week after the violent eviction of the Adivasis in Muthanga, Kerala Congress Committee President K. Muraleedharan stated that the government could not uphold its promises to its citizens. The five-acre agreement, he said, was an erroneous calculation; if Adivasi families did receive land, less than an acre would be allotted to each family.[59]

Public demand for inquiries into the 2003 police attack on the Adivasis led to official investigations by the National Commission for Women (NCW), the National Human Rights Commission (NHRC), and the Scheduled Caste and Scheduled Tribes Commission (SC/ST Commission). All three committees interviewed the Adivasis and categorically concluded that the government and the police grossly violated laws and human rights. The NCW found that the police molested Adivasi women and tortured women and children and it indicted "the Kerala government for failing to provide adequate security and basic amenities to tribals in Wayanad."[60] Despite its findings, the NCW did not call for legal action against the police; its recommendations included providing counseling for "the mentally-shattered tribal women" and rehabilitating the Adivasis, "who have lost everything."[61] The People's Judicial Enquiry Commission (PJEC) set up by the NHRC took a stronger stance against the atrocities committed by the police. Denouncing the Kerala government's claim that the police had no alternative except to use armed force, the committee pointed out that "there is no provision in the Constitution giving any right to the police to kill any person" and there is no provision in the law that permits illegal custody of men, women and children. By denying the Adivasis their land rights, the PJEC stated that the government had violated human rights and Constitutional privileges. "In the absence of any genuine attempt on the part of the government" to implement the land agreement, the Commission said, "the Adivasis had no choice but to continue their struggle and remain on the land." The Commission also held all successive governments liable "If the Adivasis had to continue their struggle for their land rights, whatever be the mode of struggle – violent or otherwise." The Kerala government refused to comply with the PJEC's explicit recommendation that "every killing by the police . . . [must] be subject to a judicial enquiry."[62] What followed was an inquiry by the Central Bureau of Investigation (CBI), and contrary to the Adivasis' expectations, the CBI investigations did not result in justice.

Despite the reports submitted by the three commissions, testimonies of the Adivasis, statements by television reporters who witnessed the attack, and video footage and photographs that clearly demonstrate the brutality of the police, the CBI not only absolved the police and the government of any and all wrongdoings, but it also filed criminal charges against the Adivasis.[63] The Forest Department followed suit and 12 cases were registered against the Adivasis under various sections of the Wildlife Protection Act and the Indian Penal Code.[64] The charges included murder, trespass, arson, destruction of property, and bearing weapons.[65] For nearly a decade, says C.K. Janu, leader of the AGMS, the 183 Adivasis accused of various charges have been forced to appear in court every month in order to avoid further warrants issued against them.[66] However, the Adivasis from Muthanga who returned to Challigada colony refute Janu's claim that the AGMS pays for court fees and transport costs from Wayanad in Kozhikode to the High Court at Ernakulum. Talking to India Vision's news reporter, the Adivasi women point out that they had to sell paddy and livestock in order to attend court proceedings. Most of the men are bed-ridden and unable to work due to the injuries they suffered during the attack. Whatever money the women make from daily labour wages goes towards legal fees. Little money is left for food, and none for medical expenses or other costs.[67] The Adivasis feel betrayed, yet again, by civil society, by the government, and by their leaders.

The Challigada colony families moved to Muthanga because their houses are built on the banks of the Kabini River and every time the water rises, the people have to relocate to dryer areas.[68] While many reports state that no one really knows what happened at Muthanga, the residents of Challigada colony clearly remember what took place. They have not forgotten the violence, beatings, gunshots, interrogations, abuse, and endless trials. But when all their testimonies have only led to a cruel denial of justice, the Challigada Adivasis rightly refuse to answer any more questions. They want answers. They want to know why a public and a media that is still interested in Muthanga never enquired about their lives earlier. Each year, the anniversary of the Muthanga occupation is marked by commemorations, gatherings, and news stories, but for the Adivasis who live in Challigada colony, the emotions that Muthanga evokes are fear, grief, and despair. The CBI and the police grilled all the Adivasis who lived in Muthanga and those who supported them and branded them as terrorists. Since 2004, the Kerala government has distributed more land to other Adivasi families, but none of the people involved with Muthanga were allotted any land.[69] This group has to prove first that they are not criminals, ivory smugglers, have no links with extremist organizations, and are not enemies of the state. In August 2011, the state acquitted 70 of the 114 Adivasis held on charges of encroachment and destruction of wildlife and forests.[70] The Adivasis point out that the forest is also their home, but they are treated like intruders and criminals because civil society and governments operate under the notion that only animals live in forests.[71]

What kind of a judicial verdict would fail to consider even investigative reports submitted by official government commissions? Can there be justice when criminal acts by law and order personnel are not viewed as culpable offences? Why did it take violent government evictions in Muthanga for the struggles of the Adivasis of Challigada colony to become the centerpiece of a television news story? These legal, constitutional, and ethical questions that Muthanga raises do not pertain to the Adivasis of Kerala alone. The violence against Wayanad's indigenous communities started long before they decided to move to Muthanga; it is rooted in a long and ruthless history of oppression and unjust laws, in centuries of betrayals and brutalities. Although ownership of land is a central issue in the Muthanga struggle, Muthanga, like other indigenous struggles, is more than a land rights battle. It is a battle for a just world.

> As I went walking I saw a sign there
> And on the sign it said "No Trespassing."
> But on the other side it didn't say nothing,
> That side was made for you and me.
> In the shadow of the steeple I saw my people,
> By the relief office I seen my people;
> As they stood there hungry, I stood there asking
> Is this land made for you and me?
> (Woody Guthrie, *This Land is Your Land*)[72]

Notes

1 Banerjee, Subhabrata, 'Whose Mine is it Anyway? National Interest, Indigenous Stakeholders and Colonial Discourses: The Case of the Jabiluka Uranium Mine', *Organisation Environment 13*(3) (2000), p. 13. Accessed October 10, 2013. doi: 10.11 77/1086026600131001

2 *Ibid.*

3 Although officially classified as "scheduled tribes," India's indigenous communities are ethnically and culturally distinct and different from each other. The term "Adivasi" most commonly refers to communities living in forest or hill regions. Despite legislative and Constitutional measures to safeguard the rights of Adivasis, policies instituted by various governments over the years have only served to disenfranchise and displace these communities.

4 Tom Hayden, 'Introduction', in: idem (ed.), *The Zapatista Reader* (New York: Nation Books, 2002), p. 2.

5 'Opening two-week session, UN Indigenous forum tackles land, resource issues,' *UN News Centre*, May 14, 2007. Accessed June 9, 2016.
    www.un.org/apps/news/story.asp?NewsID=22531&Cr=Indigenous&Cr1

6 UNPO, *Unrepresented Nations and Peoples Organisation (UNPO) Yearbook* (Netherlands: Kluwer Law International, 1997), pp. 70–71.
7 Leopoldo Jose Bartolome, Chris de Wet, Harsh Mander, and Vijay Kumar Nagraj, 'Displacement, Resettlement, Rehabilitation, Reparation and Development', *WCD Thematic Review 1*(3) (2000), p. 16. (Prepared as an input to the World Commission on Dams, Cape Town). Accessed March 3, 2012.
     http://siteresources.worldbank.org/INTINVRES/Resources/DisplaceResettle RehabilitationReparationDevFinal13main.pdf
8 Desiree Hellegers, 'From Poisson Road to Poison Road: Mapping the Toxic Effects of Capital in Linda Hogan's *Solar Storms*', *Studies in American Indian Literature 27*(2) (2015), p. 2. Accessed June 28, 2016. doi: 10.5250/studamerindilite.27.2.0001
9 Marcelo Diversi, 'Damming the Amazon: The Postcolonial March of the Wicked West', *Cultural Studies ↔ Critical Methodologies 14*(June 2014), p. 242, p. 246. Accessed July 15, 2015. doi: 10.1177/1532708614527557
10 Schwartz, Brendan and Valery Nodem. 'A Humanitarian Disaster in the Making along the Chad-Cameroon Oil Pipeline: Who's Watching?' *AlterNet*, 1 December 2009, n. pag. Accessed November 11, 2013. www.alternet.org/
11 *Ibid.*
12 Neva Collings, 'Environment', in: United Nations, *State of the World's Indigenous Peoples* (New York: United Nations Publication, 2009), pp. 89–91. Accessed November 4, 2013. www.un.org/esa/socdev/unpfii/documents/SOWIP/en/
13 Benno Hansen, *Ecowar: Natural Resources and Conflict* (Denmark: Books on Demand, 2011), p. 63.
14 'The Cerrejón mine: Coal exploitation in Colombia.' World Development Movement, May 2013. Accessed November 4, 2013.
     www.banktrack.org/manage/ems_files/download/158e31a/cerrejon_media_briefing.pdf
15 Dan Kovalik. 'Colombia: Ethnocide and Political Violence on the Rise.' *The Huffington Post*, March 28, 2016. Accessed June 12, 2016. www.huffingtonpost.com/dan-kovalik/colombia-ethnocide--polit_b_9556570.html
16 Dip Kapoor, 'Human Rights as Paradox and Equivocation in Contexts of Adivasi (original dweller) Dispossession in India', *Journal of Asian and African Studies 47*(4) (2012), p. 405. Accessed April 4, 2012. doi: 10.1177/0021909612438092
17 Shelly Saha-Sinha, 'India's New Mineral Policy will Usher in Gloom for Adivasis', *Infochange India* (1 January 2009). Accessed March 8, 2012.
     http://infochangeindia.org/environment/analysis/indias-new-mineral-policy-will-usher-in-gloom-for-adivasis.html
18 Rita Gebert, Annie Namala, and Jayant Kumar, 'Poverty Impact Assessment Report, Chhattisgarh', in: *Report for the European Union, Berlin and Raipur* (2011), pp. 12–13, p. 15. Accessed March 12, 2012. www.giz.de/en/downloads/giz2011-en-poverty-impact-assessment-report.pdf
19 Bharadwaj, Sudha. *Gravest Displacement, Bravest Resistance: The Struggle of Adivasis of Bastar, Chhattisgarh Against Imperialist Corporate Landgrab.* Sanhati (June 1, 2009). Accessed March 19, 2012. http://sanhati.com/excerpted/1545/
20 Mathew Areeparampil, 'Displacement due to Mining in Jharkhand', *Economic and Political Weekly 31*(24) (June, 1996), p. 1524. Accessed October 10, 2013.
     www.jstor.org/stable/4404276
21 *Ibid.*
22 'Jharkhand Tribal Groups up in Arms against Projects.' *Press Trust of India* (15 November 2005). Accessed October 12, 2013. http://in.rediff.com/news/2005/nov/15 jhar.htm
23 Pavithra Narayanan, *What Are You Reading? The World Market and Literary Production in India* (New Delhi: Routledge, 2012), p. 58.

24 Peter Popham, 'A United World? Benetton and Native Indians of Patagonia Clash Over Land', *The Independent* (6 July 2004). Accessed November 14, 2013.
    www.independent.co.uk/news/world/americas/a-united-world-benetton-and-native-indians-of-patagonia-clash-over-land-552212.html

25 Jeffrey Bury, 'Mining Mountains: Neoliberalism, Land Tenure, Livelihoods, and The New Peruvian Mining Industry in Cajamarca', *Environment and Planning 37* (2005), pp. 222–23. Accessed November 10, 2013. doi: 10.1068/a371

26 *Ibid.*, pp. 222–27.

27 Narayanan, *What Are You Reading?* , p. 58.

28 Ramachandra Guha, 'Adivasis: Unacknowledged Victims', *Outlook India* (14 April 2010). Accessed September 12, 2014. www.outlookindia.com/website/story/unacknowledged-victims/265069

29 Government of India, *National Mineral Policy, 2008* (New Delhi: Ministry of Mines, 2008). Accessed March 11, 2012.
    http://mines.nic.in/writereaddata/Content/88753b05_NMP2008[1].pdf

30 'Malaysia: 8 more Indigenous Penans arrested for protesting against Murum Dam,' Indigenous Voices In Asia, November 7, 2013. Accessed December 15, 2015. http://iva.aippnet.org/malaysia-8-more-Indigenous-penans-arrested-for-protesting-against-murum-dam/

31 'Indigenous Protesters Rally against Sweden Iron Mining Plans', *United Press International* (*UPI*) (26 August 2013). Accessed November 10, 2013. www.upi.com/Indigenous-protesters-rally-against-Sweden-iron-mining-plans/37241377489780/

32 Wojtek Gwiazda, 'Tahltan First Nation Protesters of Coalminee Project Prepare for Arrests', *Radio Canada International* (20 September 2013). Accessed June 14, 2016.
    www.rcinet.ca/en/2013/09/20/tahlton-first-nation-protesters-of-coal-mine-project-prepare-for-arrests/

33 'Panama Campaign Update: Ngöbe People and Environmentalists Protest Proposed Mining Law', *Cultural Survival* (9 February 2011). Accessed November 9, 2013.
    https://www.culturalsurvival.org/news/panama/panama-campaign-update-ng-be-people-and-environmentalists-protest-proposed-mining-law

34 Mufson, Steve, 'Keystone XL Pipeline Raises Tribal Concerns', *The Washington Post* (17 September 2012). Accessed November 9, 2013.
    www.washingtonpost.com/business/economy/keystone-xl-pipeline-raises-tribal-concerns/2012/09/17/3d1ada3a-f097-11e1-adc6-87dfa8eff430_story.html

35 Dalee Sambo Dorough, 'Human Rights', in: United Nations, *State of the World's Indigenous Peoples* (New York: United Nations Publication, 2009), p. 205.

36 Rodolfo Stavenhagen, *The Emergence of Indigenous Peoples* (New York: Springer Heidelberg, 2013), p. 95.

37 Hayden, *The Zapatista Reader*, p. 2.

38 Balakrishnan Rajagopal, *International Law from Below: Development, Social Movements and Third World Resistance* (Cambridge: Cambridge University Press, 2003), p. 196.

39 *Ibid.*

40 Dorough, 'Human Rights', p. 199.

41 Olle Östensson, 'The Stakeholders: Interest and Objectives', in: James Otto and John Cordes (eds), *Sustainable Development and the Future of Mineral Investment* (Paris: UNEP, 2000), p. 3/2. Accessed November 2, 2013.
    http://apps.unep.org/redirect.php?file=/publications/pmtdocuments/-Sustainable%20Development%20and%20the%20Future%20of%20Mineral%20Investment-20001552.pdf

42 Howard Zinn, *The Zinn Reader: Writings on Disobedience and Democracy* (New York: Seven Stories Press, 1997), p. 403.

43 Stavenhagen, *The Emergence of Indigenous People*, p. 92.

44 Graeme Neate, 'Native Title Claims: Overcoming Obstacles to Achieve Real Outcomes', National Native Title Tribunal, Government of Australia (6 October 2008). Accessed September 15, 2014. http://apo.org.au/node/1280

45 G. Prabhakaran. 'No Adivasi Will Die of Hunger If We Get Our Alienated Land Back', *The Hindu* (6 June 2013). Accessed November 8, 2013.
www.thehindu.com/news/national/kerala/no-adivasi-will-die-of-hunger-if-we-get-our-alienated-land-back/article4784879.ece

46 '"Invasion Day" Protests Held', *Special Broadcasting Service* (23 August 2013). Accessed November 11, 2013. www.sbs.com.au/news/article/2011/01/26/invasion-day-protests-held

47 'Chile Protests: Indigenous Groups March against Columbus Day', *Huffington post* (14 October 2013). Accessed November 10, 2013. www.huffingtonpost.com/2013/10/14/chile-protests-columbus_n_4098351.html

48 Darley J. Kjosavik, 'Politicising Development: Re-imagining Land Rights and Identities in Highland Kerala, India', *Forum for Development Studies 37*(2) (2010), pp. 260–261. Accessed November 8, 2013. doi: 10.1080/08039410.2010.481448

49 'Muthanga Nine Years Later,' *Ini Avar Parayatte*, Indiavision news channel, Kochi, Kerala, (12 December 2011), 3:34–4:05 minutes. Video. Translated from Malayalam by Pavithra Narayanan. *Ini Avar Parayatte* was a special news program produced by Indiavision, a 24-hour Malayalam news channel that was based in Kochi, in the southern Indian state of Kerala. Indiavision shut down its operations in 2015.

50 *Ibid.*, 4:51–5:25 minutes.

51 Mari Marcel Thekaekara, 'What Really Happened: An Account of the Sequence of Events in the Muthanga Forest on February 19', *Frontline 20*(6) (March 15–28, 2003). Accessed September 12, 2013. www.frontline.in/static/html/fl2006/stories/20030328002204600.htm

52 *Ibid.*

53 R. Krishnakumar, 'The Muthanga Misadventure,' *Frontline* 20(6) (March 15–28, 2003). Accessed September 12, 2013.
www.frontline.in/static/html/fl2006/stories/20030328002504200.htm The Asian Human Rights Commission released a list of Adivasi women and children missing after the attack. Of the 65 persons named on the list, 12 have been traced (Biju Govind, 'Missing Adivasis Traced', *The Hindu* [12 March 2003]). Accessed August 5, 2013. www.thehindu.com/2003/03/12/stories/2003031204750700.htm

54 'Kerala: The Government Fails to Comply with Agreement', in: Diana Vinding (ed.), *The Indigenous World 2002–2003* (Copenhagen: IWGIA, 2003), p. 323.

55 *Ibid.*; and C.R. Bijoy and K. Ravi Raman, 'Muthanga: The Real Story – Adivasi Movement to Recover Land', *Economic and Political Weekly 38*(20) (May 17–23, 2003), p. 1976. Accessed April 4, 2012. http://www.jstor.org/stable/4413574

56 Krishnakumar, 'The Muthanga Misadventure'.

57 Vinding (ed.), 'Kerala', p. 323.

58 Bijoy and Raman, 'Muthanga', p. 1976.

59 'Tribals Cannot Be Given Even One Acre', *The Hindu* (25 February, 2003). Accessed September 4, 2013.
www.thehindu.com/thehindu/2003/02/25/stories/2003022506710400.htm

60 'Rehabilitate Tribals, NCW Tells Govt', *The Times of India* (23 March, 2003). Accessed October 4, 2013. http://timesofindia.indiatimes.com/city/thiruvananthapuram/Rehabilitate-tribals-NCW-tells-govt/articleshow/41184746.cms

61 *Ibid.*

62 People's Judicial Enquiry Commission, 'Preliminary report on Muthanga', Pucl.org (17 March, 2003). Accessed April 4, 2012. www.pucl.org/Topics/Dalit-tribal/2003/muthanga-report.htm

63 'Muthanga Probe: CBI to File Chargesheet Soon', *The Hindu* (19 February 2004). Accessed April 5, 2012. www.thehindu.com/2004/02/19/stories/2004021905550300.htm

64 '132 Tribals Remanded', *The Hindu* (22 February, 2003). Accessed March 3, 2012. www.thehindu.com/2003/02/22/stories/2003022205220400.htm

65 'CBI Report on Muthanga: Tribals Allege Conspiracy', *The Hindu* (20 October, 2004). Accessed March 2, 2012. www.thehindu.com/2004/10/20/stories/2004102005330400. htm and 'Muthanga Case: 70 Tribespeople Acquitted', *The Hindu* (3 August, 2011). Accessed April 5, 2012. www.thehindu.com/todays-paper/muthanga-case-70-tribes people-acquitted/article2318058.ece

66 Subash Gatade, 'Interview with Ms. C. K. Janu, Leader of Tribals in Kerala', Sacw. net (18 March, 2005). Accessed October 10, 2013. http://www.sacw.net/Nation/gatade 18032005.html

67 'Muthanga Nine Years Later,' 11:00–11:10 minutes; and 14:41–14:52 minutes.

68 *Ibid.*, 22.34 minutes.

69 *Ibid.*, 7:40–8:05 minutes; and 13:20–14:03 minutes.

70 'Muthanga Case', *The Hindu*.

71 'Muthanga Nine Years Later,' 16:00–16:10 minutes

72 The lyrics quoted here are from <woodyguthrie.org>.

13 Black gold and environmental enemy no. 1

Towards a visual history of coal

Stefan Siemer

Introduction

Over the past century, mining and especially coal mining has produced images of its own.[1] Here, one will find machines, workspaces and mining landscapes as well as miners with their families and neighbourhoods. These images could be regarded as a kind of visual heritage one can refer to as historical documents or pieces of art. In connection with specific media, they play an active role in forming our understanding and perception of coal mining. Images of coal have left their traces in historical textbooks as well as on websites of the coal industry or as part of anti-coal campaigns. We find mining motives on framed oil paintings in museums, on web-based photography, on posters or postcards. One can come across them even in small size: In 2013, the US postal service dedicated its series 'Forever Stamps' to America's industrial workers, among them an underground coal miner.[2]

However, in analysing and understanding these visual representations, false notions and an intrinsic Eurocentrism have to be taken into consideration. The ongoing discussion about the environmental impact of coal mining may exemplify that. While in European countries like Germany or Britain the discussion is dominated by climate change, in India or Indonesia people are far more aware of land destruction and land loss as a result of extensive opencast mining. Different experiences create different images. But this is also true the other way round. Images on the worldwide web are part of the global campaigns of coal companies designed to propagate positive images of coal as an energy resource of the future.

In my chapter, I will ask how visual representations of coal and coal mining appear in non-European contexts and especially focus on the tension between local traditions and global images. Therefore, I will draw attention to Western perceptions of mining as well as refer to examples from mining countries like China, Indonesia or India. I am very well aware of my restricted Western view and scope, knowing that this is by far no comprehensive approach. Nevertheless, my starting point will be the 19th century Western tradition with examples from Britain, Germany, the United States and France. I will try to unfold them into a more global context in accordance with the fact that late 20th-century coal mining is a global business and that environmental problems have become more and more global. I will exemplify this in three different stories or narratives, which I think are essential. In the first part, I will refer to nostalgic views of coal as 'black gold', then deal with the notion of miners and their workspaces and conclude with the notion of dirty coal and its environmental impact on landscapes, also in the context of climate change.

But at the beginning some remarks about visual history as a topic of research, especially with regard to coal mining, are in order. During the past two decades, visual culture studies have broadened our traditional notion of images and changed the ways in which to look at them.[3] From this perspective, they are no longer arranged into a hierarchy with highly esteemed artwork at the top. Instead, they are seen as part of our everyday world. Right at the beginning of his book *Visual Culture*, Richard Howell states that "we live in a visual world" and that in order to understand it, we have to learn new ways of reading and understanding images on the same level as traditional books.[4] There is an obvious need for visual literacy to analyse them with regard to history and specific techniques. Therefore, visual culture studies refer to the precarious status of film and photography in a digital age and question their traditionally assumed matter of fact and documentary character.[5] With reference to the famous slogan of Kodak from around 1900, one can say that pressing the button of the camera is only one part of the story; the practice of (digital) processing and selecting images afterwards matters as well. Last but not least, looking at images is part of a social practice with a huge impact in creating specific visions of class, race or gender.[6] Briefly put: It is not so much the reality, which produces images, but also images, which produce a reality of their own. Their crucial role in constructing our common notion of history is discussed in Gerhard Paul's comprehensive visual history of the 20th century. In his introduction, he states the important technical innovations in the field of electronic and digital reproduction, not to mention the groundbreaking technique of lithography dating from early 19th century, which made illustrations part of everyday culture.[7] Covering the whole spectrum of graphic arts and of course photography, his collection presents detailed stories of visual icons like Sam Shere's photograph of the exploding Hindenburg at Lakehurst, Joe Rosenthal's flag hoisting soldiers at Iwo Jima, Eddie Adams photograph of the shooting of a captured Vietcong or the collapsing twin towers on September 11, 2001.[8]

One could also easily put images of coal mining into this visual canon. As part of an everyday culture of work and industry, they have as well left their traces in 19th- and 20th-century history. One can find them in historical textbooks illustrating mining techniques and machines or serving as historical evidence of workspaces and neighbourhoods.[9] In a recent publication about industrial heritage in the Ruhr area and in Upper Silesia, photographs and postcards are shown as means of documentation of architecture from a comparative point of view.[10] But otherwise they often appear as anonymous illustrations without reference to their ambiguous role as documents and individual statements within a given historical context.

But what is the purpose of these visual representations and in which context do they appear? As early as 1958, Heinrich Winkelmann presented coal mining as precious artwork without a broader context of different visual and social practices.[11] Quite in contrast to him and with an unconventional Marxist approach, Francis D. Klingender discusses the influence of modern industry upon the work of 18th- and

19th-century British artists like Joseph Wright of Derby or William Turner.[12] He states that around 1800 mines and machines harmoniously fitted into sublime landscape paintings, while later on a negative perception of modern industry prevailed and for many artists, industry increasingly became part of a vision from hell. But foremost photography left its traces in the common imagination of the industrial world. As an example, Klaus Tenfelde discusses the essential role of photography for the company of Krupp. It formed a positive image for a wider public and also strengthened the corporate identity of the workers as part of the Krupp Family.[13]

Especially with regard to coal, recent studies focus on photography, as well. In 1989, an exhibition was dedicated to 20th-century documentary photography reflecting the working conditions and everyday life of miners.[14] Later on, Michael Farrenkopf presented a selection of photographs from the archives of the Deutsches Bergbau-Museum Bochum stating the more or less official character of photography in propagating what the head of the mining companies wanted the public to see. The photographs show industrial architecture, technical equipment and workers arranged to group portraits propagating a positive image of buildings, machines and products.[15] With regard to the photographic archive of about 4000 photographs of the Appalachia-based Consolidation Coal Company, Geoffrey L. Buckley analyses in detail the practice and use of photography as part of US mining history of the early 20th century.[16]

Black gold

Until today, our popular Western notion of coal is dominated by the term of black gold, creating a positive notion of an abundant energy resource, which became crucial for the 19th-century industrialisation. Therefore, the term became a starting point for quite ambiguous notions of coal mining between documentary realism and coal nostalgia referring to hard, dirty and dangerous labour as well as the memory of mining communities long-since vanished. On a more global scale, positive notions of coal as a worldwide commodity appear in the logos and advertisement campaigns of mining companies.

With the rise of documentary photography around 1900, photographers like Jacob A. Riis and Lewis W. Hine focused on the urban and industrial poor and, taking up the title of Riis famous essay on New York from 1889, *How the Other Half Lives*. But it was Hines who, around 1910, with his moving portraits of breaker boys and trapper boys working in the coal mines of West Virginia and Pennsylvania, created some of the icons of documentary photography.[17] Later on, Bill Brandt in his *English at Home* from 1937 and Chris Killip with his sea-coaler photographs from the 1980s continued this tradition of social photography up to the present time. In the second half of the 20th century, most of these photographs were distributed through newspapers and magazines focusing on social life and historical events in order to describe *things as they are*, creating a universal language of photojournalism.[18] Here they appeared on an equal level with the text and were arranged in series telling a story of their own, with only some marginal addition by written captions.[19]

But outside their original context, photographs live a life of their own. As reproductions shown i.e. on posters, calendars, reports by coal companies or as photographical histories without any relation to a written narrative, they provoke, as Raphael Samuel has pointed out, an unspecific "aura of pastness".[20] At least to the casual observer they seem like a peephole into another world, which he could set in relation to his own. On a more local scale these visual recollections of things past allow the close observer to compare in detail his own modern living conditions with those of former residents and get an idea how his neighbourhood has changed over time. It is an ambiguous experience. Photographs feed our nostalgia and notions of a better past and also make us believe that living conditions have significantly improved. Moreover, they allow us to participate in historical events we never experienced and therefore play such an essential role in forming a collective memory culture.

Popular photographical histories, for example, covering the golden era of the Ruhr industry from 1900 to the 1960s, have been an important factor in creating a regional identity of the '*Ruhrgebiet*'.[21] With the growing interest in the industrial heritage and everyday experiences of miners, photographs also became part of web-based collections, like the collection of the University of Pittsburgh. Here one can find photographs of coal miners and their workplaces from the 1890s onwards.[22]

The photographic history of Pottsville/Pennsylvania, published by Leo L. Ward and Mark T. Major in 1995, may illustrate popular nostalgic notions of coal and the practice of rearranging them in new contexts. As both authors state in their introduction: "The views in this book are scenes of a bygone era. They are views of anthracite mining and the coal miners who worked in the mines".[23] The cover shows a group of young miners in their working suits but without any working tools leaning casually against a wooden shed. The photograph suggests an atmosphere of rest and calm after a shift. The sepia hue of the black-and-white softens the impression of blackness of their faces, creating an air of privacy, despite of their obviously hard working conditions.

The book is part of the series *Images of America* by the US publisher Arcadia. Since the early 1990s, Arcadia has been one of the main publishers in the field of popular books on local history. In 2003, there were about 2000 titles in ten series available and it has been estimated that Arcadia sold 1.5 million books in 2006. Referring to the *Images of America*, Mark Rice points at the standardisation of history: Each book has the same length and same size and presents its photographs in the same sepia coloration.[24] Quite solemnly, the publisher characterises the series as follows:

> Local authors transform dusty albums and artefacts into meaningful walks down memory lane. Millions of vintage images become tiny time capsules, re-establishing memories of the formerly familiar, introducing generations to what once was, and reminding us all of what has been (and can be) in every corner of our nation.[25]

As a time machine, photography virtually takes the observer back into a world of a hundred years ago. But coal nostalgia could also be interpreted as an attempt of former coal-based communities to deal with the decline of the coal industry. From that point of view, historical retrospection by photography serves as a means of developing new identities in the context of the new post-industrial landscapes.

There is also a photography-based memory culture in those countries with a mining industry still active. Since 2007, the web-based *Queensland Historical Atlas* is dedicated to the history of North Western Australia. As an ongoing project, this virtual museum presents and comments historical objects as well as photographs and other images. Preserving the memories of a landscape deeply changed by the exploitation of resources, the *Queensland Historical Atlas* focuses especially on mining in photographs from the early 20th century up to the 1970s.[26]

One can find photographical records in other more specialised contexts, as well. Alan Murray's *On the Edge* tells the story of Australian coal mining unions during the 1990s, referring to the archives and the photographical record of Australia's main trade union, the Construction, Forestry, Mining and Energy Union (CFMEU).[27] In documenting various campaigns, its photographs address those who have experienced the struggle against the coal companies. With the names of the participants given in the captions, the book also works like a family album, remembering those who played an active role in the historical events. Here, images are not part of a sentimental journey back into history but rather evidence of an ongoing history of strong unions.

The examples discussed so far refer to the Western tradition of coal mining and a visual history deeply rooted in 19th-century perceptions of the golden age of coal. However, while this story addresses a more or less local public, today's story of the black gold is part of a global coal industry. It is told by photography as well as by graphic media creating new images quite different from the black-and-white nostalgia of former times.

Logos may serve as very fitting starting points. Consisting mainly of graphic elements, they are 'written' in a universal language that is easily understood and play an important part in the corporate identity of companies, as well as advertisement campaigns, which can be understood by a wider public and which are not restricted to local traditions and languages. Otto Neurath, a pioneer in establishing visual symbols as means of visual education in the 1920s and 1930s, has pointed out that their origins could be traced far back to early modern history as part of scientific and educational textbooks.[28] Instead of extensive written explanations, the publishers of 18th- and 19th-century illustrations tried to reduce information to a basic level to convey clear messages. They were forerunners of today's abstract graphic pictures and pictograms which have become part of our everyday life and, thanks to modern electronic media, have been disseminated all over the world. We find them on our computer desktop, on traffic signs, on statistical charts as well as on company websites, annual reports, magazines and newsletters.

This universal language also took hold of coal. The logo of the state-owned Coal India Limited (CIL), the biggest mining company worldwide founded in 1974, shows a piece of coal in the shape of a diamond referring not only to its chemical quality as carbon but foremost as a precious mineral.[29]

Three captions strengthen this symbolic notion and allow for a more precise interpretation of the specific role of coal in the history of India. First of all, the slogan "Empowering India" underlines the central role of coal for the Indian economy as "a national asset".[30] The remainder deals with the company's social responsibility for environmental and social rehabilitation programs. The message is quite clear: Indian coal is part of a nationwide collaborative effort and a source of wealth for all people without environmental damage. Other logos communicate different messages. Apart from the written label, the logo of the Chinese mining company and energy provider Shenhua shows layers of coal seams which pile up to a black mountain with a top piece of coal broken out, ready for burning.[31] It is a clear-cut reminder of the abundance of easily accessible coal in China.

The rhetoric of abundance is also found on the cover of the German magazine *Der Spiegel* from 1973 that shows a huge black dump of coal leaving out a small speck of blue sky with a tiny pithead frame in the background (Figure 13.1). In the context of the caption "Energy crisis: Rescue through coal?", the looming pile of coal seems like an ironic allusion to the black landscapes of 19th-century Coketown. But the cover has to be read quite differently. At the climax of the first oil crisis, the fear of running out of oil and rising costs put coal on the agenda again after it had

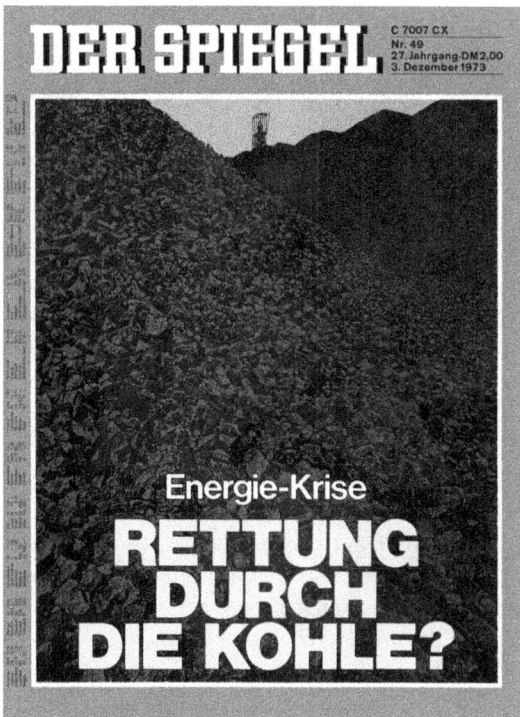

Figure 13.1 "Energy crisis: Rescue through coal?" (cover illustration), 1973 (*Der Spiegel* 49/1973)

previously been strongly challenged by oil, gas and nuclear energy. Therefore, the cover illustration visualises cheap and abundant coal as salvation for the Western industries. Apart from the historical context, coal is also visualised here as black matter instead of showing any reference to miners at work with their machines or to a clearly identifiable workplace in this image. The pithead in the background is not part of a colliery but an abstract icon and contrary to understandable expectations, there is no allusion to the Ruhr area as the main coal production site in Germany.

Leaving out the local context, images of coal became more and more international and it seems that by the 1970s, the visualisation of coal mining in a clearly identifiable national context had been replaced by a kind of more abstract and international style. This style reflected the fact that coal had become a global commodity. The conveyor belt of the Australian coal producer Whitehaven might be seen as an appropriate emblem of this global availability.[32] There is no visible beginning and no end to the belt. In an ever-running stream, coal is delivered from one end of the world to the other without any solemn notion of coal as salvation to Western industries. Instead, we see coal as a part of a complicated and highly mechanised production process which begins with mining and ends with incineration in a power plant in order to produce energy for domestic and industrial use. The black gold of former times is a commodity, black matter ready for sale to customers everywhere in the world.

Workplaces

This ambiguity of coal between local and global images is also present with regard to the miners' workplaces. Since the subterranean part of a coal mine is not easily accessible, rather than gaining first-hand experience, the public has to rely on illustrations in novels and textbooks with images of hard-working men, as well as technical drawings giving an abstract impression of machines, tunnels and galleys. But today, the working coal miner seems quite absent from the images disseminated by coal companies via the internet, advertisement brochures and annual reports. He has been replaced by machines and working tools which seem to be the real agents of coal production.

I will begin with a late 19th-century fiction in which realistic and fantastic elements merge.[33] In his novel *Les Indes noires* from 1877, the French novelist Jules Verne describes a hidden world of coal deep underneath the surface of Scotland. Curiously enough, Verne transfers the mining village from the surface to the underground. In one of his illustrations, Jules Férat shows us this fantastic setting as a vast cavern in which tiny cottages are dispersed under a sky of stone supported by gigantic columns. This Coal City is supplied with fresh air by ingenious shafts and lit by powerful electric lamps. Here, Dante's vision of an underground place as hell has been turned into a secure place, where industrious miners and their families exploit the mineral resources easily accessible for the general benefit of mankind. For Verne and Férat, the underground workspace has lost its terror and is no longer a dangerous place. Eight years later, Zola published his novel *Germinal*, describing his experiences among coal miners in the Pas-de-Calais area in northern France. Quite in contrast to Verne, for Zola the place underground was a dangerous place, the miners threatened by the catastrophes of collapsing shafts, tunnels and galleries.

Nevertheless, up to late 19th century, realistic pictures of underground workspaces were scarce. Before the 20th century, the narrow shafts and tunnels had not been places in which artists and photographers could work properly.[34] For a long time, the imagination of the subterranean was therefore restricted to more or less abstract technical drawings, which show mining devices like machinery or cut sections. One has only to remember the iconic images in Georg Agricolas' *De Re Metallica* from 1556 or the detailed copperplates illustrating the article *Minéralogie* in Diderot and D'Alemberts *Encylopédie* from the 18th century. One has also to notice that within these mechanised workspaces, the miner himself plays a minor role.

Around 1900, images of mining workspaces found their way into the exhibitions of the newly established Museums of Science and Industry. For the first time, visitors could experience the underground without any danger. As an example, the Deutsches Museum at Munich, founded in 1903, not only dedicated 3600 square metres to its visitors' mine that formed one of the biggest departments of the museum, but also commissioned large-sized paintings in order to describe various machines and production processes.[35] According to the basic idea of explaining machines and artefacts from an engineer's point of view, these images were part of an educational program designed to explain science and industry as well as to describe technical inventions in line with European culture.[36]

Mainly focusing on machinery, these images leave no room for miners' working experience. But even the early engravings of the Royal Commissions on Children's Employment in the Mines and Mills from the 1830s and 1840s, showing a child hauling coal in narrow underground galleries, one of the most frequently reproduced images in mining history, describe hard working conditions in a quite technical way.[37] More personal images of miners are only those that depict settings above ground. Curiously enough, Andrew Roy's *A History of the Coal Miner of the United States* from 1907 shows idyllic scenes with miners departing from wife and children or returning from the mine to their tiny houses with garden.[38] The first photographs with underground miners also retained this artificial technical style. In the early times, most of them were staged due to the poor technical equipment and orders of the mining companies which preferred a positive public image.[39] But during the 20th century things changed. Photographs of the underground workspace focusing on the hard and dangerous work of the mines became an essential part of a social realism. In 1948, the British magazine *Coal*, published by the National Coal Board, displayed a photograph on its cover that showed miners below ground. The exhausted faces of the shaft sinkers speak for themselves. Far from any hidden heroism, the photograph depicts the work as it is: Dirty and dangerous.

This documentary style found its counterpart in the working-class hero of 20th-century political propaganda.[40] His extraordinary qualities of physical strength and power are combined with dexterity in the use of his working tools. Like a soldier, he fights not for himself but for higher purposes to achieve industrial and technical progress for a better life. Quite in contrast to this 19th-century technological optimism and despite its totalitarian attitude, the working-class hero could also be seen as an example of man's dignity as a tool-bearing animal set against modern anonymous and mechanised workspaces. It very much is this heroism that propagates mining as a masculine domain quite opposite to the evidence of the important role of women as mineworkers all over the world.[41]

The propagation of this ideal is a central part of a visual history of coal. It emerged foremost in the context of political propaganda of the 20th century.[42] A Chinese poster from 1970 shows a typical working-class hero with a pick-axe in front of an invisible seam of coal (Figure 13.2). He virtually attacks the coal like a soldier would his enemy. Quite in contrast to the giant-like miner in the foreground, in the background one can distinguish workers teeming like ants around the buildings of a coal mine. The message of the poster is quite clear and is emphasised in the title: "The engagement for coal mining means great support for the communist revolution". There is no advanced mechanical equipment visible. The miner performs his role in a quite archaic manner domi-nated by physical strength. Further examples of this pictorial style meant to propagate a political ideology could be found during the Soviet era, especially in case of the Stakhanovite movement of the 1930s.

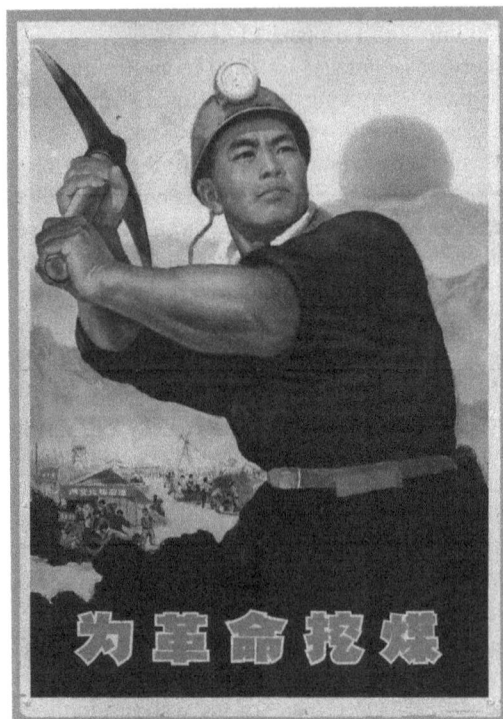

Figure 13.2 "The engagement for coal mining means great support for the communist revolution" (poster), 1970 (International Institute of Social History/Stefan R. Landsberger Collection)

But beyond pure political ideology there is a hidden appeal of moral behaviour in this coal-based propaganda. Even clear-cut posters have to be read in a specific regional or national context. As in the case of India, coal has been described by the ruling colonial British "as the great engine of moral improvement, the greatest instrument of civilisation for the people".[43] This education by coal mining also has a clear nationalistic bias. In India, coal achieved an iconic status as a national symbol after independence, which led to the Coal Mines Nationalisation Act of 1973 and two years later to the founding of Coal India Limited as a state-owned company.[44] According to this narrative, the picture of the working-class hero could be seen as part of a national propaganda designed to show the labour-intensive sector of mining as an essential part in the development of the country.

Today, those images of male heroism, which are deeply rooted in a certain national and political context, seem outdated. One can find their counterpart on the websites of the global mining companies where huge machinery has replaced working men, dwarfing them to mere technical operators. With regard to the Australian opencast mines in the Bowen Basin, Luke Keogh points out: "everything is 'big' in the coal industry: big trucks, big buckets, big holes, big production, big profit, big landscape change; it is this 'big' vernacular that dominates the industry".[45] At the same time, these machines became part of a global image of

Figure 13.3 Campaign poster for ILO Convention 176 (English version), 1995 (IndustriAll)

the miner's workplace. On a global scale, working men seem to be replaceable. As giant working tools, the machinery of an opencast mine bears the connotation of an overwhelming power over nature, shaping whole new landscapes. The huge shovel of a dragline, which the US coal mining company Arch Coal presents on its website, places the miner and his working device within a moon-like landscape of an opencast mine.[46] The machine is clearly the main focus, smooth and free from dirt and dust. Strikingly enough, no trace of mined coal is visible. If we were not aware that this is a coal company's website, the shovel could be regarded as an abstract piece of artwork. The worker himself is placed here to give reference of scale to the dimension of the shovel. Despite his passiveness, he is proud of his power, too. He has turned his physical strength over to the machine, which he operates and controls. But this picture bears another important message: Coal is clean, it is a product mined by mighty clean machines in clean workspaces.

The globalisation of images also takes place in the campaigns of the International Labour Organization (ILO). The poster featuring the *Safety and Health in Mines Convention ILO 176* shows a stylised mining tunnel lit by a bright light at its end in the shape of the number 176 (Figure 13.3). Walking towards this light is a miner with a pickaxe. He turns his back to the beholder. But far from any documentary attitude or even heroism, the picture is also a representation of a mining workspace. It is the underground tunnel translated into a global visual style, addressing miners all over the globe.

Deadly coal

The visual history of coal also deals with the disastrous environmental impact of producing and burning it. In 2002, the British magazine *The Economist*, in a cover story, termed coal the "Environmental Enemy No 1", describing the disastrous environmental costs of coal mining, especially with regard to carbon dioxide.[47] Therefore, a crucial factor in the public notion of coal is the ongoing discussion about global warming and climate change, which has shifted public attention from mining to burning. During the last three decades, coal has not only become a global commodity but also a decisive factor in environmental health on a global scale.

Before coal is burned, it has to be mined and it is this process, which even today leaves deep visible traces in natural environments all over the world. But not only have the production of fossil resources and the related industries been agents of change. One has to consider other energy sources like dams for hydroelectricity or, more recently, windmills or solar panels. Mike J. Pasqualetti has termed these old and new landscapes, "energy landscapes" and pointed out that they could be "windows in the past", containing different historical layers of former activities in gaining energy and natural resources.[48] Pasqualetti also makes us aware of the lasting impact of large-scale mining activities shaping completely new kinds of landscapes. Therefore, I will reflect on recent coalscapes as further examples of the influence of local notions and traditions on the creation of images of coal and coal mining.

In India, the share of opencast mining in the coal sector was at about 80 per cent in 2000/1.[49] The consequences of these activities for the residents were dramatic. As Nesar Ahmad and Kuntala Lahiri-Dutt have pointed out:

When communities are forced to leave the land that they have lived on for generations, they not only lose farming land but are also deprived of the forests, waters in ponds, streams and springs, and grazing lands on which their life was dependent. The social bonds that existed between individuals are ruined.[50]

The affected communities have no legal instruments to wield against the power of the state-owned coal mining company. It is a dispossession of the poor, often indigenous people, which come from the lowest strata of society.[51] On the internet, one can see photographs showing these devastated landscapes and the disastrous change from agriculture to mining. But in the tradition of Western documentary photography, they are first of all produced for an international news market addressing foremost a non-Indian public.[52] Therefore, it would be interesting to look at different images used for environmental campaigns by residents and to examine them with regard to local visual cultures.

In Indonesia, the mining advocacy network Jaringan Adovaksi Tambang (JATAM) dedicates its work to the disastrous impact of mining on the natural landscape and residents. Since the Indonesian resources boom of the 1970s, attention has shifted from ore to coal in order to supply the Asian market and foremost China. Due to a lack of regulation in recent years, coal companies could acquire permits to exploit new coalfields without regard for environmental damage. In its brochure *Deadly Coal*, the Indonesian non-governmental organisation (NGO) Jaringan Adovaksi Tambang (JATAM) sums up the dramatic impacts of opencast coal mining for Borneo.[53] Its residents are affected by lack of food production, clean water supply and above all the displacement of indigenous people. The transport of coal by river barges from mines to the coast also contaminates the water and forest destruction leads to catastrophic flooding in cities like Samarinda. The cover illustration of *Deadly Coal* summarises these environmental threats into a suggestive picture. It shows a close-up of an opencast mine as part of a bleak and dehumanised landscape. At the margin of a small pond, a meagre tree gives evidence of a forest that had existed before and which has been replaced by broad roads built for huge trucks transporting coal. Interestingly enough, miners and their work are absent from this picture. However, one has to ask at whom this English-language brochure is addressed. It seems that environmental issues and their images are less part of local campaigns but rather much more targeted at a global public and one has to ask whether a specifically Indonesian view on landscape and environmental change exists.

Since the 1970s, new coalscapes also emerged in the Appalachian Mountains in the eastern parts of the United States. Here, coal is mined by blasting up whole mountains, a method known as mountaintop removal. It is practised by the mining companies as the most efficient way of gaining access to the coal underneath. As a result, it leaves deep traces in a rural and sparsely populated region. Since the 1980s, this practice has provoked massive protests by residents and NGOs in Kentucky and West Virginia.[54] In line with the protest of local environmentalists Beehive Design Collective, a local grassroots organisation from Maine, specialising in educational graphics campaigns, has developed a large canvas entitled

Figure 13.4 The True Cost of Coal (poster) (Beehive Design Collective)

True Cost of Coal (Figure 13.4). Teeming and swarming with hundreds of different scenes, the canvas works as a kind of graphic novel describing the disastrous impact of opencast coal mining on natural landscapes. But the picture is not only about mining in detail. It works as a comprehensive narrative that describes a coal-based industry together with the production of everyday consumer products. The message of this coalscape leads far beyond the ambivalent notion of good versus bad coal. On a basic level, it puts into question our modern and industrialised way of living and asks whether wealth and progress are worth the deep impact of coal on our natural habitat.

But quite in contrast to the local impact of mining through devastation of entire landscapes, the burning of coal has created a new image on a global scale. To understand this new image, one has to begin with early representations of the smokestack as a symbol for the coal-based industries. Even this everyday piece of architecture has a history of its own and, quite surprisingly, it once was a symbol of industrial progress and wealth: Keep the smokestacks smoking in order to keep the industry running and create new jobs. As a mighty piece of architecture towering over most other buildings, the smokestack with its trail of ashes and dust is a landmark of the traditional industrial landscape, indicating workshops, residential areas and infrastructures. It is an essential part of 19th-century Coketown, a new industrial type of city described by Charles Dickens in his novel *Hard Times*. Later on, for Lewis Mumford, the main features of this new urban landscape are destruction and disorder, an archetype and model for all further cities in the Western world.[55] Therefore, the notion of the smokestack as an emblem of the coal-related industrial world is quite ambivalent: It stood for wealth and progress as well as for the destructive power of the new industries.

In the course of the 20th century, this notion of the smokestack changed significantly. A crucial part in this change of image was the public discussion of climate change and carbon dioxide emissions.[56] In 1986, the term climate catastrophe was on the public agenda for the first time and it was soon recognised as a new kind of global environmental threat. It could be seen as a turning point in environmental history. Founded by the United Nations in 1988, an Intergovernmental Panel on Climate Change (IPCC) started working to gain objective scientific expertise. Led by the scientists of the IPCC, the following decade saw a debate on human-made greenhouse gases, foremost carbon dioxide. They set politics in action. The international conferences at Rio in 1992 and Kyoto in 1997 established a world-wide institutional framework to reduce greenhouse gases. At this time, the fossil resources of coal and oil were regarded as the main producers of carbon dioxide. Therefore, one could ask how the ongoing discussion on climate change and the burning of fossil resources has created and continues to create a new image of coal. Other than the black skies of Coketown, the threat of global warming is quite invisible and lurks behind blue skies. The discussion on climate change also marks a turning point in the visual history of coal, also indicating its public perception as environmental enemy no. 1.

In 2006, Davis Guggenheim's documentary film *An Inconvenient Truth: A Global Warning* about Al Gore's campaign against climate change got the attention of a worldwide public. This is no place for discussing the film in the context of a visual history of coal. Instead, I will focus on the poster promoting the film.[57] It is a powerful visual expression of widespread fears of climate change. Beyond that the picture also bears evidence to the hidden visual and rhetorical traditions behind actual debates.

In an apocalyptical scene one can discern several power plants emitting smoke and vapour towards a dark sky. Quite in contrast to this, a group of smokestacks emits a white cloud of vapour, which turns into the typical shape of a hurricane seen from above. Obviously, the scene points at mankind's bleak future, in which it will be facing the consequences of uncontrolled industrialisation and burning of fossil resources. Related to the discussion on climate change one can discern two dominant motives. One of them, the coal-fired power plant, has become the major target of the environmentalists and their campaigns during the past decade. Referring to the traditional image of Coketown, its characteristic smokestacks or cooling towers emerged as an epitome of climate change. In contrast to this, the other motive, the aerial photography of the hurricane, alludes to a recent event. One year before the release of the film in August 2005, hurricane Katrina had devastated the Gulf of Mexico region with almost 2000 casualties and flooded and destroyed parts of New Orleans. Therefore, the picture links the burning of fossil resources as cause of a global change of climate with a catastrophe, which experts think to be a result of this change. Thus, the message of this photo collage is quite clear: To avoid a catastrophic future, immediate action is necessary. But beyond linking the burning of fossil resources and natural catastrophes, the picture also offers a deeper level of meaning. Like the dark world beneath our feet, the image of black skies still lives in our collective memory as an archetypal emblem of threat and punishment. Accordingly, the burning of coal is set in a kind of doomsday scene, threatening humankind with collective destruction. As a moral statement, this vision of coal also reflects on humankind's sins against the environment. The ambiguity of the local and the global could not have been expressed better.

Conclusion

In my chapter, I have tried to discuss a visual history of coal mining from a comparative point of view regarding the intrinsic tensions and ambiguities of local and global images. On the one hand, most of the visual representations mentioned were created within a local tradition referring to a certain community or nation, i.e. nostalgic photographs of American mining communities, paintings of Chinese working-class heroes or brochures dealing with the environmental impact of coal mining on Borneo. On the other hand, they emerged as global icons since coal is a global commodity, as in the case of the logos of coal companies operating worldwide, or the smokestacks of coal-burning power plants in campaigns against climate change. However, it is not only a specific visual style or a visual culture that makes images local or global. Their status also depends on specific media as means of a local or worldwide dissemination as well as on a specific historical context like the Western tradition of good versus bad coal, the working-class hero in China or coal as national asset in India.

However, from this comparative point of view, a visual history of mining has still to be written. It could be an interesting new field of future research and could sharpen our perception of different kinds of mining cultures. The examples presented so far are a possible starting point for further discussion on a more elaborate scale.

Acknowledgements

I would like to thank Kuntala Lahiri-Dutt, Peter Colley, Jie Mao and Manfred Warda for valuable hints on the understanding of coal in other national contexts.

Notes

1   In the following, I prefer 'image' to 'picture' as a comprehensive term reflecting more the reception than the production of visual representations.
2   <http://about.usps.com/news/national-releases/2013/pr13_062.htm> (accessed 30 May 2016).
3   The discussion on photography as a historical source dates back into the 1960s at the latest. See the lively account about photography and new social history by Raphael Samuel, *Theatres of Memory: Past and Present in Contemporary Culture* (London: Verso, 1994), pp. 315–336.
4   Richard Howell, *Visual Culture* (Cambridge: Polity, 2003), p. 1.
5   Howell, *Visual Culture*, pp. 164–166.
6   Gillian Rose, *Visual Methodologies: An introduction to researching with visual materials*, 3rd ed. (London: Sage, 2012), pp. 11–13.
7   Gerhard Paul, *Das Jahrhundert der Bilder*, 2 vols. (Göttingen: Vandenhoeck & Ruprecht, 2009), p. 27.
8   Paul, *Das Jahrhundert der Bilder*, vol. 1: p. 412, p. 674; vol. 2: p. 354, p. 686.
9   See for example Roy Anthony Church, *The History of the British Coal Industry, Vol. 3: Victorian pre-eminence 1830–1913* (Oxford: Clarendon, 1986); William Ashworth, *The History of the British Coal Industry, Vol. 5: The nationalized industry 1946–1982*, (Oxford: Calrendon, 1986); Barry Supple, *The History of the British coal Industry, Vol. 4: The political economy of decline 1913–1946* (Oxford: Clarendon, 1987); Klaus Tenfelde (ed.), *Geschichte des deutschen Bergbaus, vol. 4: Rohstoffgewinnung im Strukturwandel. Der deutsche Bergbau im 20. Jahrhundert* (Münster: Aschendorff Verlag, 2013).

10 Thomas Parent (ed.), *Montanrevier: Bilder aus dem Ruhrgebiet und Oberschlesien* (Essen: Klartext, 2008).
11 Heinrich Winkelmann, *Der Bergbau in der Kunst* (Essen: Glückauf, 1958).
12 Francis D. Klingender, *Art and the Industrial Revolution*, ed. and rev. by Arthur Elton (London: Evelyn, Adams & Mackay, 1968). The first edition of Klingenders study appeared in 1947.
13 Klaus Tenfelde (ed.), *Bilder von Krupp: Fotografie und Geschichte im Industriezeitalter* (München: CH Beck, 1994), p. 313.
14 Ulrich Borsdorf and Rudolf Kania (eds), *Ausbeute: Bergbau und Bergarbeit in der Fotografie* (Essen: Klartext Verlag, 1989). Just recently an exhibition featuring global coal mining from today's point of view heavily draws on photojournalism to get a picture of living and working conditions of miners: Ulrike Stottrop (ed.), *Kohle Global. Eine Reise in die Reviere der anderen* (Essen: Klartext Verlag, 2013). With English version on CD.
15 Michael Farrenkopf, *Mythos Kohle: Der Ruhrbergbau in historischen Fotografien aus dem Bergbau-Archiv Bochum* (Münster: Aschendorff Verlag, 2009).
16 Geoffrey L. Buckley, *Extracting Appalachia: Images of the Consolidation Coal Company 1910–1945* (Athens, OH: Ohio University Press, 2004).
17 Thilo Koenig, 'Die andere Seite der Gesellschaft. Die Erforschung des Sozialen', in: Michel Frizot (ed.), *Neue Geschichte der Fotografie* (Köln: Könemann, 1998), pp. 347–357.
18 Mary Panzer, *Things as they Are: Photojournalism in context since 1955* (London: Aperture, 2006), pp. 9–33.
19 Robert Lebeck and Bodo von Dewitz, *Kiosk: A history of photojournalism 1839–1973* (Göttingen: Steidl, 2001).
20 Samuel, *Theatres of Memory*, p. 322.
21 Sigrid Schneider, 'Images: Bilder vom Ruhrgebiet', in: idem (ed.), *Schwarzweiss und Farbe: Das Ruhrgebiet in der Fotografie* (Essen: Pomp, 2000), p. 25.
22 <https://images.library.pitt.edu/> (accessed 30 August 2018).
23 Leo L. Ward and Mark T. Major, *Pottsville* (Charleston, SC: Arcadia Publishing Books, 1995).
24 Mark Rice, 'Arcadian Visions of the Past',' *Columbia Journal of America Studies 9* (2009), p. 8. Arcadia developed as a subsidiary of the UK-based Tempus Publishing, founded by Alan Sutton in 1993. In 1997 Tempus also founded a successful subsidiary in Germany (Sutton Verlag) which has a design quite identical to *Images of America*.
25 <www.arcadiapublishing.com/mm5/merchant.mvc?> (accessed 15 May 2014).
26 <www.qhatlas.com.au/content/mining> (accessed 6 June 2016). See also Luke Keogh, 'The Storied Landscape: A Queensland collection' (unpublished PhD thesis) (The University of Queensland, 2011).
27 Alan Murray, *On the Edge: A narrative history of Australian coal miners and their Union in the 1990s* (Sydney: CFMEU, 2012). Another book supported by CFMEU collects personal portraits of former coal miners recorded during an oral history project of the late 1980s and early 1990s, illustrated with historical photographs of miners and workspaces dating back to early 20th century. See Fred Moore, Paddy Gorman and Ray Harrison, *At the Coalface: The human face of coal miners and their communities: An oral history of the early days* (Sydney: CFMEU, 1998).
28 Otto Neurath, *From Hieroglyphics to Isotype: A visual autobiography*, Matthew Eve and Christopher Burke (eds) (London: Hyphen Press, 2010).
29 <www.coalindia.in/> (accessed 14 April 2014).
30 Kuntala Lahiri-Dutt, Radhika Krishnan and Nesar Ahmad, 'Land acquisition and dispossession: private coal companies in Jharkhand', *Economic & Political Weekly 47* (2012), p. 40.
31 <www.csec.com/> (accessed 29 August 2014).
32 <www.whitehavencoal.com.au/> (accessed 1 October 2013).
33 See Eckhard Schinkel (ed.), *Über Unterwelten. Zeichen und Zauber des anderen Raums. Ausstellungskatalog* (Essen: Klartext, 2014).

34 There are early exceptions like Pehr Hilleström's (1732–1816) Piranesi-like pictures of the Falun mineworks. For the 20th century, one could find several examples like Henry Moore's impressive series of drawings showing Yorkshire coal miners underground. See Winkelmann, *Der Bergbau in der Kunst*, pp. 328, 422–425.

35 Conrad Matschoss, *Das Deutsche Museum. Geschichte, Aufgaben, Ziele* (Berlin: VDI-Verlag, 1933), pp. 75–83.

36 Manfred Döbereiner, 'Der Gemäldebestand des Deutschen Museums', in: Klaus Türk (ed.), *Arbeit und Industrie in der bildenden Kunst* (Stuttgart: Franz Steiner, 1997), pp. 187–204; Hans-Liudger Dienel, 'Bilder und Leitbilder der Technik', in: Türk (ed.), *Arbeit und Industrie in der bildenden Kunst*, pp. 160–179.

37 Samuel, *Theatres of Memory*, p. 327.

38 Andrew Roy, *A History of the Coal Miner of the United States from the Development of the Mines to the Close of the Anthracite Strike of 1902 Including a Brief Sketch of Early British Miners*, 2nd ed. (Columbus, OH: Press of JL Trauger Printing Comany, 1907), p. 87, p. 121.

39 Farrenkopf, *Mythos Kohle*, p. 12.

40 See Willi Kulke, 'Für Fortschritt und Planerfüllung. Helden der Arbeit', in: LWL-Industriemuseum (ed.), *Helden. Von der Sehnsucht nach dem Besonderen* (Essen: Klartext, 2010), pp. 273–288.

41 Gill Burke, 'Women miners: here and there, now and then', in: Kuntala Lahiri-Dutt and Martha Macintyre (eds), *Women Miners in Developing Countries: Pit women and others* (Farnham: Routledge, 2006).

42 Marien van der Heijden, Stefan R. Landsberger and Kuiyi Shen, *Chinese Posters: The IISH-Landsberger Collections* (München: Prestel Pub, 2009).

43 Kuntala Lahiri-Dutt (ed.), *The Coal Nation: Histories, ecologies and politics of coal in India* (Farnham: Ashgate, 2014), p. 1.

44 Lahiri-Dutt (ed.), *The Coal Nation*, pp. 11–17.

45 <www.qhatlas.com.au/content/coal> (accessed 29 May 2014).

46 <www.archcoal.com/> (accessed 14 April 2014).

47 The Economist, 'Environmental Enemy No 1', *The Economist*, July 6, 2002.

48 Martin J. Pasqualetti, 'Reading the changing energy landscape', in: Sven Stremke and Andy van den Dobbelsteen (eds), *Sustainable Energy Landscapes* (Boca Raton, FL: CRC Press, 2013), p. 12.

49 Urmila Jha-Thakur and Thomas Fischer, 'Are open-cast coal mines casting a shadow on the Indian environment?', *International Development Planning Review 4* (2008), p. 241.

50 Nesar Ahmad and Kuntala Lahiri-Dutt, 'Engendering mining communities: examining the missing gender concerns in coal mining displacement and rehabilitation in India', *Gender, Technology and Development 10* (2006), p. 315.

51 Ahmad and Lahiri-Dutt, 'Engendering mining communities', p. 319.

52 See examples in Stottrop (ed.), *Kohle Global*, p. 130, 133, 143, 149. Unfortunately I did not find any evidence of images of opencast mining as part of environmental campaigns in India.

53 Arief Wicaksono (ed.), *Deadly Coal: Coal extraction and Borneo Dark generation* (Jakarta 2010), <https://english.jatam.org/dmdocuments/DC%20ingg02.pdf> (accessed 12 May 2014).

54 Penny Loeb, *Moving Mountains* (Lexington, KY: University Press of Kentucky, 2007).

55 Lewis Mumford, *The City in History: Its origins, its transformations, and its Prospects* (New York: Houghton Mifflin Harcourt, 1961), p. 447.

56 See for the following discussion Walter Hauser (ed.), *Klima: Das Experiment mit dem Planeten Erde* (Stuttgart: Theiss, 2002).

57 <https://en.wikipedia.org/wiki/An_Inconvenient_Truth> (accessed 29 May 2014).

14 Environmental history and global mining

Towards a neo-materialist approach

Timothy James LeCain

Introduction

From the moment we are born to the day we die, many of us are never more than a few yards away from a nearly pure piece of copper metal. In a hyper-technological and hyper-consumerist nation like the United States, there are 250 kilograms of copper for every inhabitant, a total of some 74 million metric tons of copper in wires, pipes, radiators, and countless other things dispersed throughout their built environment.[1] The per capita numbers are less striking for most other nations, but copper is still omnipresent wherever humans have most fully embraced the complex mix of social and material patterns of existence that we call modernity. That we do not typically perceive the copper around us is a product of both familiarity and obfuscation. Where tangled webs of copper power and phone lines can still be found above ground, we quickly learn to see past them. In their homes and buildings, most people demand (and building codes typically require) that the copper wire and pipes be hidden away behind walls and floors—for safety reasons, to be sure, though for aesthetic as well. With some notable exceptions,[2] we humans generally prefer not to be reminded of the metallic stuff that makes our built environment function. Much the same can be said for the hundreds of other metals, from aluminum to zinc, that we share our homes and cities with, as well as the coal, oil, and gas and other mineral treasures of the earth that constantly flow around and through us and our machines.

Given how ubiquitous minerals and metals are in our everyday environment, it is striking how little attention we give to them. This is echoed in an equally odd neglect of the ways these metals are extracted and refined, as well as to what comes of them once they cease to be useful and are thrown aside. Whatever the stage of use, the metallic and mineral fabric of the world around us rarely attracts much human interest. Among environmental historians, most scholars have focused on the extraction and processing end of the metallic life cycle, and to a lesser degree, recycling, leaving the actual use of metals to historians of technology, architecture, and art. This bias is even implicit in the working title of this chapter: an environmental history of global mining, rather than an environmental history of global minerals and metals.

In recent years, however, new analytical and methodological approaches to the environmental history of mining and other subjects have begun to suggest how we can move beyond earlier conceptual limitations by paying more attention to the role of material things like copper, coal, and oil. Elsewhere, I have called this efflorescence of new ideas the 'neo-materialist flip,' a term meant to suggest a departure from the dizzying series of 'turns' in recent decades, most of which have preserved the traditional academic focus on human sociocultural phenomena.[3] Regardless of terminology, the essential idea is that a growing number of neo-materialist scholars are in essence proposing that scholars need to flip our standard way of understanding and analyzing the world on its head. Instead of a world in which humans stand above and manipulate a separate material environment, neo-materialism argues that humans and their sociocultural systems are inextricably embedded in and a product of their material world. Fundamentally, neo-materialist theory suggests that it no longer makes sense to draw a clear conceptual line between humans and matter (or nature, as some prefer to call it), but rather that humans and their cultures are made of and from matter and cannot logically be understood in isolation from it. Put simply, the neo-materialist flip asks us to consider the many ways in which the material world creates us as much as we create it.

At first blush this may seem to be one of those ideas that seems so obvious as to be banal: who would deny, for example, that Thomas Edison needed the material powers of copper, electrons, and magnetized iron to invent the electric light bulb? Yet for most historians, the material side of that and most other stories literally goes without saying. It is irrelevant. They lavish all their attention on telling the human side of the story but treat the non-human material world as little more than a passive stage. If we all recognize that copper and electrons mattered in the invention of electric light bulbs, why then do we pay so little attention to them? Part of the answer, I argue, is that everything in the conventional humanistic worldview has conspired against paying attention to material things, from our reflexive anthropocentrism to the long dominance of social and cultural constructivism, and most fundamentally, our belief that we control and manipulate a separate natural material world rather than emerge with and from it.

In this chapter, my intent is neither a comprehensive survey of the sizeable global literature on mining and the environment nor of the rapidly growing body of neo-materialist literature, but rather to use a neo-materialist frame to suggest both how the environmental history of mining has evolved in recent decades and where it might yet go. I argue that neo-materialist theory offers a potentially powerful means to bring the history of mining out of its status as a somewhat marginal specialized historical subfield and to instead make it a far more central part of our understanding of history broadly construed. By avoiding the earlier tendency to limit mining history to clearly bounded topical fields like labor history, the history of technology, economic and business history, or even an environmental history defined as a focus on the natural destruction caused by mineral extraction, a neo-materialist approach suggests that historians reframe their analysis to give greater attention to minerals and metals themselves and both their human and non-human interactions. Most radically, scholars are now beginning to develop the analytical tools needed to ask deeper question about how mining and its products not only helped to create the environments humans live in, but also to shape human sociocultural systems. By moving from an environmental history of mining to something more like a *material history* of minerals and metals, we can better identify the essential role that the many extraordinary treasures of this earth have played in transforming us into the unusual mammals that we call humans.

Global mining and the environment

From the earliest days of the field, historians of mining around the world have often at least mentioned its environmental effects. Lou and Herbert Hoover's influential English translation of the classic sixteenth-century Saxon mining text, Georgius Agricola's *De Re Metallica*, called attention to Agricola's own discussion of the adverse environmental effects of mining.[4] Subsequent historians of mining followed a similar path, assuming that to tell the environmental history of mining was to discuss the damages (or, for those wishing to avoid declensionist narratives, the changes), the extraction, and processing of minerals caused to the surrounding landscape and its organisms. Since environmental history as a distinct scholarly field had its origins primarily in the United States, much of the early work on mining and the environment focused on the American West where the effects of large-scale mineral extraction were most evident. Indeed, one of the foundational early scholars of American history, Frederick Jackson Turner, had identified both the influence of the western 'frontier' environment and its exploitation through mining as central forces in the creation of a supposedly unique American character and politics.[5] Many of the specifics of Turner's famous 'Frontier Thesis' were later generally discredited. However, some of the pioneering generation of environmental historians, such as William Cronon, saw much of value in Turner's insistence on the influence of the environment on pioneers, even if they disagreed with most of the details of his argument.[6] Likewise, reinforced by the undeniable historical importance of western mining, Turner's ideas gave the history of mining an unusual prominence in the American historiography for a time. Much of this early work focused on topics like the California Gold Rush, the creation of the great western mining corporations and fortunes, and the labor conflicts that ensued as powerful mining capitalists sought to deskill miners and destroy nascent unions. At times these scholars would make at least some mention of the environmental damages caused by mining, typically focusing on tailings dumped into creeks and rivers that damaged downstream farmers and toxic smelter smoke that poisoned crops and animals.[7] In the case of the immense destructiveness of hydraulic mining in California, the topic even inspired an early exemplar of what would later come to be called environmental history with Robert Kelley's pioneering *Gold vs. Grain*.[8]

As environmental history emerged as a distinct field in the late 1970s, a few American mining historians began to reexamine their topic through this powerful new lens.[9] One of the earliest monographs was Duane Smith's 1994 survey, *Mining America*, a policy-oriented history that chronicled the mining industry's relative lack of concern for environmental issues until public and governmental pressures began to force changes in the 1970s.[10] However, much of the richest new work came not from within the field of mining history, but rather from environmental historians, historical geographers, cultural historians, and other scholars bringing new questions and methods to bear on the subject. Tailings and smelter smoke pollution continued to attract the most attention, though scholars increasingly analyzed these issues in ways that probed broader human ideas about the natural world and their relation to social history issues like community, class, and gender.[11] Smoke pollution offered a particularly useful means of broadening the environmental history of mining to encompass the urban and even domestic spheres, as smoke did not respect conventional human boundaries between country and city, factory and home. Coal, for example, was not only mined in some distant countryside colliery but also subsequently burned for heat in both urban factories and homes. The increasing use of soft sulfur-laden bituminous coal in place of cleaner-burning anthracite in the late nineteenth and early twentieth centuries, would eventually spark a new recognition of the complex interconnections between cities, homes, and the mineral products of mines. As the environmental historian of England, Peter Thorsheim, argues that this smoke pollution undermined earlier beliefs that a 'natural' substance like coal could not possibly be toxic, thus beginning to blur previously clear distinctions between the natural and artificial.[12] Environmental historians coming out of a public health and medicine background further questioned these divisions by pointing out that workers in mine smelters were often exposed to the same toxins as non-workers living downwind from the smelters. On a biological level, mining workers, farmers, and towns people were all linked by the powerful effects that sulfur, arsenic, lead, and other byproducts of mineral smelting had on the health of highly porous human bodies.[13]

In the 1990s, the influential American environmental historian Richard White provided another potentially powerful tool for rethinking the place of mining and minerals in history. White's short 1996 monograph, *The Organic Machine*, suggested that human technologies like dams always remain integrated into natural systems, becoming the 'organic machines' of the title. White also encouraged environmental historians to think about how humans have historically come to know nature through their work, sharply departing from the earlier tendency to see any use of the natural world solely as exploitation. While White did not discuss mining in his own work, his doctoral student Kathryn Morse drew on these and other ideas to rethink one of the classic topics of American mining history, the Klondike Gold Rush. In her 2003 book, *The Nature of Gold*, Morse analyzes the hordes of Alaskan gold seekers in terms of their bodily engagement with the harsh northern environment. Morse argues that the miners not only came to know the nature of the Klondike through their bodies, but also depended on organic machines like coal-fired ships and factory-made food to reach and develop their gold mines. For Morse, the Klondike Rush is not so much an example of the human exploitation of a pristine natural world as a product of a profound engagement with nature in its many other anthropogenic forms.[14]

However, while Morse devotes a good deal of attention to the human interactions with material things like steamboats, rivers, and canned pork and beans, when she turns to the topic of the gold itself, her analysis largely reverts to a more traditional approach in which a highly mutable human culture exists independently of and shapes a largely passive material world. The value of gold, Morse argues, 'can be an intangible, cultural, human creation,' one that 'takes concrete, material form only when human beings act on it.' Likewise, Morse asserts that, 'stripped of its profound cultural meaning, gold was not much different from any other part of the earth.'[15] Somewhat incongruously for a book with the title *The Nature of Gold*, Morse devotes only a brief passage to discussing the biogeochemical nature of gold:

> Separate from the question of its value, gold, of course, *was* the product of nature. It did and does have a nature, a set of chemical, physical, tangible attributes. Gold is bright yellow, the color of the sun. It is heavy, extraordinarily malleable, and profoundly enduring; it never decays. As some human societies discovered centuries ago, those natural characteristics made gold useful in particular ways. Gold was well suited to be forged into jewelry and other ornaments as a symbol of enduring power and wealth. It also worked well as a representative and store of value.[16]

Having spent all of seven sentences on the gold itself, however, Morse quickly reminds the reader that none of this really matters, as 'the nature of gold, its bundle of physical characteristics, did not determine how human beings would use it.' Of course, it is certainly true that the nature of gold did not *determine* its use, but neither was gold infinitely fungible or merely an empty signifier whose actual materiality was irrelevant. As is often the case in contemporary humanistic thinking, Morse seems to suggest that since the historical role of material things like gold is less readily evident than that of humans, it does not need to be discussed at all.

This eagerness to shift the focus from the material to the cultural, as well as the implicit assumption that the two are clearly different, is also evident in the book's foreword by the prominent environmental historian, William Cronon. Almost as if an afterthought, Cronon admits that, 'Sure, gold has certain attractive features that set it apart from other elements on the periodic table,' and he goes on to note the metal's malleability, resistance to rust, and 'seductive color.' Given this, he suggests that many might conclude that, 'Surely the human attraction for this magical stuff was ordained by nature and maybe even hard-wired into our very genes?' By reducing the issue at hand to the idea that humans have a natural or even genetic lust for gold, Cronon seems only to raise the topic to suggest its essential foolishness, dismissing his rhetorical query with a curt one-word answer: 'Perhaps.' Like Morse, Cronon seems eager to leave such trivial material matters behind, to insist that whatever the attraction gold might hold for humans, it 'always exists within a web of cultural relationships that change over time to reflect the historical epoch in which they occur.' The gold itself is far less important than the 'different lenses' through which Morse will ask the reader to view the Klondike Gold Rush.[17]

Despite their somewhat misleading rhetorical tactics, Morse and Cronon are obviously correct that one of the key ways in which gold or other minerals change history is through their historical intersection with humans. However, it is not the *only* way. As I have argued elsewhere, other animals and organisms inter-act with minerals as well, providing other histories and stories worth telling.[18] More importantly, while the human–mineral intersection is clearly critical, among many humanistic scholars like Morse it too often becomes an excuse to subse-quently ignore the material side of the interaction all together and focus solely on a human cultural sphere, which largely constructs reality. In Morse's analytical frame, humans think about and assign cultural values to gold, but they do not think with or through gold. Morse's approach thus leaves little room to consider, for example, how the very material nature of gold might have helped to create human cultural and social phenomena, including perhaps even the very idea of money, capital, and transferable wealth.

Given the early dominance of environmental history by American scholars, the methods and ideas pioneered there have been highly influential as the field became increasingly international in scope. In terms of mining and the environ-ment, scholarship from and about Great Britain and Germany has been especially productive in recent years. As already noted, the problem of coal smoke pollution in England inspired a rich environmental history literature, and the same can be said about copper smelting pollution in Cornwall and Wales.[19] Recently, British copper mining and smelting technology has been the subject of an international collaborative research project called 'A World of Copper,' led by the British his-torian Chris Evans, although environmental aspects are not the primary focus.[20] One of the leaders of a new generation of German environmental historians, Frank Uekoetter, adopts a transnational perspective in *The Age of Smoke*, his comparative history of American and German responses to coal, copper, and other forms of smoke pollution.[21] Uekoetter also edited a recent collection of arti-cles on the environmental history of mining in central Europe, where he rightly notes that the topic is inherently international, as mining 'shows little respect for national boundaries.'[22]

Almost anywhere around the globe where sustained mineral exploitation has taken place has begun to attract at least some attention from environmental histo-rians. The literature on the Australian mining industry has grown significantly in recent years, as has work on the important gold, silver, and copper mines of South America.[23] The environmental history of Asian mining has also begun to flourish, most notably with Brett Walker's perceptive analysis of the tailings and smoke pol-lution problems caused by the Ashio copper mining operations in Japan. Walker's history of the devastating sulfur and arsenic pollution at Ashio offers a model for understanding how many non-Western nations raced to modernize their extrac-tive industries while paying little heed to the environmental costs. Drawing on some of recent scholarship incorporating the history of medicine and bodies into environmental history (discussed in more detail below), Walker also adopts an analytical approach that resists traditional divisions between mining and a broader sociocultural sphere. By focusing on human bodies and health, Walker suggests how the process of extracting and purifying minerals had profound effects on Japanese sociocultural phenomena of all sorts, from rice cultivation to the deeply traditional practice of silkworm raising.[24]

In sum, while there is still much work to be done around the world, this brief overview suggests that the global environmental history of mining has begun to mature in recent years. However, despite some of the promising new ideas noted above that are pushing the topic beyond its traditional focus on tailings and smelter smoke, scholars have only just begun to better incorporate mining and the environment into our broader understanding of global human history. Indeed, the environmental history of global mining is particularly well suited for a deep reassessment, precisely because mining continues to be widely seen as one of the most inherently unnatural of human activities. Some cutting-edge scholars notwithstanding, most would still likely agree with the American scholar Lewis Mumford's assertion some 80 years ago that the development of deep underground mining constituted the first fully artificial technological environment. In Mumford's view, mining marked the moment when humans decisively stepped out of the natural world and became the creators of their own environments and destinies.[25]

Likewise, most scholars of mining and the environment still embrace the broader conceptual and sociocultural divisions that draw and enforce clear lines between the extraction of minerals and their subsequent use and disposal. Miners who extracted copper, for example, are typically seen as clearly distinct from the engineers who designed the copper mines, who in turn are to be distinguished from the factory workers who fabricated copper pipes, the plumbers who installed the pipe, the homeowners who used the copper plumbing, and the scrap metal dealer who melted it down into copper ingots ready to be turned to some new purpose. In many ways, these functional categories are entirely logical: the work of a miner extracting copper bears greater resemblance to that of a miner extracting silver than it does to a mining engineer, plumber, or homeowner working with copper. However, these categories also reflect the human propensity to measure all things primarily in terms of themselves, a propensity that can obscure other equally insightful ways of understanding the world. Rather than focusing solely on the human social patterns that emerge from the various stages of a mineral's life cycle, what new insights might emerge from organizing our analysis around the common material thread that unites them: the mineral itself?

The neo-materialist flip

Most of us rarely think about the metallic environment we live in, encouraging the belief that who we are as humans has little or nothing to do with the stuff with which either our 'natural' or human-built environment are made. To be sure, we readily recognize that the functional operation of air conditioners, cell phones, and cars affects how we live. Yet the material things that make these technologies possible—copper wires and radiators, steel frames and bodies, rare earth mineral batteries and semiconductors—are of little interest, save perhaps when their failure disrupts the seamless operation of the devices we increasingly depend upon.

Our neglect of this built environment in which humans live the vast majority of their lives stems from a number of flawed but deeply held beliefs. First,
most contemporary and even environmentally sensitive humans assume that the
'environment' is essentially someplace other than where they are. Humans and
their cities and technologies might affect a 'natural' environment—generally by
damaging it—but they are not themselves a part of that environment. Second,
since humans are commonly understood as the unnatural users and exploiters of a
distinctly separate environment, it follows that this environment cannot possibly
shape them in any significant sense. Humans increasingly recognize, of course,
that altering this supposedly exterior natural environment can result in pollution,
climate change, and other forms of ecological feedback that can adversely affect
them as individuals and societies. But merely recognizing some ecological connection between humans and their environment is not the same as understanding
the human animal as an inextricable part—and hence to some degree a product
of—that environment. Indeed, the belief that humans are merely 'connected' to
an environment simply reinforces the idea that there is a gulf between humans
and their environment that is only occasionally bridged. By contrast, consider that
humans do not feel the need to point out that bears or alligators are 'connected' to
their environment. Third, humans typically assume that even if they are somehow
linked to their environment biologically, they remain largely or entirely independent of it culturally. Culture, we often believe, is an abstract product of our
brains in isolation from our material surroundings. We might think about things,
or assign things great cultural meaning and value, but we don't think through or
with things, or even *as* things. Most humans believe that they, in contrast to every
other animal on the planet, are predominantly creatures of ideas rather than of
matter, of spirit rather than stuff.[26]

In the academic world, this division between humans and nature—or culture and matter—and the associated marginalization of materialism reached
something of an apotheosis with the so-called cultural or linguistic turns that
began in the 1980s and dominated academic thinking well into the new millennium. While only the most extreme of constructivists ever doubted the reality
of a separate material world that existed outside of human language and ideas,
the post-modern focus on the centrality of discourse strongly favored idealist rather than materialist historical investigations and explanations.[27] Still, the
penetration of post-modern theory into the academy was always uneven, and
many historians either ignored it or embraced only its more moderate forms.
Environmental historians in particularly retained their long-standing allegiance
to the agency of the environment, even though such a stance was out of step
with post-modernist thought. At the same time, though, environmental historians were radically expanding their definition of 'nature' and matter to include
technologies and the bodies of human themselves, as well as the bodies of animals like cows and horses that humans have bred to be living technologies.
The work of Richard White, Christopher Sellers, and Linda Nash all suggest
the ways in which humans know and experience their environments through
their bodies, pointing towards what Sellers terms an 'embodied history.'[28] In her

examination of late-nineteenth century American views of their environment, Nash convincingly argues that the idea of a 'bounded body' was yet another illusory conceit of the modernist worldview with its dreams of human liberation from and mastery over nature.[29] Historians of the environment and technology have made similar points in arguing that some animals can best be understood as technologies. From a neo-materialist perspective, the expansion of the concept of technology to include animals and other organisms has radical implications, as it poses a direct challenge to the conventional culture–matter boundary. If, as Edmund Russell and other scholars have now convincingly argued, a cow deliberately bred by humans to serve a specific instrumental purpose is indeed a technology, where precisely is the line between the technological and the natural, and hence the cultural and the material?[30]

This process, Edmund Russell argues, constitutes what he terms 'evolutionary history.' The genetic structures of organisms like dogs, cattle, and cotton have coevolved with human sociocultural phenomena, frequently without any conscious intention on the part of humans. Russell even challenges conventional accounts of the British Industrial Revolution that emphasize the central role of human social and technological forces, such as capitalism or the invention of steam-powered spinning. While not denying the significance of these factors, Russell argues that they all overlook the centrality of a specific type of cotton that had only recently begun to be imported from the New World. The product of a long and largely unintentional co-evolutionary history with indigenous peoples, New World cotton had unusually long and strong fibers that could withstand the harsh handling meted out by early mechanical spinning. This co-evolved cotton fiber, Russell argues, was at least as important to the rise of the British Industrial Revolution as new machines and social relations of production, yet historians have largely ignored its role in favor of narratives that emphasize the centrality of human creativity and initiative.[31]

As with new ideas framing human bodies and culture as embedded in nature, the concept of evolutionary history suggests that human sociocultural practices not only affect other organisms, but are often embedded into their genetic structure, which is to say, embedded in what modernist Western thinkers have often defined as 'nature.' Yet, if human culture and technology become part of the 'natural' material world, it obviously no longer makes sense to argue that there is some sort of dialectical dance between distinct cultural and material spheres— rather, we must strive to understand both how culture is a material thing and matter itself can be a type of distilled culture.

Environmental historians have also made great strides in breaking down the conventional modernist distinctions between the city and nature, in part by pioneering the analysis of the material and ecological flows that sustain cities.[32] The belief that the city is the antithesis of the natural is perhaps even older and more powerful than the idea that technology is inherently unnatural. Yet as historians increasingly embrace a materialist analytical position, they can begin to perceive the city as a place of material flows and processes in which the human and non-human are inextricably linked.[33] As one of the founders of urban environmental history, Martin Melosi, argues in a recent article that historians must move beyond the idea, that cities and the human-built environment are somehow unnatural, and that humans constitute 'a separate category from the rest of living things.'[34]

Historians of technology and the environment have thus mounted attacks on multiple fronts against the traditional understanding of human history. However, these ideas have been developed largely in isolation from each other and the broader historical and humanistic disciplines, their shared characteristics and radical significance remaining largely unrecognized. One key theme that unites all of these disparate approaches is their shared challenge to the belief that humans and their sociocultural systems are distinct from the material world around them. That the material world matters, not just in shaping human affairs, but also in constituting human bodies and the stuff of their sociocultural existence.

Surprisingly, however, while historians of environment and technology were busily defending and extending their foundational interests in materiality over the past few decades, the wider academic world's infatuation with the cultural turn was beginning to wane. Over the past decade, humanists in a wide variety of disciplines have shown a renewed interest in the material world. Although its roots can be traced back to at least the 1990s, the rise of a so-called 'new materialism' is now beginning to attract wide scholarly attention. In part, the movement was clearly a reaction against the domination of social and cultural constructivist theories in the academy, although materialist theory is not necessarily antithetical to constructivism. Most new materialist thinkers would not deny the importance of human cultural constructs of the world, but would rather argue that a narrow focus on this aspect has encouraged a dangerous neglect of the complex ways in which humans interact with a dynamic and creative material world. New materialist ideas have also derived in part from the array of approaches sometimes called 'post-humanist,' a term that suggests a shift away from the anthropocentrism of conventional humanistic approaches.[35]

A recent attempt to explore the possibilities of a new materialist approach is the 2010 collection edited by Diana Coole and Samantha Frost, *The New Materialisms*.[36] In their insightful introduction, Coole and Frost note that human beings 'inhabit an ineluctably material world,' yet that this essential materiality has been marginalized in recent decades. 'We share the feeling current among many researches,' they write, 'that the dominant constructivist orientation to social analysis is inadequate for thinking about matter, materiality, and politics in ways that do justice to the contemporary context of bio politics and global political economy.'[37] While their new materialism need not be antithetical to older constructivist methods, Coole and Frost call for a far more vibrant role for matter in its interaction with humans and their social systems. Theirs is a matter that is 'active, self-creative, productive, unpredictable,' a matter that 'becomes' rather than simply 'is.' Matter recognized in this way requires that we rethink conventional concepts of causation and agency that have long been simplistically anthropocentric. The human species, they argue, must be 'relocated within a natural environment whose material forces themselves manifest certain agentic capacities and in which the domain of unintended or unanticipated effects is con-siderably broadened.'[38]

Jane Bennett's influential 2009 work, *Vibrant Matter*, takes a somewhat simi-lar stance, though she is more successful in applying her theoretical ideas to actual material things. Under the banner of what she calls 'vital materialism,' Bennett strives to strip away both anthropocentrism and biocentrism in order to arrive at a concept of an environment that is more than merely a passive or sometimes recalcitrant stage for human action. Bennett argues that it is illogi-cal to conceive of humans as solely acting within and influencing a separate environment. Humans affect nature, but the nonhuman also affects culture. If, as she insists, matter is an active force in creating the human, then we also need to 'readjust the status of human actants: not by denying humanity's awesome, awful powers, but by presenting these powers as evidence of our own constitutions as vital materiality.' Making perhaps the most succinct statement of the analytical potential of neo-materialist theory, Bennett observes that 'human power itself is a kind of thing power.'[39]

These and the ideas of other neo-materialist thinkers hold great promise. Ironically, though, many neo-materialists (including more than a few of the authors collected in Coole and Frost's edited volume) have often been more inter-ested in discussing human ideas about matter rather than matter itself. Theories sprout vigorously, but concrete historical examples remain fallow. The same is often the case in other supposedly materialist approaches, suggesting that even when we deliberately set out to discuss the material world, it is exceedingly diffi-cult to avoid being pulled back into talking mostly about ourselves.[40] Nonetheless, if the intriguing theoretical insights of the new materialist are brought together with the recent empirically grounded insights of historians of the environment and technology noted above, we can begin to understand how even seemingly passive material things like metals and minerals can shape history and human culture.

The matter of minerals and metals

What might a neo-materialist history of mining and minerals look like? A detailed answer will demand many books and articles as yet unwritten. But drawing on some pioneering literature from mining history itself, let me briefly suggest some of the rich possibilities. We do best, of course, to begin with the minerals and metals themselves. As the philosopher Manuel De Landa reminds us, humans themselves are made of minerals. For eons, soft watery creatures dominated life on earth, fluid animals akin to modern-day jellyfish who could only move and survive in water. About 500 million years ago, a few of these gelatinous ani-mals literally got some backbone by incorporating minerals like calcium (or more precisely, calcite, a molecule of calcium and carbon united with three of oxy-gen, $CaCO_3$) into their bodies. These hard, rigid calcareous skeletons would, in time, make it possible for humans and other vertebrates to occupy almost every ecosystem on the planet. Calcium skeletons would even enable humans to mine limestone rocks, which despite being made of the very same stuff as their bones, have conventionally been understood as the antithesis of living things.[41] Human bodies and brains also depend on hundreds of other metals and minerals to function. Copper not only surrounds us, it is inside us, working closely with iron to constantly form new red blood cells.[42] Indeed, it probably did not take long for the humans who first began to work with iron to notice that the stuff that provided them with so many useful things had the same unique taste as their own blood—a taste that we can only describe with the tautological term 'metallic.'

Given how deeply embedded copper and other metals and minerals are in our bodies, it is perhaps slightly odd that humans are so ill-equipped to perceive their presence around them. Evolutionary biology and psychology offer a reasonable explanation: humans evolved to perceive the things necessary to their immediate biological survival, which were primarily other organisms that might provide food or present a threat. Eating these plants and animals usually provided the trace amounts of minerals and metals essential to human metabolism, leaving little need to evolve the ability to smell or see them more acutely. As a result, perhaps there is no 'mineralphilia' equivalent to E. O. Wilson's concept of 'biophilia,' which posits that humans have an evolutionarily embedded affinity for other organic things.[43] Regardless, were it possible for humans to smell the presence of copper, as can some species of fish, we would constantly be reminded of its unusual concentration in our built environment in comparison to its relatively dispersed geological state elsewhere.[44] Likewise, if humans could perceive electromagnetic fields the way that some reptiles can, we would not only sense the copper around us, but would recognize the constant dynamic changes in these fields as the electrical power the copper wire carries shifts.

Perhaps the most obviously useful neo-materialist approach is to take seriously Jane Bennett's observation that 'human power is a type of thing power,' potentially giving a whole new significance to mining and minerals. Indeed, before Bennett articulated the concept of 'thing power,' a few environmental historians were already beginning to think along similar lines. In his 2008 history of the famous American Ludlow Massacre of 1914, *Killing for Coal*, Thomas Andrews argues that the material properties of Colorado coal deposits played a central role in creating the sociopolitical power of both Rockefeller's coal company and the worker's union that attempted to resist it. In very much a neo-materialist fashion (though he does not use the term), Andrews asks us to think about the role played by both the energetic and geological nature of the coal. The accumulated solar energy stored in the coal was the original source of the power wielded by the Rockefeller Corporation and, more broadly, the United States itself.[45] Material energy, Andrews suggests, is intimately associated with sociocultural power. Indeed, this intimate connection between energy and power has recently found a more general theoretical and methodological statement. In an important 2011 article, Edmund Russell and several colleagues persuasively argue that material and social power are intimately linked, that energy power is the ultimate source of the social power that allows some humans to dominate others. Put in terms of Bennett's analysis, human power can be understood as a type of thing power, which in turn is sometimes a type of energy power. 'All power, social as well as physical,' Russell and his co-authors conclude, 'derives from energy.'[46] In this, Russell and Andrew suggest the role of ecology, with its embrace of all manner of material phenomena, as the key means of understanding the creation of human social powers. As Patrick Joyce and Tony Bennett recently noted, human agency and intention clearly mattered in the creation of this kind of power, but 'this agency gained its powers only to the extent that it was able to stabilise and capture material agencies' which 'operated according to logics of their own, often independent of human intention and indeed awareness.'[47] Or, as the urban historian Chris Otter notes, 'Materiality is not "outside" power, any more than it is "outside" the economic.'[48]

Although less avowedly materialist in its analytical approach, Andrew Isenberg's 2006 study, *Mining California: An Ecological Study*, also suggests how human social power was extracted from the material world in the hydraulic mining of gold and silver deposits in the American state of California. In this Isenberg means more than the obvious point that the wealth derived from the gold and silver could translate into human social power. Rather, he demonstrates that gaining wealth and power from mining also came from exploiting the ability of the environment to absorb the accompanying pollution and destruction.[49] Thus, social power may be a product not only of material energy, as Russell and his colleagues suggest, but also of material entropy—that is, the successful management of the disorder inevitably created by the *use* of any physical energy source. I have developed this concept in some detail through a close analysis of the conflicting entropic effects that underlay the so-called 'Smoke Wars' of the early twentieth century that pitted American ranchers against the a powerful copper mining and smelting corporation in Montana. The sociocultural power of the ranchers, I argue, derived from the ability of their cattle to reduce the entropic disorder of the environment by efficiently concentrating the energy of the widely dispersed plants of the 'open range' in their own bodies. As with Isenberg's hydraulic mines, however, the mining company's power depended on a contrary entropic process of dispersing copper smelter pollutants like sulfur and arsenic over the same open range. When the cattle reconcentrated this dispersed arsenic in their own bodies, they died, thus undermining the energetic basis of the ranchers' power with disastrous effects.[50]

The material energetic basis of human sociocultural power, however, is in some sense the low hanging fruit for a neo-materialist theory and method. If we are to more fully understand the materiality of human power or other sociocultural phenomena, we cannot limit our analysis to the energetic, entropic, or other mathematically quantifiable properties of matter. Rather, an even more interesting question is how the material world in all of its complex biogeochemical nature can help to create or even constitute the human sociocultural world in ways that go well beyond ideal or discourse. As the archaeologist Nicole Boivin notes, 'the material world, by virtue of its physical nature and the complementary sensory awareness of the human body, can impact people in very different ways that have nothing to do with the communication of abstract notions.'[51] Historians and other humanists must develop new methods, perhaps taking their lead from the British anthropologist Tim Ingold, who argues we must learn to engage 'directly with the materials themselves, following what happens to them as they circulate, mix with one another, solidify and dissolve in the formation of more or less enduring things.'[52]

In this, Thomas Andrews' insightful analysis of the physical properties of Colorado coal deposits again offers a powerful empirical example. The emergent nature of the interactions between miners and coal came not primarily from the material's energy content, but rather from other geological and physical properties that cannot easily be quantified. Coal miner solidarity, Andrews argues, emerged in no small part from their experience of working with the material properties of the coal itself, such as the coal's relative softness and its association with methane gas. These properties made dangerous roof falls and explosions a constant risk, deeply influencing the social interactions of the human beings working with the coal. Rather than assuming that the cultural idea of worker solidarity was primarily a discursive one transmitted by pamphlets, speakers, and conversations, Andrews illustrates instead how solidarity was also a physically embodied phenomena. Instead of learning just from Marx, the miners could in some very real sense learn from the coal, extracting a material basis for an alternative social organization.

The political scientist Timothy Mitchell makes this point explicit in his intriguing 2011 analysis of the sociocultural effects of human efforts to extract coal and oil, *Carbon Democracy*. Mitchell persuasively argues that the material nature of coal deposits forces states to rely on large numbers of workers to extract it. This sociotechnical environment was more conducive to democratic control than that of oil, he argues, which could be obtained with far fewer workers and thus encouraged more centralized authoritarian regimes.[53] On a broader analytical level, Mitchell insists that scholars must move beyond the traditional belief that coal and oil create their greatest effect mostly through the money they generate and that the spread of democratic practices is mostly the result of the spread of an abstract idea. To actually transform coal and oil into useful and profitable commodities, Mitchell instead argues, involves 'establishing connections and building alliances—connections and alliances that do not respect any divide between material and ideal' or even between the 'human and nonhuman.'[54] It is in these complex webs of social and material interconnections that the possibilities for democracy are facilitated or foreclosed. Indeed, his analysis suggests that these material structures of energy extraction, processing, and flow, constitute the stuff of 'democracy' every bit as much as do abstract ideas about the theory and procedures of a democratic system. 'Workers were gradually connected together not so much by the weak ties of a class culture, collective ideology or political organisation,' Mitchell argues, 'but by the increasing and highly concentrated quantities of carbon energy they mined, loaded, carried, stoked and put to work.'[55]

Conclusion

Although it seems obvious that humans can only act and exist through the use of material things and phenomena, it is extraordinary how little attention historians and other humanists have given to this fundamental reality of our existence. As the feminist-theorist Elizabeth Grosz argues, 'Life is the growing accommodation of matter, the adaptation of the needs of life to the exigencies of matter.'[56] Second only to the organic things that feed, house, and clothe us, mineral and metallic things are the essential materials that help create human societies and their diverse sociocultural systems. One possible future of the environmental history of mining, then, lies in moving beyond the modernist dichotomies that separate mine and city, nature and technology, biology and culture, and the 'natural' and built environment. Instead, historians should strive to analyze the many ways metals and minerals flow through our societies and our bodies, helping to create human power and culture even as they entangle us every more deeply in their own material demands. In this new environmental history of mining, copper, iron, gold, and countless other treasures of the earth will take on new importance and meaning as we come to better appreciate that these are not merely things that humans use. Rather, these are the very things that make us human.

Notes

1 R. B. Gordon, M. Bertram, and T. E. Graedel, 'Metal Stocks and Sustainability', *Proceedings of the National Academy of Sciences 103* (2006), pp. 1209–1214.
2 The Italian architect Renzo Piano made the exposed internal workings of the George Pompidou Centre in Paris part of the building's external design. However, the copper electrical wires and plumbing are still encased in yellow and blue plastic tubes, so the metal itself cannot be seen. Nor has this design and aesthetic approach been widely imitated.
3 These ideas are examined in detail in my book in the Cambridge University Press series, Studies in Environment and History, *The Matter of History: How Things Create the Past* (Cambridge: Cambridge University Press, 2017). For a briefer recent introduction, see: Timothy James LeCain, 'Against the Anthropocene: A Neo-Materialist Perspective', *History, Culture, and Modernity 3* (2015), pp. 1–28.
4 Georgius Agricola, Herbert Clark Hoover, and Lou Henry Hoover, trans., *De Re Metallica* (New York: Dover, 1950).
5 Frederick Jackson Turner, 'The Significance of the Frontier in American History', originally presented as a paper at the 1893 meeting of the American Historical Association in Chicago.
6 William Cronon, 'Revisiting the Vanishing Frontier: The Legacy of Frederick Jackson Turner', *The Western Historical Quarterly 18* (1987), pp. 157–176.
7 One of the early leaders of American mining history, Clark Spence, noted the tailings pollution caused by large-scale gold dredging: Clark C. Spence, 'The Golden Age of Dredging', *Western Historical Quarterly 11* (1980), pp. 401–414. An early discussion of smelter pollution, more self-consciously environmental in focus, is: Donald MacMillan, 'A History of the Struggle to Abate Air Pollution from Copper Smelters in the Far West, 1885–1933' (PhD thesis) (University of Montana, 1973).
8 Robert Kelley, *Gold vs. Grain: The Hydraulic Mining Controversy in California's Sacramento Valley* (Glendale, CA: A. H. Clark Co., 1959).
9 Two valuable surveys of this literature are: Katherine G. Morrissey, 'Rich Crevices of Inquiry: Mining and Environmental History', in: Douglas C. Sackman (ed.), *A Companion to American Environmental History* (Chichester: Wiley-Blackwell, 2010); and, Gavin Bridge, 'Contested Terrian: Mining and the Environment', *Annual Review Environmental Resources 29* (2004), pp. 205–259.

10 Duane A. Smith, *Mining America: The Industry and the Environment, 1800–1980*, (Denver, CO: University of Colorado Press, 1994).

11 Richard V. Francaviglia, *Hard Places: Reading the Landscape of America's Historic Mining Districts*, (Iowa City, IA: University of Iowa Press, 1991); Katherine G. Morrissey, 'Mining, Environment and Historical Change in the Inland Northwest', in: Dale Goble and Paul Hirt (eds), *Northwestern Lands and People: An Environmental History* (Seattle, WA: University of Washington Press, 1999), pp. 479–501; Katherine G. Aiken, '"Not Long Ago a Smoking Chimney was a Sign of Prosperity", Corporate and Community Response to Pollution at the Bunker Hill Smelter in Kellogg, Idaho', *Environmental History Review 18* (1994), pp. 67–86; and Kathryn Morse, *The Nature of Gold: An Environmental History of the Klondike Gold Rush* (Seattle, WA: University of Washington Press, 2003).

12 Peter Thorsheim, *Inventing Pollution: Coal, Smoke, and Culture in Britain Since 1800* (Athens, OH: University of Ohio Press, 2006). Morrissey identifies the historiographical significance of this shift: Morrissey, 'Rich Crevices of Inquiry'. See also: David Stradling, *Smokestacks and Progressives: Environmentalists, Engineers, and Air Quality in America, 1881–1951* (Baltimore, MD: Johns Hopkins University Press, 1999); Lynne Page Snyder, 'The Death-Dealing Smog over Donora, Pennsylvania: Industrial Air Pollution, Public Health, and Federal Policy, 1915–1963,' (PhD thesis) (University of Pennsylvania, 1994); and, Angela Gugliotta, 'Class, Gender, and Coal Smoke: Gender Ideology and Environmental Injustice in Pittsburgh, 1868–1914', *Environmental History 5* (2000), pp. 165–193.

13 Morrissey, 'Rich Crevices'; Christopher C. Sellers, 'Factory as Environment: Industrial Hygiene, Professional Collaboration and the Modern Sciences of Pollution', *Environmental History Review 18* (1994), pp. 55–83; and Joel A. Tarr, *The Search for the Ultimate Sink* (Akron, OH: University of Akron Press, 1996).

14 Kathryn Morse, *The Nature of Gold: An Environmental History of the Klondike Gold Rush* (Seattle, WA: University of Washington Press, 2003).

15 Morse, *The Nature of Gold*, pp. 8, 192.

16 Morse, *The Nature of Gold*, p. 22.

17 William Cronon, 'Foreword: All That Glitters', in: Morse, *The Nature of Gold*, pp. ix–xiii, quote on xi. To be fair, it is not entirely clear if these are Cronon's views or if he is simply summarizing Morse's ideas.

18 Timothy James LeCain, 'An Impure Nature: Memory, Geese, and Neo-Materialism at America's Biggest Toxic Superfund Site', *Global Environment 11* (2013), pp. 16–41.

19 See for example: Thorsheim, *The Invention of Pollution*; Ann Beck, 'Some Aspects of the History of Anti-Pollution Legislation in England, 1819–1954', *Journal of the History of Medicine 14* (1959), pp. 475–489; Edmund Newell, 'Atmospheric Pollution and the British Copper Industry, 1690–1920', *Technology and Culture 38* (1997), pp. 655–689; Edmund Newell and Simon Watts, 'The Environmental Impact of Industrialisation in South Wales in the Nineteenth Century: "Copper Smoke" and the Llanelli Copper Company', *Environment and History 2* (1996), pp. 309–336; and, D. A. Radcliffe, 'Ecological Effects of Mineral Exploitation in the United Kingdom and their Significance to Nature Conservation', *Proceedings of the Royal Society of London 339* (1974), pp. 355–372.

20 Chris Evans and Olivia Saunders, 'A World of Copper: Globalizing the Industrial Revolution, 1830–1970', *Journal of Global History 10* (2015), pp. 3–26.

21 Frank Ueokoetter, *The Age of Smoke: Environmental Policy in Germany and the United States, 1880–1970* (Pittsburg, PA: University of Pittsburgh Press, 2009). The American environmental historian, Andrew Isenberg, also compares the two nations: Andrew C. Isenberg, 'Mercurial Nature: The California Gold Country and the Coalfields of the Ruhr Basin, 1850–1900', in: Ursula Lehmkuhl and Hermann Wellenreuther (eds), *Historians and Nature: Comparative Approaches to Environmental History* (Oxford: Berg Publishers, 2007), pp. 124–145.

22 Frank Uekoetter, 'Introduction: Mining and the Environment', in: Frank Uekoetter (ed.), *Mining in Central Europe: Perspectives from Environmental History* (Munich: Rachel Carson Center Perspectives, 2012), p. 5. A good example of this is cross-border smoke pollution. See: John D. Wirth, *Smelter Smoke in North America: The Politics of Transborder Pollution* (Lawrence, KS: University Press of Kansas, 2000).

23 Robin Libby and Tom Griffiths, 'Environmental History in Australasia', *Environment and History 10* (2004), pp. 439–474; Elizabeth Dore, 'Environment and Society: Long-Term Trends in Latin American Mining', *Environment and History 6* (2000), pp. 1–29; Nicholas A. Robins, *Mercury, Mining, and Empire: The Human and Ecological Cost of Colonial Silver Mining in the Andes* (Bloomington, IN: Indiana University Press, 2011); Jason Moore, '"This Lofty Mountain of Silver Could Conquer the Whole World": Potosi and the Political Ecology of Underdevelopment, 1545–1800', *The Journal of Philosophical Economics 4*(1) (2010), pp. 58–103.

24 Brett L. Walker, *The Toxic Archipelago: A History of Industrial Disease in Japan* (Seattle, WA: University of Washington Press, Weyerhaeuser Series, 2009). Another excellent recent monograph on the Ashio pollution story that focuses more on its political effects is Robert Stolz, *Bad Water: Nature, Pollution, and Politics in Japan, 1870–1950* (Durham, NC: Duke University Press, 2014).

25 Lewis Mumford, *Technics and Civilisation*, 1934 (New York: Harcourt Brace, 1963), pp. 65–77.

26 This idealist view of culture is insightfully explained and critiqued in, Nicole Boivin, *Material Cultures, Material Minds: The Impact of Things on Human Thought, Society, and Evolution* (Cambridge, MA: Cambridge University Press, 2010).

27 For a recent overview and critique of this succession of intellectual turns, see the collection of essays in the American Historical Review Forum: 'Historiographical "Turns" in Critical Perspective', *American Historical Review 117* (2012). In her perceptive comment on the forum, Julia Adeney Thomas rightly points out that the authors largely failed to take note of what is arguably the most important recent theoretical shift in the discipline: the rise of the new materialism and the 'environmental turn.' See: Julia Adeney Thomas, 'Comment: Not Yet Far Enough', *American Historical Review 117* (2012), pp. 794–803.

28 Richard White, '"Are You an Environmentalist or Do you Work for a Living?": Work and Nature', in: William Cronon (ed.), *Uncommon Ground: Rethinking the Human Place in Nature* (New York: W. W. Norton, 1996), and, Richard White, *The Organic Machine* (New York: Hill and Wang, 1995). Also see: Christopher Sellers, 'Thoreau's Body: Towards an Embodied Environmental History', *Environmental History 4* (1999), pp. 486–514.

29 Linda Lorraine Nash, *Inescapable Ecologies: A History of Environment, Eisease, and Knowledge* (Berkeley, CA: University of California Press, 2006). See also, Nancy Langston, *Toxic Bodies: Hormone Disruptors and the Legacy of DES* (New Haven, CT: Yale University Press, 2010).

30 Philip Scranton and Susan R. Schrepfer, *Industrializing Organisms: Introducing Evolutionary History* (New York: Routledge, 2004).

31 Edmund Russell, *Evolutionary History: Uniting History and Biology to Understand Life on Earth* (Cambridge, MA: Cambridge University Press, 2011).

32 Maria Kaika, *City of Flows: Modernity, Nature, and the City* (New York: Routledge, 2005). A somewhat similar idea can be found in the recent development of commodity chain analysis. See: Jennifer Bair, *Frontiers of Commodity Chain Research* (Stanford, CA: Stanford University Press, 2009); Alex Hughes and Suzanne Reimer (eds), *Geographies of Commodity Chains* (New York: Routledge, 2004); and, Matthew Evenden, 'Aluminum, Commodity Chains, and the Environmental History of the Second World War', *Environmental History 16* (2011), pp. 69–93.

33 Joel Tarr, *The Search for the Ultimate Sink* (Akron, OH: University of Akron Press, 1996); Martin Melosi, *The Sanitary City: Urban Infrastructure in America from Colonial*

Times to the Present (Baltimore, CT: Johns Hopkins University Press, 2000); Craig Colten, *Transforming New Orleans and Its Environs: Centuries of Change* (Pittsburgh, PA: University of Pittsburgh Press, 2000); Craig E. Colten, *An Unnatural Metropolis: Wresting New Orleans from Nature* (Baton Rouge, LA: Louisiana State University, 2005); Matthew Gandy, *Concrete and Clay: Reworking Nature in New York* (Cambridge, MA: MIT Press, 2002); and, Matthew Klingle, *Emerald City: An Environmental History of Seattle* (New Haven, CT: Yale University Press, 2007).

34 Martin V. Melosi, 'Humans, Cities, and Nature: How Do Cities Fit in the Material World', *Journal of Urban History 36* (2009), p. 7.

35 Good introductions are, Cary Wolfe, *What Is Posthumanism?* (Minneapolis, MN: University of Minnesota Press, 2009), and Rossi Braidotti, *The Posthuman* (Cambridge: Polity Press, 2013).

36 Diana H. Coole and Samantha Frost, *New Materialisms: Ontology, Agency, and Politics* (Durham, NC: Duke University Press, 2010).

37 Coole and Frost, *New Materialisms*, p. 6.

38 Coole and Frost, *New Materialisms*, p. 9.

39 Jane Bennett, *Vibrant Matter: A Political Ecology of Things* (Durham, NC: Duke University Press, 2010), p. 10.

40 This critique is raised in both, Tim Ingold, 'Materials Against Materiality', *Archaeological Dialogues 14*(1) (2007), p. 2, and Bjørnar Olsen, 'Material Culture After Text: Re-Membering Things', *Norwegian Archaeological Review 36* (2003), p. 88.

41 Manuel De Landa, *A Thousand Years of Nonlinear History* (New York: Zone Books, Swerve Editions, 2000), pp. 26–27. Of course, most large limestone deposits were formed from the skeletons and shells of billions of ancient sea creatures, so geologists—if not the broader public—recognize that both human skeletons and limestone are the products of organic processes.

42 Larry E. Johnson, 'Copper Deficiency and Toxicity', *Merck Manual*, online version, <www.merckmanuals.com/home/disorders-of-nutrition/minerals/copper-deficiency-and-toxicity>, accessed 24 September 2015.

43 Edward O. Wilson, *Biophilia: The Human Bond With Other Species* (Cambridge, MA: Harvard University Press, 1984). Which is not to say that there is *no* evolutionary or biological explanation for why almost all humans are fascinated by the appearance and texture of metals and gems like copper, gold, rubies, diamonds, etc.

44 Researchers have found that some salmonid species find their way up a maze of branching stream courses to return to their birthplace in part by sensing the unique mixture of minerals dissolved in the water from different watersheds. Unusual high levels of copper produced by mining, though, can actually harm the ability of fish to sense these subtle differences in mineral concentration. See, David H. Baldwin, et al., 'Sublethal Effects of Copper on Coho Salmon: Impacts on Nonoverlapping Receptor Pathways in the Peripheral Olafactory Nervous System', *Environmental Toxicology and Chemistry 22* (2003), pp. 2266–2274.

45 Thomas G. Andrews, *Killing for Coal: America's Deadliest Labor War* (Cambridge, MA: Harvard University Press, 2008).

46 Edmund Russell, James Allison, Thomas Finger, John K. Brown, Brian Balogh, and W. Bernard Carlson, 'The Nature of Power: Synthesizing the History of Technology and Environmental History', *Technology & Culture 52* (2011), p. 248.

47 Patrick Joyce and Tony Bennett, 'Material Powers: Introduction', in: idem (eds), *Material Powers: Cultural Studies, History and the Material Turn* (London: Routledge, 2010), p. 14.

48 Chris Otter, 'Locating Matter: The Place of Materiality in Urban History', in: Joyce and Benett, *Material Powers*, p. 55.

49 Andrew Isenberg, *Mining California: An Ecological History* (New York: Hill & Wang, 2006).

50 Timothy James LeCain, 'Copper and Longhorns: Material and Human Power in Montana's Smelter Smoke War, 1860–1910', in: John McNeill and George Vrtis (eds), *Mining North American: An Environmental History since 1522* (Berkeley, CA: University of California Press, 2017), pp. 166–190.
51 Boivin, *Material Cultures, Material Minds*, pp. 96–97.
52 Tim Ingold, *Being Alive: Essays on Movement, Knowledge, and Description* (Abingdon: Routledge, 2011), p. 14.
53 Timothy Mitchell, *Carbon Democracy: Political Power in the Age of Oil* (London: Verso, 2011).
54 Mitchell, *Carbon Democracy*, p. 7.
55 Mitchell, *Carbon Democracy*, p. 27.
56 Elizabeth Groz, *Time Travels: Feminism, Nature, Power* (Durham, NC: Duke University Press, 2005), p. 132.

15 Mining heritage

Comparative perspectives

Stefan Berger

Introduction

In many parts of the world mining was once an important economic activity and is so no more. Those areas have, in many cases, undergone deindustrialisation, or, at the very least, structural transformation that has been accompanied by debates on what to do with the material and immaterial remnants of a now defunct mining industry. In this context, narratives of the past have been developed that fitted the interests of diverse actors in deindustrialisation and structural transformation processes and that could justify their particular visions of the future. Among those actors, we find politicians, administrators, businessmen, trade unionists, academics, preservationists, artists, architects, engineers, tourism specialists and former workers in the industry. As these actors have rarely been in agreement over which strategies would lead to successful structural transformation, the visions of the past that underpinned those strategies have usually been highly contested, especially as they often implied different visions of the future.

This chapter will review, in comparative perspective, a select number of former mining regions, all of them coal-mining regions, in the capitalist West, the post-communist East European states, and in communist China. In particular we will focus on the Ruhr area in Germany, the South Wales coalfield in Britain, Nord Pas de Calais in France, Asturias in Spain, Upper Silesia in Poland, the Jiu valley in Romania, the Appalachian coalfield in the US and Shenyang province in northeast China. These regions have been chosen as they are part of a comparative research project on mining heritage that I have been heading since 2012 and that seeks to compare the Ruhr with other coal-mining regions, where mining has either come to an end or where the industry is in crisis.[1] In the first part of this chapter, we shall all-too-briefly review different conceptualisations and theorisations of mining heritage. In the second part, we then examine the preconditions for heritage narratives to emerge, concentrating on socio-economic, political and cultural factors and actors. In the third and final part, we then analyse the narratives themselves, highlighting particular narrative tropes and strategies and relating them to the generation of different futures for those mining regions. The chapter will end with some preliminary and tentative conclusions on why mining heritage initiatives have been more successful in some regions than in others and what such success means for the framing of this heritage.

My comparison here has to be provisional, as much more research on additional former mining regions is needed and indeed more research on those regions that have been selected here is needed too, before we can say something more definite about reasons for the emergence and the shape of mining heritage in a global perspective. However, what follows should be seen as a first attempt to provide answers to questions that are surely worth answering, as they affect millions of people around the globe: what do different forms of heritigisation of mining mean for local/regional and national identities? How are former class identities preserved or obliterated? And there are questions that I cannot deal with here in a brief chapter but that would be equally crucial to include in future work, such as: how is the gendering of former mining places affected by heritage discourses? What perspectives are developed in heritage on cultural, religious, ethnic and racial cleavages and solidarities that once characterised mining regions? Overall, this chapter aims to provide glimpses into meaning making through heritage and the ways in which such constructions of the past open up vistas onto possible futures of former mining regions.

Conceptualisations and theorisations of mining heritage

Mining heritage is part and parcel of industrial heritage. In the English language, the term 'industrial archaeology' emerged in the 1950s to describe a discipline that was to concern itself with industrial heritage. Michael Rix, in *The Amateur Historian* in 1955, called for greater preservation of the industrial heritage of eighteenth- and nineteenth-century industrial Britain, which he argued was threatened by extinction.[2] Four years later, in 1959, an industrial archaeology committee was founded within the Council of British Archaeology. By the mid-1960s a National Record of Industrial Monuments was created. An Association for Industrial Archaeology was founded in 1973, two years after the North American Society for Industrial Archaeology had come into being.[3]

While industrial archaeology was foundational to the study of industrial heritage, the latter quickly developed in extraordinarily multi-disciplinary ways. Urban planners, architects, art historians, historians, preservationists, archaeologists, engineers, museologists, sociologists, economists, geographers and political scientists have all contributed to this field over recent decades, and it has been a great challenge to combine and draw together the different methodological, theoretical and empirical insights provided by so many different disciplines.[4]

If the professionalisation of the study of industrial heritage took place earliest in the English-speaking world, developments here were being watched closely elsewhere. In Germany, for example, the 1970s saw a string of publications by experts seeking to transport the concept of industrial archaeology into Germany, and the word '*Industriearchäologie*' is in use, but industrial culture (*Industriekultur*) has become the far more popular term. It was arguably first used by Tilman Buddensieg, who, in 1979, published a book on the industrial design of the electronics giant AEG with the term '*Industriekultur*' in its title. Hermann Glaser then significantly extended the meaning of the term in the 1980s to include not just the aesthetics of industrial design and enterprise culture but the historical life worlds of industrial modernity *tout court*. Ulrich Borsdorf, the founding director of the Ruhrmuseum, sought to extend the notion of industrial culture to include the everyday of the life worlds of workers in regions of heavy industry.[5] We still have to investigate the conceptual history of the field 'industrial archaeology/industrial culture/industrial heritage' in comparative perspective more fully to see what systems of meaning have been associated with terms that have also been used to describe mining heritage.

If the meaning of concepts and terms are by no means the same in different parts of the world, we have nevertheless witnessed an impressive internationalisation of research in industrial heritage, of which mining heritage is one part. The International Committee for the Conservation of the Industrial Heritage was founded in 1973 and it currently has representatives from over 40 countries in the world and organises regular conferences.[6] It has been important in developing standards for industrial heritage conservation and in publishing calls to support such conservation, e.g. the Charter for Industrial Heritage of 2003 and a range of thematic reports and studies. Much of its work, like that of other heritage bodies, is conservation-oriented and quite technical. The mounting criticism of an alleged lack of critical perspectives on heritage and a one-sided concern with the materialities of conservation has led to the formation of 'critical heritage studies', which now has thousands of adherents around the globe and also organises a thriving bi-annual conference.[7] Critical heritage studies has been more concerned with meaning making and representation through heritage rather than documentation and preservation, and it has also attempted to intervene politically in heritage debates arguing that heritage is, in various ways, an important resource for a politics of the left.

If we define industrial heritage as that heritage which emerges in industrial spaces and which impacts, through worlds of work, urban spaces and transportation, on the everyday of the people living in such industrial spaces, then industrial heritage is part and parcel of a historical culture of regions of industry, including mining regions. Historical culture, according to Jörn Rüsen, finds expression through the articulation of historical consciousness in society.[8] Both historical culture and historical consciousness is always contested and pluralistic and yet it is also always used in order to stress overarching commonalities in societies, e.g. of class, religion, ethnicity, region etc. In this sense, historical culture is characterised by a paradox: it strives towards unity and produces plurality and difference.

Mining heritage as historical culture is therefore intensely political as it influences historical consciousness and constructs identities. It is part and parcel of what Hayden White has termed 'the practical past', i.e. a past that is meaningful in the present.[9] It can also be related to Friedrich Nietzsche's 'critical past' that he juxtaposed to an 'antiquarian past'.[10] Whereas the former is relevant for life in the present, the latter is a dead weight around the neck of the living. The concept of nostalgia has often been related to Nietzsche's antiquarian past, but as a whole variety of authors have pointed out more recently, there is also a 'critical' and 'practical' edge to nostalgia, which relates the passions of nostalgia to the retention of forms of historical consciousness that have political meaning in the present.[11]

Historical consciousness is, of course, closely related to memory. In memory studies, Anna Cento Bull and Hans Lauge Hansen have adapted the concept of 'agonism' in the political theory of Chantal Mouffe in order to differentiate an agonistic memory culture from a cosmopolitan and an antagonistic one.[12] Agonistic memory cultures are characterised by attempts to contextualise memory socially and historically rather than locate it on a normative scale of good and evil. It also incorporates a radical multi-perspectivity that allows for contestation and the representation of voices on the left, otherwise easily written out of an allegedly universal cosmopolitan memory culture. Furthermore, it is open-endedly dialogic (in a Bakthinian sense) rather than consensually dialogic (in a Habermasian sense). And finally, it seeks to harness the emotions and passions of solidarity that have been underpinning all political projects of the left. Their plea is to develop more agonistic memory cultures as they see this as one way of repoliticising memory culture. I would argue that these ideas can usefully be introduced to heritage studies. In the case of mining heritage, it would mean looking for an agonistic memory culture of mining that would preserve a specific left-wing politics and specific left-wing values that neoliberal forms of deindustrialisation have pushed to the margins of debate or eliminated altogether.

When we discuss mining heritage in the following pages, it should be clear that we do not refer to the remnants of former mines alone. Instead we refer to landscapes of mining which also always have been 'mindscapes'.[13] Landscapes imply, as Juan Manuel Wagner has argued, that we have to maintain physical objects signifying a web of spatial relations. These include many objects of the everyday, which are in no way special or extraordinary, but which contain authenticity and emotional value.[14] With regard to the mining heritage, Hans-Werner Wehling has argued that its industrial landscape consists not only of the mines themselves, but also of the slagheaps, the networks of regional communication and transportation and the housing estates where miners lived.[15] Landscapes of industrial culture thus represent a regional ensemble of objects that preserve an emotive tie of the region's inhabitants to their past. And such landscapes imply mindscapes that include the immaterial heritage of mining, from food to leisure pursuits to communication and many other ways of thinking, acting and behaving that cannot be memorialised in an object.

While, on one level, the material objects of mining heritage – mine buildings and towers, shafts, housing estates, slagheaps and railway lines – are arguably similar in mining regions all over the world, on another level, there are important differences, for example in the functionality of their architecture. While some mine owners were content with the cheapest and most functional design, others were keen to represent in the mine their own social status and ambition. The architectural value of the mine buildings has often been important for the decision to preserve them. More representative buildings have often been regarded as more worthy of preservation in comparison with their simpler and less spectacular variants. It is no coincidence that the first mining building that was ever put under preservation in Germany was an art deco machine hall at Zeche Zollern in Dortmund.[16] In 1969, this marked the beginning of the industrial heritage boom in the Ruhr area of Germany. But many of the buildings of this industry in this part of Germany looked more like castles and palaces than like mine buildings. The aesthetics of the buildings were and are an important argument for their preservation.

The decision for or against preservation is also often related to plans for re-use.[17] Proposals on how to deal with these objects have also been similar in many parts of the world: they can be preserved as technical and architectural monuments, or they can be used to house mining, technical or regional museums. They can be re-used in a public or commercial sense. They can be turned into housing or office space. They can be adorned with light installations and turned into platforms for art and cultural spectacles.[18] The many different uses of mining heritage have been strongly related to the needs of the housing industry, the tourism industry and to business and commercial interests.[19] These uses have also had an impact on the dominant narrativisations that underpin the meaning making of that heritage. Hence we shall now first look at what socio-economic, political and cultural preconditions had to be in place in order to result in the preservation of mining heritage.

Preconditions for mining heritigisation

The consequences of deindustrialisation and the success of the structural transformation of mining regions play a vital role in determining the attractiveness of the option to develop mining heritage initiatives. Arguably the most developed mining heritage landscape worldwide can be found in the Ruhr region of Germany, home of one of the most important coalfields in Europe between the last third of the nineteenth century and the first two thirds of the twentieth.[20] Here the coal crisis started in the late 1950s. Ten years later, in 1969, all coal companies were put under the roof of a quasi-state-supported company, the Ruhrkohle AG (RAG), and over the next 50 years, coal was softly phased out, with the help of moderate subsidies paid by the German taxpayer. These 50 years were also characterised by the structural transformation of the region away from coal (and eventually also from steel) towards other industries and the ever-growing service sector. While the social and economic problems of the region during this period and in the present have been considerable, deindustrialiation did not leave whole towns and areas devastated.[21] Social immiseration was not entirely prevented but much cushioned by a comprehensive welfare state and a culture of solidarity rooted in the German idea of 'social partnership'.[22] Such relative socio-economic success of the process of structural transformation of the Ruhr has nurtured a positive look at the mining past and a desire to preserve its heritage for generations to come.

Other mining regions in Western Europe have done less well than the Ruhr in coping with the structural transformation brought by the end of mining. Many of the mining regions of Britain have been rapidly deindustrialised after the epic miners' strike of 1984/5 that could not prevent the government's determined effort to close down the industry. As a result of rapid and brutal deindustrialisation, without any concern for the miners, the mining regions of Britain were devastated.

When I first visited the South Wales coalfield in 1987, many of the villages and towns resembled ghostly places. Many former miners and their families had left, and the few attempts to think about economic revival strategies, like building dual carriageways into the valleys and trying to attract companies to newly set-up industrial estates did not have the desired effect of energising former mining communities. The failed structural transformation of former mining regions, like South Wales, left little room for the preservation of mining heritage.[23]

The situation in the French mining regions, especially the Nord Pas de Calais, was not as disastrous as it was in Britain. Welfare plans were better and strategies of economic revival were more successful, although not on a par with the Ruhr. Nevertheless the mining regions were not as devastated by the closure of the mines as was the case in Britain, so that there was also more scope for the creation of mining heritage.[24] All other mining regions in Western Europe can be said to be somewhere in between the extremes of the German example on the one side and the British on the other. A particularly interesting case is that of Spain, where mining in Asturias was protected by the Franco dictatorship as a vital part of the national economy. The crisis hence only hit the Asturian mining industry after the transition to democracy and Spain's entry into the European Union. The winding down of the industry here took place over a longer period of time and was accompanied by generous social plans for miners made redundant, thereby preventing social hardship. However, the structural transformation of the Asturian coalfields has not been very successful so that the young in particular have few opportunities and little choice than to emigrate to other parts of Spain or abroad, if they want to find jobs. This mixture of a cushioned decline of the mining industry in combination with a failure to develop successful strategies for structural transformation has also produced some attempts to develop a mining heritage.[25]

The fate of mining regions in Western Europe did not depend on forms of ownership. The industry had been nationalised in Britain, France and Spain whereas it had remained in private hands in West Germany after 1945. Yet the most brutal form of deindustrialisation with the most devastating social consequences took place in Britain, whereas West Germany was most successful in securing a future for its former mining region at the Ruhr. If we look to Eastern Europe, here the mining industry was devastated by the move from a state-controlled command economy under communism to a free-market economy under capitalism after the communist dictatorships came to an end in the late 1980s and early 1990s. The relatively sudden crisis and collapse of mining in Eastern Europe and the subsequent social and economic crises in mining regions formed a problematic background for heritage initiatives, although some attempts were made, in particular in the Polish region of Upper Silesia.[26] By comparison, the Jiu valley in Romania saw much less successful attempts to preserve some of its mining heritage.[27]

If we move to mining regions outside of Europe, we observe in the US and Australia developments that are more akin to Britain than they are to continental Western Europe. Australia is a somewhat special case as mining has been mainly booming here over recent decades. Hence, unlike the European scenario, there is no end to the mining industry in sight. The question of heritigisation is far less urgent and prominent. Where mining has come to an end, there is little effort to construct any form of heritage around it, partly because of the geographical remoteness of some of those mining regions, partly because of the pressures on real estate in the more booming coastal areas of the continent.[28] In the US, mining is also still an ongoing economic activity, although there are many areas where mining now belongs to the past. Under neoliberal economic conditions we have also seen few attempts to cushion the social effects of deindustrialisation. Many of the former mining regions have experienced massive social crises and immiseration as well as a very visible decline of towns and regions. Structural transformation has not been successful and once again, the economics of failure have proven detrimental to attempts to preserve the mining heritage.[29] A final intriguing case is that of communist China, where many regions, including the northeast of the country, have witnessed the decline of a mining industry. The structural transformation of those regions is ongoing, but the closure of the mining industry has not seen massive social hardship, as strategies for economic restructuring have been put in place and are at least partly working. Thus, the northeast of China is still home to a booming motor industry and it has become the place where the new high-speed trains of China are being built. Yet the concept of industrial heritage seems alien to many in China. There have been tentative attempts to move to concepts of industrial heritage and there is undoubtedly much interest in ideas of industrial heritage. The economics would certainly allow the move to heritigisation in a grand style, but it seems unclear to date whether such a route will be taken.[30]

If we have found a clear correlation between the economic success of structural transformation in former mining regions and the emergence of industrial heritage, we also have to admit that economic success alone is not a sufficient reason to explain the prominence of industrial heritage initiatives. In addition to economic preconditions for heritigisation, we also have to consider political conditions. The politics of structural transformation in the Ruhr took place within a corporatist framework, where the state, the employers and the trade unions all shared the same premises about the desire of a long-term strategy for social change. Political support (which included both main political parties, the Christian Democrats and the Social Democrats) for such change was crucial in ensuring its success. The mining union, today's IGBCE, was willing to manage economic decline responsibly, ensuring an excellent deal for its members in exchange for agreeing to the managed end of the industry. And the employers were willing to come together to form one big company, the RAG, that was to deliver the socially responsible long-term end of mining in the Ruhr. An active macro-economic politics of the German state also played an important role in attracting new investment to the region and in helping the region to make the transition to a service economy, especially around a new and extensive landscape of higher education, built from the 1960s onwards. Thus we can say that the relative economic success of the Ruhr region was politically grounded.[31] Those involved in the political management of the long decline of mining developed a close emotional relationship to the industry and were among the stalwarts of a booming heritigisation of mining from the 1960s onwards.

The strong corporatist consensus in the Ruhr region has found no real equivalent elsewhere. In the Nord Pas de Calais, the politics of the mining region was badly divided between communism and social democracy. Whereas the former for a long time sought to defend the industry in a militant fashion, the latter attempted to move towards forms of structural transformation. However, the political rivalry meant that both dominant political forces were initially not keen to opt for heritage initiatives. The Communists saw them as one way of agreeing to deindustrialisation, where the Social Democrats did not want to invest in reminders of a past that, in their view, had to be transformed. It was therefore only after the collapse of communism and the definitive end of the mining industry in the 1990s that more heritage initiatives were started.[32]

The politics of the end of mining also had a major impact on British mining heritage initiatives. The Conservative government under Margaret Thatcher, who had crushed the industry alongside its unions, wanted to get rid of anything reminiscent of the bitter civil war like strife that had accompanied the final death struggle of the industry. Heritage was an unwelcome reminder of a past that was at best embarrassing and at worst traumatic. Hence, any heritage initiatives in the former coalfields of Britain tended to come 'from below'. In South Wales' local, often Labour-run, councils, former miners who had often also been active trade unionists and a whole range of left-of-centre intellectuals with deep sympathy for the culture of the mining regions were crucial in setting up what remained of the mining heritage of that region.[33]

In Asturias, the politics of the demise of mining was complicated by the collusion of the mining unions and left-wing political parties, above all the Communists and the Socialists, in the end of the industry. Their relatively conciliatory stance vis-à-vis the state as employer in a nationalised industry ensured generous redundancy packages for miners, but many of them and even more so their descendants, who had been left without a future perspective, felt betrayed by these deals which stood to them in marked contrast to the militant and revolutionary traditions of the region. Hence heritage initiatives in Asturias have to cope with the dissonances produced by a clash of memory – the memory of that militant tradition versus the memory of the conciliatory end of a proud tradition.[34]

In Eastern Europe the politics of the transition from communism to post-communism was not favourable to mining heritage initiatives. After all, the communist regimes had turned the miner into the iconic proletarian for whom those regimes allegedly worked and who legitimated the 'dictatorship of the proletariat'. The representation of the miner as pillar of the communist regime (regardless of how truthful it was) meant that after the fall of communism, miners were still associated with the regime and therefore there was little support for mining heritage under post-communist regimes.[35] Poland can be seen as an exception to this politics of post-communism, as the workers had an important hand in bringing down communism. After all, it was the independent trade union Solidarnosc that mobilised against communism in the 1980s and was eventually successful in removing it from Poland. The working class, including the miners, were therefore not one-sidely associated with communism in post-communist politics. Hence, a mining region, like Upper Silesia and a mining town like Kattowice, can remember their mining past through massive state-of-the art museums and heritage initiatives.[36]

In the Anglo-world outside of Europe, a more neoliberal political frame in the US and elsewhere meant that there was at best modest political support for mining heritage initiatives. If we look, for example, at the mining in and around Appalachians, it is clear that community-based initiatives, often supported by local politicians, were most prominent, whereas larger state initiatives were mostly absent.[37]

In communist China, the politics of the Communist Party has been to unleash the forces of a turbo-capitalism under the label of the 'Chinese way to communism'. This 'Chinese way' amounts to a stern belief in the fast modernisation of Chinese society allowing for rapid social mobility. Deindustrialisation has to be met with active policies of structural transformation that return 'failing regions' to modernising ones. Interestingly, the 'Chinese way to communism' does not seem to be much interested in the working-class experience, concentrating instead on the Party and on the social climbers in Chinese society (to which most successful Party officials belong). Hence the traditions of the working class as symbolised in mining heritage are of no great interest to the Party, which might explain why so far there has at best been limited efforts to preserve the mining heritage in those Chinese regions in which mining is a phenomenon of the past. The politics of the Communist Party of China is not supportive of a memory culture of failure.[38]

If economics is the basis of mining heritage initiatives, politics is the key to their success. It can open doors and provide the necessary political support for heritage initiatives. How these initiatives ultimately look like and what forms of institutionalisation of mining heritage are chosen depends largely on the respective cultures of deindustrialisation and the cultural representatives in charge of framing the mining heritage in different parts of the globe. Hence cultural preconditions for mining heritage also cannot be neglected.

In our most successful mining heritage landscape, the Ruhr, we encounter a dense and rich network of cultural institutions and cultural actors who, in close collaboration with economic and political elites, shape the outlook of mining heritage in the region. At the beginning of the mining heritage movement in the Ruhr in the 1960s, we find a social movement from below seeking to preserve this heritage against the odds. At that time the powerful alliance of industry, unions and state sought to get rid of the remnants of the mining industry as quickly as possible. By contrast, a motley alliance of social activists, including art historians, historians, urban planners, preservationists, architects, teachers and representatives of a range of other professions, as well as mining trade unionists and ordinary miners, campaigned to keep the remnants of the industry, not just including former workplaces but also housing estates, slagheaps, rail tracks, canals and other features of the industrial landscape associated with mining. In the course of the 1970s and 1980s, this social movement was coopted by the powerful corporatist elites in the Ruhr who put their influence, financial muscle and power behind preservationist attempts. The International Building Exhibition Emscher Park during the 1990s marked the highpoint of attempts to anchor mining heritage in the region and make it the anchor point of regional identity.[39]

Elsewhere in Western Europe, we also encounter attempts to preserve the mining heritage from the 1990s onwards. If we look to the Nord Pas de Calais and to Asturias, we see that those attempts were far less based on the mobilisation of civil society in a broad social movement. Instead many of the initiatives were initiated by local and regional politics in the hope that they might contribute to the structural transformation of those regions and help their population to reconcile the past with the present and future. However, in the absence of accompanying successes for the structural transformation of those regions, heritage initiatives appear to be somewhat disconnected from the dominant cultures of despair, disappointment and bitterness which can often be encountered among the former mining milieu in Nord Pas de Calais and Asturias. In this sense, heritage initiatives seem often to ignore the lives of ordinary workers in the industry.[40]

In communist Eastern Europe representatives from a cultural/academic background often play an influential role in supporting mining heritage initiatives from below. This certainly is the case in the Jiu valley in Romania, but here, as elsewhere, these movements from below often struggle to win support among the economic and political elites of post-communist countries. Hence they remain relatively small and insignificant.[41] The exception here again is Poland, where there has been significant support for mining museum and mining heritage. Museum and heritage professionals have been key in promoting mining heritage. The mine Guido in Zabrze in Upper Silesia or the galleries devoted to mining in the brand new museum of Silesia are good examples of the strong institutionalisation of mining heritage in Upper Silesia that is also keen to learn from the Ruhr and regularly invites individuals and delegations from the cultural scene in the Ruhr to hear how they have been representing their mining past to their audience.[42]

Outside of Europe, the lack of official political and economic support has still led to the emergence of a social activism around mining heritage, where individual groups form social movements keen to work towards preservation. Oral history, the formation of small community-based museums and an active social memory, resulting in memorial walks and other forms of remembering the mining past, abound here. Artists, intellectuals, academics as well as trade unionists and former workers all play an important role in these 'heritage from below' movements.[43] In China the Communist Party tightly controls all attempts to mobilise civil society in favour of mining heritage. Therefore all cultural representatives and institutions are working in close collaboration with and under the guidance of the Communist Party. Academics, artists, preservationists and museologists are also active here, as elsewhere in mining regions, but they have to frame their work according to the overall interpretation of the Communist Party. Given that the idea of mining heritage is still relatively new in China, the actual forms it might take are also at present difficult to gauge. However, the vast resources of the Chinese state could make mining heritage into an impressive showcase of mining heritage initiatives worldwide. The narrativisation of the meaning of such heritage for communist China still needs to be worked out. In 2006 the 'Wuxi Recommendation on Protecting Industrial Heritage During Fast Economic Development' was practically a copy of the 2003 TICCIH Charter for the Industrial Heritage. The party-state is interested in heritage discourses as it perceives heritage as one way of proclaiming and underlining its international status.[44]

Narrativisations of mining history

So far we have discussed the socio-economic, political and cultural preconditions for the emergence of mining heritage initiatives in the different parts of the world. We can summarise that mining heritage is strong where the economic transformation of former mining regions has been relatively successful, where there is broad political support for mining heritage and where there is strong cultural engagement from civil society in mining heritage initiatives. This, however, does not yet say much about the specific narrativisations of mining heritage that are associated with that heritage in different parts of the world. Hence this third part of the chapter will ask about the diversity of the heritage narratives associated with mining.

If we start again with our model region for mining heritage, the Ruhr, we find a remarkably uniform and homogeneous narrative. It is structured around the idea that mining forms the basis for the region's identity. As a region, the Ruhr did not exist before mining. Four towns in the Ruhr, Duisburg, Essen, Bochum and Dortmund have a prominent medieval and early modern history but the region only developed rapidly with mining and industrialisation from the middle of the nineteenth century onwards. Especially in the north of the Ruhr, villages grew exponentially to 'industrial village' status of 100,000 people and more within the period of one generation. The dominant heritage discourse starts from the assumption that life in the region was centred on mining (and coal and steel production that accompanied mining). Other industries that also had been a prominent part of the Ruhr industry, at various times, including textiles, chemicals and retail industries, are marginalised by the memory discourse of the mining heritage. The construction of a distinct miners' culture through mining heritage is extended to become a characteristic of the region as a whole. The solidarity engendered by working in dangerous conditions underground thus is portrayed, by extension, as a hallmark of the entire region, as is an alleged openness to foreigners (due to a long history of migration) and a supposed penchant for modernity (in technological and cultural terms). Furthermore the narrativisations of mining heritage in the Ruhr relates in intriguing ways to the environmental damage for which mining was responsible. Like structural change has overcome the economic crisis, so it has also dealt successfully with environmental damage and today what is framed as 'industrial nature' (*Industrienatur*) is an even better and more diverse form of nature than could be found at the Ruhr before industrialisation.[45] It fits with an overall narrative that focuses heavily on the long history of successful structural change and pride in the achievements of how this was managed. As the Ruhr was once the preeminent mining region in Germany, it is today the preeminent region for successful structural change. Overall this is a proud narrative seeking legitimation and re-assurance from the past that what has been done has been good and is going to be even better in the future.[46]

Similarly upbeat narrativisations are harder to construct for other former mining regions. If the Ruhr narrative can be described as success narrative, the dominant narrative of mining heritage in former British coalfields is one of heroic and defiant defeat. Given the oppositional nature of much of the mining heritage in Britain, its vanishing point is the 1984/5 miners' strike. It is constructed as heroic attempt to save an industry and a way of life from the evil machinations of Margaret Thatcher. The recognition that now mining communities belong firmly to the past is only occasionally broken by ironic defiance, which serves as rhetorical reminder of the defiance of the miners in 1984/5. The visitor to the museum at the UNESCO world heritage site 'Bit Pit' in Blaenavon is faced, at the end of the exhibition, with a listing of all mines ever active in the South Wales coalfield, underneath which he will find the words: 'We Will Be Back'.[47] The mining landscapes in South Wales and elsewhere in Britain have been largely demolished. It is hard nowadays even to recognise these regions as former mining regions. What has been preserved are individual buildings and sites, which stand in the landscape in a slightly forlorn and lost way, and the stories presented at these sites are stories firmly contained in the past, despite the striking reference to the future at the end of the Blaenavon exhibition. Many sites offer tours by former miners who are portrayed as an authentic voice from that past, telling visitors about their former work underground, their social relations above ground and their common hatred for Margaret Thatcher. In former mining communities, as well as in London and Glasgow, the celebrations on the announcement of her death were particularly vigorous – with people dancing around makeshift fires, shouting 'the witch is dead, the witch is dead'.[48]

If we go across the Channel, to the Nord Pas de Calais region, narrativisations of its mining heritage are neither emphasising the successes of ongoing structural transformation nor any defeats leading to the end of the industry. Rather, we encounter here a strangely technical endeavour to tell those born into a post-mining era about the lost world of the miner. Political actualisations to the present are rare, and instead we find authentic sites, also crowned with the world-heritage label, that depict life underground and above ground, explaining the technology and using a language that distances the visitor from a world that has disappeared, just like the former dominant political orientation of the miners, communism, has disappeared. It is interesting in this respect to reflect on the fact that attempts to revitalise the region culturally have led to the introduction of high culture to the region – thus we now have a section of the Louvre set up in a new architectural home in Lens. Hence other forms of culture rival the depoliticised memory of the once dominant miners' culture. The recent successes of the right-wing populist Front National in the Nord Pas de Calais region indicate that this memory can be successfully re-politicised – as memory of loss, of dispossession and discrimination.[49]

If the official heritigisation of the Nord Pas de Calais seems rather apolitical, its equivalent in Asturias seems aimless. The problematic end of the industry, referred to above, means that an institutionalised mining heritage is not very developed. Some sites, like the spectacularly located Arnao mine museum, are associated strongly with the mining company and here a technological perspective on the past dominates.[50] Some sites, such as the Mining and Industry Museum in El Entrego, are created by local and regional politics.[51] Here one looks in vain for a coherent attempt to narrativise the mining past of the region. The exhibition design seems haphazard, presenting in kaleidoscopic fashion unrelated parts of a heritage that do not fit into a structured narrative. Other parts of such mining heritage, such as the youth culture in the coalfields of Asturias, have developed a politicised oppositional narrative accusing government, mining companies, unions and established left-wing parties of betraying the militant heritage of mining. It can be described as a narrative of despair as it is based on heroic and positive depictions of the past that has come to an end. There is no way from that past to the present and the future; the future is only made up of a lack of opportunities and the choice to leave the coalfields to find one's luck and fortune elsewhere.[52]

In communist Eastern Europe, we find prominent narrativisations of mining heritage only in and around Katowice in Poland. Not unlike in the Ruhr, they are partly functionalised to underpin the construction of a regionalist narrative.[53] Mining heritage is one part of a Silesian identity that somehow is different from a Polish identity. The story of mining thus fits into the renaissance of regionalisms in the post-communist Poland. Again, like in the Ruhr, the regionalist story is bound up with a greater national one. If the Ruhr is narrativised as powerhouse for the German nation and eventually for a united Europe, Upper Silesia is narrativised in terms of the struggle of the Polish nation for sovereignty between the late eighteenth and early twentieth centuries.[54]

In other post-communist parts of Eastern Europe, mining heritage does not fare so well and narrativisations of the mining past are extremely rudimentary and deficient. In the Jiu valley in Romania, the memory of that past may well be lost in future, as the likelihood of it getting institutionalised seems so small. Yet there is a community-based activism, involving artists, architects and local inhabitants who are struggling against the coal mining company and the Romanian government to institutionalise forms of industrial heritage. The outcome is uncertain, but it seems as though many of the conditions for a successful establishment of industrial heritage are missing here.[55]

Mining regions in the US are also characterised by strong local community-based activities. They have been unable to prevent the destruction of most mines and much of the accompanying mining landscape, but through oral history and community museums and commemorations they have preserved the communicative memory of mining and have also partly been successful in institutionalising it. An impressive example is the documentary film *After Coal: Stories of Survival in Appalachia and Wales*, by documentary filmmaker Tom Hansell. It amounts to an impressive portrayal of the devastation coal brought to the environment of those communities, but it also celebrates the strong cultural identities and feelings of solidarity prevalent in both coalfields. The film is testimony to the desire of this community-based activism to retain values and ideas that are perceived as positive in the past of those mining regions and that are often in stark contrast to the dominant neoliberal scenarios of the future for those regions.[56]

In China, mining regions like the one in Shenyang in the northeast of the country are memorialising the industry of the region in ways that celebrate its past achievements without discussing its contemporary problems. The Shenyang industrial museum is housed in a huge and now defunct industrial complex with an impressive modernist facade. Inside we learn about the industrialisation of the region – under Japanese and Russian tutelage. The complex stories of imperialism and communist internationalisation that are part and parcel of that story are not dealt with in any depth. Instead we get huge doses of proud heroism – the achievements of Chinese workers, under the tutelage of the Communist Party, have contributed to making China a great nation. This storyline combines pride in class with pride in region and pride in nation. The museum is strong on presenting objects and artefacts from the industrial area and it also recreates parts of the factory, including a working canteen, and some of the 1950s-style working-class housing to give visitors a physical experience of the industrial past. The sheer size of the museum and the wealth of its displays are no doubt impressive, but the storyline avoids problematic areas and is streamlined into a narrative of heroic success that is, however, firmly anchored in the past. There is no discussion of how this relates to the present or future of Shenyang.[57]

Overall, we have attempted to present here all-too-briefly different kinds of narrativisations of mining heritage in different former mining regions of the world. Narratives of success stand next to narratives of heroic defeat, depoliticised narratives, narratives of despair and narratives of nostalgia. The narrativisations have been shown to be strongly related to which set of heritage actors are prominent and to how the economic, political and cultural preconditions for the emergence of mining heritage have been fulfilled.

Conclusions

By way of conclusion I would like to return to the theoretical points made in the first part of this chapter. Some of the dominant narratives that we encountered are part and parcel of the construction of a 'practical past'. In other words, they seek to make a political point in the present in order to shape the future. This 'practical past' approach to mining heritage can be part of the official culture, like in the Ruhr, or it can also be related to oppositional cultures seeking to preserve something that is seen as not particularly developed in the official culture of mining heritage, like in South Wales, the Appalachians and Asturias. Here, in their different ways, we have encountered a mining heritage, which celebrates and adores what has gone for good – with no possibility of actualising it, but with the desire to preserve values and ideas that are portrayed as having characterised those mining communities of the past. We have also found, especially in the Nord Pas de Calais, examples of an 'antiquarian past' which is heavily depoliticised and seems to speak of mining heritage for heritage's sake. In post-communist Eastern Europe, mining heritage is largely struggling against the odds to find ways of establishing itself against the background of economic decline, political hostility and cultural elites with little sympathy for industrial heritage. The Jiu valley in Romania is a good example of this. However, there are exceptions, such as Katowice and Upper Silesia, where mining heritage has been related to a growing regionalism. Its institutionalisation through museums and monuments is impressive and has taken place through forms of transregional transfer, especially from the Ruhr. A strongly official mining heritage in Shenyang is narrativised around the heroic national deeds of the people of a particular region – under the leadership of the Communist Party.

While meaning making through narrativisations thus differs quite substantially in different parts of the world, I would like to propose the introduction of greater doses of agonism in debates surrounding mining heritage. Agonistic memory, as introduced above, has, in my view, the power to repoliticise mining heritage from the standpoint of the political left. Where it has been depoliticised or repoliticised by the political right, such as in France, it would result in attempts to recapture that heritage for a politics of the left. Where elements of a practical past are already in place, like in the Ruhr, it would have the power to debate the underpinning values and interests and engender a much-needed debate if that heritage is not going to be ossified around a set of prescribed assumptions and normative values that silences and omits many other stories of that past.

In other words, an agonistic memory of mining would be capable of recapturing a vision for the future of former mining regions that would build on their pasts – politically and culturally. An agonistic memory of mining would strengthen those oppositional memories that struggle to find institutionalisation and that are in danger of being lost between an official memory culture and forgetting. It would introduce contestation and political struggle over meaning making into heritage debates and move the latter away from a *l'art pour l'art* fascination with technologies of the past or an antiquarian concern with the past record. Ultimately, an agonistic memory of mining is working towards a 'practical past' that seeks to shape the political future of those regions and give room to a political left seeking to overcome the political dominance of neoliberalism.

Notes

1 I am extremely grateful to the Regionalverband Ruhr, and the Land North-Rhine Westphalia for providing the funding to carry out this research.
2 Michael M. Rix, 'Industrial Archaeology', *The Amateur Historian* 2(8) (1955), pp. 225–229.
3 <https://industrial-archaeology.org/> [accessed 9 August 2018]; <www.sia-web.org/> [accessed 9 August 2018]; see also Marilyn Palmer and Peter Neaverson, *Industrial Archaeology: Principles and Practice* (London: Routledge, 1998).

4 The challenges of the multi-disciplinarity of the field have often been recognised. See, for example, Henrik Wagner (ed.), *Industrial Heritage in the Nordic and Baltic Countries: Seminar on Cooperation in Strategies, Research and Training* (Copenhagen: Tema publishers, 2000), pp. 7, 20, 111.

5 Ulrich Borsdorf, 'Industriekultur und Geschichte', *Forum Industriedenkmalpflege und Geschichtskultur* (Jan. 2015), pp. 16–19;

6 http://ticcih.org/ [accessed 9 August 2018]; see also James Douet (ed.), *Industrial Heritage Re-Tooled: the TICCIH Guide to Industrial Heritage Conservation* (London: Routledge, 2012).

7 On the Association for Critical Heritage Studies see, <www.criticalheritagestudies. org/> [accessed 9 August 2018]. One of its founding members and key representatives has been Laurajane Smith, whose *Uses of Heritage* (London: Routledge, 2006) has become an extraordinarily popular reference point for critical heritage studies.

8 Jörn Rüsen, *History: Narration, Interpretation, Orientation* (Oxford: Berghahn Books, 2005).

9 Hayden White, *The Practical Past* (Evanston, IL: Northwestern University Press, 2014).

10 Friedrich Nietzsche, *Unzeitgemäße Betrachtungen. Zweites Stück: Vom Nutzen und Nachtheil der Historie für das Leben* (Leipzig, 1874).

11 For a review of this growing body of literature see Stefan Berger, 'Industrial Heritage and the Ambiguities of Nostalgia for an Industrial Past in the Ruhr Valley in Germany', *Labor 16*(1) (2019), pp. 36–64.

12 Anna Cento Bull and Hans Lauge Hansen, 'On Agonistic Memory', *Memory Studies 9*(4) (2016), pp. 390–404.

13 On the concept of 'mindscape' see Dai Smith, *Aneurin Bevan and the World of South Wales* (Cardiff: University of Wales Press, 1993), p. 92 f.

14 Juan Manuel Wagner, 'Kulturlandschaftliche Identitätserhaltung in industriell geprägten Räumen durch Umnutzung und Inwertsetzung industriekultureller Relikte', in: Hans-Walter Herrmann, Rainer Hudemann and Eva Kell (eds), *Forschungsaufgabe Industriekultur. Die Saarregion im Vergleich* (Saarbrücken: Kommission für Saarländische Landesgeschichte und Volksforschung, 2004), pp. 345–360.

15 Hans-Werner Wehling, 'Montanindustrielle Kulturlandschaft Ruhrgebiet. Raumzeitliche Entwicklung im regionalen und europäischen Kontext', in: K. Fehn (ed.), *Siedlungsforschung: Archäologie – Geschichte – Geographie* (Bonn: Verlag Siedlungsforschung, 1998), pp. 167–189.

16 Thomas Parent, 'Schloss der Arbeit, Familienpütt und Ikone der Industriekultur', in: <www.lwl.org/industriemuseum/standorte/zeche-zollern/geschichte> [accessed 9 August 2018].

17 Judith Alfrey and Tim Putnam, *The Industrial Heritage: Managing Resources and Uses* (London: Routledge, 1992).

18 For the fascinating topic of lighting and industrial heritage see the intriguing research topic of Hilary Orange, *Lighting the Ruhr*, <https://hilaryorange.wordpress. com/2017/02/15/lighting-the-ruhr/> [accessed 9 August 2018]

19 Michael Stratton (ed.), *Industrial Buildings: Conservation and Regeneration* (London: E&FN Spon, 2000).

20 For an in-depth analysis see Stefan Berger, 'Industriekultur und Strukturwandel in deutschen Bergbauregionen nach 1945', in: Dieter Ziegler (ed.), *Geschichte des deutschen Bergbaus, vol. 4: Rohstoffgewinnung im Strukturwandel. Der deutsche Bergbau im 20. Jahrhundert* (Münster: Aschendorff, 2013), pp. 571–602.

21 Stefan Goch, *Eine Region im Kampf mit dem Strukturwandel: Strukturpolitik und Bewältigung von Strukturwandel im Ruhrgebiet* (Essen: Klartext, 2002); Jörg Bogumil, Rolf G. Heinze, Franz Lehner and Klaus-Peter Strohmeier, *Viel erreicht, wenig gewonnen: ein realistischer Blick auf das Ruhrgebiet* (Essen: Klartext, 2012).

22 Stefan Berger, 'Germany in Historical Perspective: The Gap Between Theory and Practice', in: Stefan Berger and Hugh Compston (eds), *Policy Concertation and Social*

Partnership in Western Europe. Lessons for the 21st Century (Oxford: Berghahn, 2002), pp. 125–138.

23 Leighton James, 'Mining Memories: Big Pit and Industrial Heritage in South Wales', in: Christian Wicke, Stefan Berger and Jana Golombek (eds), *Industrial Heritage and Regional Identities* (London: Routledge, 2018), pp. 13–31.

24 Marion Fontaine, 'Regional Identity and Industrial Heritage in the Mining Area of Nord-Pas-de-Calais', in: Wicke, Berger and Golombek (eds), *Industrial Heritage*, pp. 56–73.

25 Ruben Vega Garcia, 'Looking Back: Representations of the Industrial Past in Asturias', in: Wicke, Berger and Golombek (eds), *Industrial Heritage*, pp. 32–55.

26 Juliane Tomann, *Geschichtskultur im Strukturwandel: Öffentliche Geschichte in Kattowice nach 1989* (Berlin: de Gruyter, 2017).

27 David A. Kideckel, 'Identity and Mining Heritage in Romania's Jiu Valley Coal Region: Commodification, Alienation, Renaissance', in: Wicke, Berger and Golombek (eds), *Industrial Heritage*, pp. 119–135.

28 Erik Eklund, '"There Needs to be Something There for People to Remember": Industrial Heritage in Newcastle and the Hunter Valley', in: Wicke, Berger and Golombek (eds), *Industrial Heritage*, pp. 168–189.

29 Allen Dieterich-Ward, 'From Mills to Malls: Industrial Heritage and Regional Identity in Metropolitan Pittsburgh', in: Wicke, Berger and Golombek (eds), *Industrial Heritage*, pp. 190–213.

30 Zhengdong Li is currently undertaking doctoral research into the comparison of mining heritage in Shenyang province in North-East China and the Ruhr in Germany. I am grateful to him discussing the issue of industrial heritage in Shenyang with me. For the general context see also the contributions of Tong Lam, 'Ruins for Politics: Selling Industrial Heritage in Postsocialist China's Rustbelt', in: Stefan Berger (ed.), *Constructing Industrial Pasts: Industrial Heritage-Making in Britain, the West and Post-Socialist Countries* (Oxford: Berghahn Books, 2020), pp. 251–269.

31 Stefan Goch, 'Tief im Westen ist es besser als man glaubt? Strukturpolitik und Strukturwandel im Ruhrgebiet', in: Stefan Grüner und Sabine Mecking (eds), *Wirtschaftsräume und Lebenschancen: Wahrnehmung und Steuerung von sozio-ökonomischem Wandel in Deutschland, 1945–2000* (Berlin: De Gruyter Oldenbourg, 2017), S. 93–115.

32 Marion Fontaine, 'Between Dream and Nightmare: Political Conventions of the Industrial Past in the North of France', in: Berger (ed.), *Constructing Industrial Pasts*, pp. 184–198.

33 Stefan Berger, 'Von 'Landschaften des Geistes' zu 'Geisterlandschaften': Identitätsbildungen und der Umgang mit dem industriekulturellen Erbe im südwalisischen Kohlerevier', in: *Mitteilungsblatt des Instituts für soziale Bewegungen 39* (2008), pp. 49–66.

34 Holm Köhler, 'Industriekultur und Raumbewusstsein in Asturien', in: *Mitteilungsblatt des Institus für soziale Bewegungen 39* (2008), pp. 77–98.

35 Tibor Valuch, 'A Special Kind of Cultural Heritage: The Remembrance of Workers' Lives in Contemporary Hungary: The Case of Ózd', in: Berger (ed.), *Constructing Industrial Pasts*, pp. 242–250.

36 See the special issue on Upper Silesia of the journal *IndustrieKultur*, February 2017, especially Thomas Parent, 'Industriegeschichte in Oberschlesien im Überblick', pp. 2–5, <http://industrie-kultur.de/ik/2017/06/14/essen-heft-2-17-der-industriekultur-ist-erschienen-schwerpunktthema-ist-industrieregion-oberschlesien/> [accessed 9 August 2018]

37 Mary B. Lalone, 'Walking the Line Between Alternative Interpretations in Heritage Education and Tourism: A Demonstration of the Complexities with an Appalachian

Coal-Mining Example', in: Celeste Ray and Luke Eric Lassiter (eds), *Signifying Serpents and Mardi Gras Runners: Representing Identity in Selected Souths* (Athens, GA: the University of Georgia Press, 2003), pp. 72–92.

38 An example of how those in favour of protecting and developing industrial heritage have to relate the discussion to the policies of the Communist Party can be found in Zhi Quiao, Hao Zhang and Yan Wei, 'Introduction and Forecast of Creative Industry Parks of Industrial Heritage Culture in China', in: Yuang-Ming Liu, Dong Fu, Zhen-Xin Tong, Zhi-Quing Bao and Bin Tang (eds), *Civil Engineering and Urban Planning IV* (Boca Raton, FL: CRC Press, 2015), pp. 185–188.

39 Stefan Berger, Jana Golombek and Christian Wicke, 'A Post-Industrial Mindscape? The Mainstreaming and Touristification of Industrial Heritage in the Ruhr', in: Wicke, Berger and Golombek (eds), *Industrial Heritage*, pp. 74–94.

40 Marion Fontaine, 'From Myth to Stigma: The Political Uses of Mining Identity in the North of France', *Labor 16*(1) (2019), pp. 65–80; Ruben Vega Garcia, 'Sounds of Decline: Industrial Echoes in Asturian Music', in: Berger (ed.), *Constructing Industrial Pasts*, pp. 216–227.

41 David Kideckel, 'The Coal-Environment Nexus: How Nostalgic Identity Burdens Heritage in Romania's Jiu Valley', in: Berger (ed.), *Constructing Industrial Pasts*, pp. 228–241.

42 On the Silesian museum see Tomann, *Geschichtskultur*, pp. 256–272; on the mine Guido in Zabrze see <www.sztolnialuiza.pl/index.php?option=com_content&vie w=article&id=740:history-of-the-guido-mine&catid=144:angielski&Itemid=927> [accessed 9 August 2018]. I am grateful to Ulrich Borsdorf for sharing with me the interest from Upper Silesia in museums and heritage of the Ruhr. Borsdorf, former director of the Ruhr museum, himself has been an advisor on the Guido project in Upper Silesia.

43 An outstanding example of the merger between university-based study and community-based activism around issues of industrial heritage is provided by the Centre for Oral History and Digital Storytelling at Concordia University in Montreal. See <http://storytelling.concordia.ca/> [accessed 10 August 2018].

44 Haiming Yan, *World Heritage Craze in China: Universal Discourse, National Culture, and Local Memory* (Oxford: Berghahn Books, 2018), p. 51.

45 Pia Eiringhaus, *Industrie wird Natur – postindustrielle Repräsentationen von Region und Umwelt im Ruhrgebiet* (Bochum: SGR, 2018).

46 I have dealt with this at greater length in Berger, 'Industrial Heritage and the Ambiguities of Nostalgia'.

47 C. Stephen Briggs (ed.), *Welsh Industrial Heritage: A Review Based Upon the Proceedings of a Joint Cadw/CBA Conference Held in Cardiff, 5 Dec. 1986* (Cardiff: Cadw., 1992), pp. 104 ff.

48 'Margaret Thatcher's Death Greeted With Street Parties in Brixton and Glasgow', in: *The Guardian*, 8 April 2013, < www.theguardian.com/politics/2013/apr/08/margaret-thatcher-death-party-brixton-glasgow> [accessed 10 August 2018]

49 On the inability of the mining heritage to construct regional identity in the Nord Pas-de-Calais region see Anne-Sophie Forbras, 'Que reste t'il de l'activité charbonnière?', in: Jean-Claude Rabier (ed.), *La remonte. Le basin minier du Nord-Pas-de-Calais entre passé et avenir, Villeneuve d'Ascq*, 2002, pp. 51–63. On the attempts of the FN to mobilise the memory of mining in its campaigns, see Béatrice Giblin, 'Comment le Pas-de-Calais a basculé du PS au FN', *Libération*, 11 May 2007.

50 <http://museominadearnao.es/> [accessed 10 August 2018].

51 <www.mumi.es/media/Default%20Files/MUMI/folleto_MUMI_EN_web.pdf> [accessed 10 August 2018]

52 Ruben Vega Garcia, 'The Memory of Youth: The Industrial Past as a Factor of Identity Among Asturian Coalfield Young Adults', unpublished paper. The author is grateful to Ruben Vega Garcia for letting him read the paper.

53 Sebastian M. Büttner, *Mobilizing Regions, Mobilizing Europe: Expert Knowledge and Scientific Planning in European Regional Development* (London: Routledge, 2012), Chapter 6: 'Region-Building and Regional Mobilisation in Poland'.

54 Peter Polak-Springer, *Recovered Territory: A Polish-German Conflict over Land and Culture, 1918–1989* (Oxford: Berghahn Books, 2015).

55 Ilinca Paun Constantinescu, Dragos Dascalu and Cristina Sucala, 'An Activist Perspective on Industrial Heritage in Petrila, a Romanian Mining City', *The Public Historian 39*(4) (2017), pp. 114–141.

56 Tom Hansell, *After Coal: Stories of Survival in Appalachia and Wales*, <http://after coal.com/> [accessed 10 August 2018].

57 Matthew Felix Sun, 'Industrial Museum of China in Manchurian Shenyang', <www.artslant.com/ny/articles/show/37680-industrial-museum-of-china-in-manchurian-shenyang> [accessed 10 August 2018].

Index

This Index was compiled by Alessandra Exter and Jannik Keindorf

For Product Safety Concerns and Information please contact our EU
representative GPSR@taylorandfrancis.com
Taylor & Francis Verlag GmbH, Kaufingerstraße 24, 80331 München, Germany

www.ingramcontent.com/pod-product-compliance
Lightning Source LLC
Chambersburg PA
CBHW060808220326
41598CB00022B/2570